Fungal Omics

Fungal Omics: Methods and Applications provides a comprehensive account of trends and developments in fungal "omics" approaches. It presents literature on various "omics" applications and includes information on plant taxonomy, genetics, epigenetics, and biotechnological and industrial applications. It begins with basic concepts, extending to complex and in-depth knowledge, and finally describing the fungal genome and all "omics" comprehensively.

Key Features:

- Covers almost every aspect of fungal omics in a single source
- Features trends in phytomycology and fungal biology studies using modern molecular tools
- Contains information on improved methods in genetics, genomics, and metabolomics in different fungal species

Highlights include integrative "omics" strategies as well as trends in phytomycology and fungal biology using modern molecular tools. With detailed information on improved methods in genetics, genomics, and metabolomics, *Fungal Omics: Methods and Applications* is an invaluable resource for scientists, researchers, and students in the evolving fields of fungal omics and life sciences.

Fungal Omics
Methods and Applications

Edited by Imran Ul Haq, Siddra Ijaz,
and Hayssam Mohamad Ali

CRC Press
Taylor & Francis Group
Boca Raton London New York

CRC Press is an imprint of the
Taylor & Francis Group, an **informa** business

Designed cover image: Shutterstock

First edition published 2026
by CRC Press
2385 NW Executive Center Drive, Suite 320, Boca Raton FL 33431

and by CRC Press
4 Park Square, Milton Park, Abingdon, Oxon, OX14 4RN

CRC Press is an imprint of Taylor & Francis Group, LLC

ISBN: 978-1-032-58742-4 (hbk)
ISBN: 978-1-032-58968-8 (pbk)
ISBN: 978-1-003-45236-2 (ebk)

DOI: 10.1201/9781003452362

Typeset in Times
by Apex CoVantage, LLC

Contents

About the Editors

Prof. Dr. Imran Ul Haq, with a bright career in agriculture, plant pathology, and fungal molecular biology, has a postdoc degree from University of California Davis, USA. He is a professor in the Department of Plant Pathology, University of Agriculture Faisalabad, Pakistan. He has supervised more than 40 graduate students and established the Fungal Molecular Biology Laboratory Culture Collection (FMB-CC-UAF). He is an affiliated member of the World Federation for Culture Collections (WFCC). He has published more than 60 articles, nine books, four laboratory manuals, and several book chapters. He has made colossal contributions to fungal taxonomy by reporting novel species of fungal pathogens in plants. His research interests are fungal molecular taxonomy and nanotechnology integration with other control strategies for sustainable plant disease management.

Prof. Dr. Siddra Ijaz, with a vibrant career in agriculture and biotechnology, has a Ph.D. from the University of Agriculture Faisalabad, Pakistan, and a postdoc degree from the Plant Reproductive Biology Laboratory, University of California Davis, USA. She is a professor at the Center of Agricultural Biochemistry and Biotechnology (CABB), University of Agriculture, Faisalabad, Pakistan. She is also serving as Managing Editor of an international high-impact-factor journal, the *Pakistan Journal of Agricultural Sciences*. She has supervised more than 60 M.Phil and Ph.D. students. She was a National Curriculum Revision Committee member from 2013 to 2014. She has published over 50 articles, ten books, one laboratory manual, and several book chapters. She is the Incharge of the Molecular Biology of Plant Disease Resistance Laboratory, CABB, UAF. She is the Curator of the Fungal Molecular Biology Laboratory Culture Collection (FMB-CC-UAF) and an affiliated member of the World Federation for Culture Collections (WFCC). Her research focuses on plant biotechnology, including genomics, genome probing for identifying disease resistance genes, genome engineering for fungal disease resistance in plants, and disease management through nanobiotechnology.

Dr. Hayssam Mohamad Ali is working at the Botany and Microbiology Department, College of Science, King Saud University. Dr. Hayssam Mohamad Ali has a "WORLD RANKING OF TOP 2% SCIENTISTS" according to the Stanford Scientists' Ranking released by Stanford University in 2021 and 2022. He has published more than 300 papers and ten books, and he has reviewed more than 1000 papers in high-impact-factor journals. He is interested in studying plant physiology, biotic and abiotic stresses, sustainability, molecular biology, environmental stresses, plant biotechnology, crop improvement, plant breeding, forestry, and plant biology.

Contributors

Amjad Abbas
Department of Plant Pathology
University of Agriculture Faisalabad
Pakistan

Syed Zain Ul Abdin
Institute of Plant Protection
MNS University of Agriculture Multan
Pakistan

Abdullah
Department of Plant Pathology
University of Agriculture
Faisalabad, Pakistan

Güliz Tepedelen Ağaner
Directorate of Plant Protection Research Institute
Republic of Türkiye Ministry of Agriculture and Forestry
Bornova-Izmir, Turkey

Munir Ahmad
State Key Laboratory for Conservation and Utilization of Bio-Resources in Yunnan
Yunnan Agricultural University
China

Waqar Ahmad
Guangdong Province Key Laboratory of Microbial Signals and Disease Control, College of Plant Protection, South China Agricultural University
Guangzhou, China

Ayesha Ahmed
State Key Laboratory for Conservation and Utilization of Bio-Resources in Yunnan
Yunnan Agricultural University
China

Faizan Ali
Department of Plant Pathology
University of Agriculture
Faisalabad, Pakistan

Muhammad Amjad Ali
Department of Plant Pathology
University of Agriculture
Faisalabad, Pakistan

Hayssam Mohamad Ali
Department of Botany and Microbiology, College of Science
King Saud University
Riyadh, Saudi Arbaia

Salim S. Al-Rejaie
Department of Pharmacology and Toxicology
College of Pharmacy, King Saud University
Riyadh, Saudi Arabia

Saman Arif
Department of Plant Pathology
University of Agriculture
Faisalabad, Pakistan

Saba Aslam
Department of Plant Pathology
University of Agriculture
Faisalabad, Pakistan

Hafiz Muhammad Usman Aslam
Department of Plant Pathology
San Luis Valley Research Center, Colorado State University
USA

Muhammad Atiq
Department of Plant Pathology
University of Agriculture
Faisalabad, Pakistan

Muniza Baig
Department of Plant Pathology
University of Agriculture
Faisalabad, Pakistan

Barbaros Çetinel
Directorate of Plant Protection Research Institute
Republic of Türkiye Ministry of Agriculture and Forestry
Bornova-Izmir, Turkey

Mohamad Chikh-Ali
Department of Plant Pathology
San Luis Valley Research Center, Colorado State University
USA

Anjum Faraz
Department of Plant Pathology
University of Agriculture
Faisalabad, Pakistan

Amer Habib
Department of Plant Pathology
University of Agriculture
Faisalabad, Pakistan

Prof. Dr. Imran Ul Haq
Department of Plant Pathology
University of Agriculture
Faisalabad, Pakistan

Prof. Dr. Siddra Ijaz
Centre of Agricultural Biochemistry and Biotechnology
University of Agriculture
Faisalabad, Pakistan

Zakia Habib
Centre of Agricultural Biochemistry and Biotechnology
University of Agriculture
Faisalabad, Pakistan

Muhammad Hassan
Institute of Plant Protection
MNS University of Agriculture
Multan, Pakistan

Abdalmenem I. M. Hawamda
Department of Agricultural Biotechnology
Faculty of Agricultural Science and Technology, Palestine Technical University—Kadoorie (PTUK)
Palestine

Zobia Jabeen
Department of Plant Pathology
Balochistan Agriculture College Quetta
Pakistan

Muhammad Jabran
State Key Laboratory for Biology of Plant Diseases
Institute of Plant Protection
Beijing, China

Nuzhat Jamal
Department of Plant Pathology
University of Agriculture
Faisalabad, Pakistan

Seema Kanwal
Institute of Plant Protection
MNS University of Agriculture
Multan, Pakistan

Sobeeqa Khalid
Department of Plant Pathology
University of Agriculture
Faisalabad, Pakistan

Babar Khan
School of Biological Science and Medical Engineering
Southeast University
Nanjing, China

Nasir Ahmad Khan
Department of Plant Pathology
University of Agriculture
Faisalabad, Pakistan

Nabeeha Aslam Khan
Department of Plant Pathology
University of Agriculture
Faisalabad, Pakistan

Ayesha Lutf
Institute of Food Science & Nutrition
University of Sargodha
Pakistan

Riffat Malik
Centre of Agricultural Biochemistry and Biotechnology
University of Agriculture
Faisalabad, Pakistan

Mohamed Mohany
Department of Pharmacology and Toxicology
College of Pharmacy, King Saud University
Riyadh, Saudi Arabia

Shahzad Munir
State Key Laboratory for Conservation and Utilization of Bio-Resources in Yunnan
Yunnan Agricultural University
China

Ghalia Nadeem
Centre of Agricultural Biochemistry and Biotechnology
University of Agriculture
Faisalabad, Pakistan

Effat Naz
Department of Botany
University of Agriculture
Faisalabad, Pakistan

Kiaenat Nazir
School of Life Sciences
Hainan Normal University
Haikou, China

Ahmad Nisar
Department of Plant Pathology
University of Agriculture
Faisalabad, Pakistan

Faiz Ur Rahman
Guangdong Provincial Key Laboratory of Postharvest Science of Fruits and Vegetables/Engineering Research Center for Postharvest Technology of Horticultural Crops in South China, Ministry of Education
College of Horticulture, South China Agricultural University
Guangzhou, China

Nasir Ahmed Rajput
Department of Plant Pathology
University of Agriculture
Faisalabad, Pakistan

Hasan Riaz
Institute of Plant Protection
MNS University of Agriculture
Multan, Pakistan

Shehbaz Sabir
Institute of Plant Protection
MNS University of Agriculture Multan

Syed Abdul Sadiq
Agriculture Research Institute
Quetta, Pakistan

Maham Sarwar
Department of Plant Pathology
University of Agriculture
Faisalabad, Pakistan

Shahzad Sabir
Institute of Plant Protection
MNS University of Agriculture
Multan, Pakistan

Akhlas Shafique
Department of Plant Pathology
University of Agriculture
Faisalabad, Pakistan

Muhammad Hamza Shahid
Department of Plant Pathology
University of Agriculture
Faisalabad, Pakistan

Muhammad Huzaifa Tanveer
Department of Plant Pathology
University of Agriculture
Faisalabad, Pakistan

İhsan Barışcan Tunçkol
Directorate of Plant Protection Research Institute
Republic of Türkiye Ministry of Agriculture and Forestry
Bornova-Izmir, Turkey

Muhammad Usman
Department of Plant Pathology
University of Agriculture
Faisalabad, Pakistan

Muhammad Wahab
Department of Plant Pathology
University of Agriculture
Faisalabad, Pakistan

Muhammad Waris
Department of Plant Pathology
Balochistan Agriculture College
Quetta, Pakistan

Yonghao Ye
Jiangsu Key Laboratory of Pesticide Science
College of Plant Protection, Nanjing Agricultural University
Nanjing, China

Jinhao Zhang
State Key Laboratory for Conservation and Utilization of Bio-Resources in Yunnan
Yunnan Agricultural University
China

Laraib Zulfiqar
Department of Plant Pathology
University of Agriculture
Faisalabad, Pakistan

Preface

The dawn of DNA-based technologies manifested a new era in fungal understanding and genomics. Fungal genomes are diverse, and they vary greatly in size, ranging from as small as 8 Mb in some yeast species to over 100 Mb in certain filamentous fungi. Omics provides a core understanding of the biology, survival, and pathogenicity of fungi and is helpful in making a layout of the genes involved in a particular function. Fungi utilize different cellular pathways to perform multiple bodily functions and communicate with the microenvironment. Significant traits related to the survival of fungi, like cell cycles, the communication of microbes and plants with the fungi, development, and differentiation, could be unraveled using "omic" technologies.

The advancements in omics really need to be compiled as a single source for the research community and professionals. This book delineates the entire philosophy of fungal "omes and omics." The book offers a comprehensive account of the latest trends and developments in omics approaches and their applications in life science. It provides a complete spectrum of fungal or fungus-related omics. The contents of the book make it an invaluable resource for students, researchers, and scientists in fungal molecular biotechnology.

This state-of-the-art book covers all areas and aspects of fungal omics, from genetics to metabolomics. This book equips readers with a deeper understanding of fungal biology by delving into fungal genetics, epigenetics, genome, transcriptome, proteome, physiome, metabolome, and their related applications. It is a must-read for anyone seeking comprehensive knowledge in the radical field of fungal omics. This book will definitely bridge the gap and provide recent and updated advances in fungal and system biology.

Imran Ul Haq

Siddra Ijaz

Hayssam Mohamad Ali

1 Overview of Fungal Communication

Siddra Ijaz, Imran Ul Haq, and
Hayssam Mohamad Ali

1.1 FUNGAL COMMUNICOME

The communication process in a cell is initiated by a message generated by a sender. It terminates when a receiver catches it. These three components are the obligatory and integral steps of any kind of communication stairway (Cottier and Mühlschlegel 2012). The communication cycle may be of varied levels of complexity in nature (Aravind et al. 2003; Pertseva and Shpakov 2009). In fungi, pheromones have been revealed to make an imperative contribution to the sexual cycle when locating and attracting a sexual mate. This group of compounds induces the fertilization process in fungi, encompassing two distinct phases, plasmogamy and karyogamy, prior to meiosis (Martínez-Soto and Ruiz-Herrera 2017). In *Saccharomyces cerevisiae*, they are diffusible peptides of the mating type (MAT) locus. When their synthesis platforms are "a-cells" (MATa), they are known as "a-factors;" they are called "α-factors" when they are synthesized by "α-cells" (MATα).

The prepro-α-factor is a pheromone precursor encoded by MFα1. Before maturation, pheromones undergo a secretory pathway process involving a cascade of proteolytic reactions (Cowan and Storey 2003). Contrary to this, "a-factors" are farnesylated and are required for ABC transporter Ste6p. The receptor Ste2p on "a-cells" interacts with "α-factors;" however, the receptor Ste3p on "α-cells" interacts with "a-factors."

Both receptors are G-protein-coupled receptors, and the binding of these receptors (Ste2p and Ste3p) with pheromones (α-factors and a-factors) stimulates the dissociation of subunits of G proteins (heterotrimeric) into α (monomeric) and βy (dimeric) subunits. This results in Ste4p activity, recruiting Ste5p to the cellular membrane. The process leads to kinase cascade activation, which phosphorylates Fus3p and Kss1p (MAP kinases) (Hoeflich and Woodgett 2001). Upon phosphorylation, Fus3p starts to travel toward the nucleus, which induces Ste12p, a transcriptional factor that ultimately causes the expression of pheromone-responsive genes (Taylor et al. 2019).

This mechanism results in a morphological response in mating cells, "shmoo," which establishes in response to pheromones (directional cell growth). When sexual mates establish contact by "shmoo," then plasmogamy occurs between them (Cottier and Mühlschlegel 2012).

DOI: 10.1201/9781003452362-1

1.1.1 Mitogen-Activated Protein Kinase Signaling Cascade

The most dynamic switch in the signal transduction process is that of mitogen-activated protein kinase (MAPK). This multistage event encompasses sequential phosphorylation-assisted signal amplification that results in a signal relay in the MAPK cascade. MAPK kinase (MAPKK) activates MAPKs, which are serine/threonine kinases (Braicu et al. 2019).

MAPKK (a "dual-specific" kinase) induces the phosphorylation of substrates at the tyrosine (Tyr) motif and serine/threonine (Ser/Thr) motif. MAPKK relays signals from pattern recognition receptors or stimulus-activated upstream effectors. These receptors are GTP-binding proteins, response regulators, and other kinases (Lopez et al. 2020).

MAPKs remain inactive until or unless they are phosphorylated by their upstream kinases, which results in conformational changes. These changes consist of the rotation of terminal lobes and the phosphorylated loop's movement toward the active site. Phosphorylation in the activation loop region stimulates the reorganization of the C-terminal lobe, triggers the rotation of the N-terminal lobe (N-lobe), and leads to the formation of an ATP-binding active site for catalysis (Pegram et al. 2021).

1.1.2 Cyclic Adenosine Monophosphate-Dependent Signaling

In *Saccharomyces cerevisiae*, cyclic adenosine monophosphate (cAMP)-dependent protein kinase A (PKA) is the principal cAMP signaling enzyme. The PKA enzyme mediates almost all effects on fungal cell physiology (Jacob et al. 2022). It is a tetrameric enzyme having two regulatory subunits and two catalytic subunits. This enzyme is inactive in a cell when the cAMP level is low (non-inducing condition).

Contrary to this, its increased level (inducing condition) results in its binding with regulatory subunits of PKA. The interaction leads to a conformational change, stimulating PKA tetramer degradation and releasing the catalytic subunit. These catalytic subunits then start to phosphorylate the downstream substrates and can stimulate or suppress transcription factors and regulatory cellular functions (Ribeiro et al. 2019; Cai et al. 2021).

REFERENCES

Aravind, L., Anantharaman, V., Iyer, L.M. Evolutionary connections between bacterial and eukaryotic signaling systems: A genomic perspective. *Curr. Opin. Microbiol.* 2003, 6, 490–497.

Braicu, C., Buse, M., Busuioc, C., Drula, R., Gulei, D., Raduly, L., Rusu, A., Irimie, A., Atanasov, A.G., Slaby, O., et al. A comprehensive review on MAPK: A promising therapeutic target in cancer. *Cancers* 2019, 11, 1618.

Cai, E., Sun, S., Deng, Y., Huang, P., Sun, X., Wang, Y., Chang, C., Jiang, Z. Histidine kinase Sln1 and cAMP/PKA signaling pathways antagonistically regulate Sporisorium scitamineum mating and virulence via transcription factor Prf1. *J. Fungi* 2021, 7, 610.

Cottier, F., Mühlschlegel, F.A. Communication in fungi. *Int. J. Microbiol.* 2012, 1, 351832.

Cowan, K.J., Storey, K.B. Mitogen-activated protein kinases: New signaling pathways functioning in cellular responses to environmental stress. *J. Exp. Biol.* 2003, 206, 1107–1115.

Hoeflich, K.P., Woodgett, J.R. Mitogen-activated protein kinases and stress. *Cell Mol. Response Stress* 2001, 2, 175–193.

Jacob, S., Bühring, S., Bersching, K. Recent advances in research on molecular mechanisms of fungal signaling. *Encyclopedia* 2022, 2(2), 840–863.

Lopez, E.D., Burastero, O., Arcon, J.P., Defelipe, L.A., Ahn, N.G., Marti, M.A., Turjanski, A.G. Kinase activation by small conformational changes. *J. Chem. Inf. Model.* 2020, 60, 821–832.

Martínez-Soto, D., Ruiz-Herrera, J. Functional analysis of the MAPK pathways in fungi. *Rev. Iberoam. Micol.* 2017, 34, 192–202.

Pegram, L.M., Anderson, J.W., Ahn, N.G. Dynamic equilibria in protein kinases. *Curr. Opin. Struct. Biol.* 2021, 71, 215–222.

Pertseva, M.N., Shpakov, A.O. The prokaryotic origin and evolution of eukaryotic chemosignaling systems. *Neurosci. Behav. Physiol.* 2009, 39, 793–804.

Ribeiro, L.F.C., Chelius, C., Boppidi, K.R., Naik, N.S., Hossain, S., Ramsey, J.J.J., Kumar, J., Ribeiro, L.F., Ostermeier, M., Tran, B., et al. Comprehensive analysis of aspergillus nidulans PKA phosphorylome identifies a novel mode of CreA regulation. *mBio* 2019, 10, e02825–18.

Taylor, S.S., Meharena, H.S., Kornev, A.P. Evolution of a dynamic molecular switch. *IUBMB Life* 2019, 71, 672–684.

2 Genomics and Metagenomics of Fungi

Shehbaz Sabir, Hafiz Muhammad Usman Aslam, Mohamad Chikh-Ali, Saba Aslam, Seema Kanwal, Muhammad Hassan, Syed Zain Ul Abdin, and Hasan Riaz

2.1 INTRODUCTION

The substantial transformation in biological and biomedical research is being influenced by two key factors: the development of technology and the remarkable growth in the amount of knowledge about DNA sequences. As a result, we are currently in an era in which novel types of experiments are practical, and previously unthinkable numbers of observations, analyses, and discoveries are being performed. Unfortunately, the billions of DNA bases do not tell us how each gene functions, how cells function, how organisms are formed, what causes disease, how we age, or how to develop a medication. This is where genomics comes into play. Genomics refers to the study of an organism's entire genome and also examines in great detail how genes and other DNA sequences that make up an organism's genome interact with one another and with each other's structures, functions, and interactions. Genomic research aims to discover the genetic underpinnings of numerous biological functions, characteristics, diseases, and evolutionary relationships. Genomics aims to understand life, not only to identify its constituent parts, using experimental and computational methods that make use of as much sequence information as possible (Lockhart and Winzeler, 2000). Another discipline that is interconnected with each genomics is metagenomics. The word "metagenomics" refers to the study of the genetic material present in the microbiota. In contrast, the terms "microbiota" and "microbiome" refer to the collection of microbial species that are present in any given environment. Metagenomics also refers to the structure and function of the whole nucleotide sequences that have been separated and examined from each organism in a bulk sample, usually microorganisms. Genomics is a multidisciplinary science that uses whole-genome cloning, molecular characterization, and identification to understand the structure, function, and evolution of genes and their interactions with one another to better understand the phenomics of an organism. Metagenomics is typically used to study a specific community of microorganisms, such as those that reside on crops, in the soil, or in a water sample (Anastassiou, 2001). Over the past few years, the focus of research has shifted from isolating species to looking at how these species interact in environments that are more similar to those found in their native

 DOI: 10.1201/9781003452362-2

habitats. This tendency has been reflected in the growing application of nucleic acid sequencing from environmental materials without any previous understanding of the species present (Ursell et al., 2012).

Fungus is a representative of an important kingdom in the eukaryotic domain of life. Because they produce antibiotics, ferment goods like bread and beer, and break down cellulose, some species have great economic worth, despite the fact that just a tiny part of the alleged millions of fungal species has been described. Living organisms serve as specimens and as a starting point for the morphological, molecular, and biochemical identification of species (Tedersoo et al., 2016). Numerous genes linked to fungal pathogenicity have been discovered and characterized as a result of the use of molecular genetics research in the study of phytopathogenic fungi. These include the genes responsible for detoxifying the antifungal substances made by plants (Soanes et al., 2002). The first significant fungal genomics milestone was the publication of *Saccharomyces cerevisiae* entire genome sequence, an organism that significantly contributed to our basic understanding of eukaryotic cell physiology, biochemistry, and genetics, as well as baking, brewing, and industrial ethanol production. The process of fermenting sugar into alcohol is one of the final steps in converting biomass into biofuel. *Trichoderma reesei* is yet another fungus that has been sequenced and used industrially (Martin et al., 2008). Microbiome research can considerably advance our understanding of the ambient reservoir of fungal microorganisms, which is a topic that is frequently understudied. The numerous fungal genomes that have already been sequenced, along with the substantial number still being analyzed, are of great importance for advancing research in various scientific fields. The field of fungal genetic research, in particular, has seen a significant impact of next-generation sequencing technology (Cuomo and Birren, 2010). The entire phylogeny of the fungal kingdom has been fundamentally rewritten in light of information obtained from enhanced DNA sequencing. Numerous times, DNA analysis has revealed that distinct fungi that were once thought to be the same species were actually quite distinct from one another (Begerow et al., 2010). The immense complexity of many fungal reproductive structures was uncovered by comparing the DNA sequences of numerous fungi that were once believed to be extremely distinct species but that actually turned out to have similar sequences. The decision of which name to choose for a particular fungus species has been complicated by this revelation and has generated some debate ("One Fungus, One Name"). The benefits of sequencing the DNA of fungi much outweigh the disadvantages. By combining whole-genome sequencing with other cutting-edge "-omics technologies" like phenomics, transcriptomics, and metabolomics, researchers have been able to comprehend fungus and their interactions with the environment like never before (Taylor, 2011). The fungi that make up a mycobiome can be identified using two sequence-based techniques. The most popular technique is sequencing after PCR amplification of the rRNA operons' internal transcribed spacer (ITS) sections, particularly ITS2 between the 5.8S and 28S genes. Fungal analyses now employ ITS2 sequences as typical fungal barcodes since fungi are so diverse (Schoch et al., 2012). The second method uses shotgun metagenomes to identify species. The majority of solutions combine proprietary databases with search techniques like BLAST,

USEARCH and UBLAST, GhostX, and DIAMOND (Buchfink et al., 2014). Some metagenomics databases host and analyze datasets simultaneously using their own procedures. Using MG-RAST, which employs a sequentially updated pipeline to perform in-depth graphical analyses of user-uploaded datasets, it was possible to identify fungi in grain dust from a swine facility. The pipeline relies on matching sequence reads to rDNA to identify eukaryotic DNA while discarding any non-rDNA readings. EBI Metagenomics, a metagenomics database and accompanying workflow hosted by the European Bioinformatics Institute (Hunter et al., 2014). Despite having over 16,000 meticulously kept datasets, EBI Metagenomics did not begin to recognize eukaryotic DNA until the release of version 4.0 on September 4, 2017. Pipeline v4.0, like MG-RAST, only employs rDNA sequences to identify fungal species, and it has only been used to analyze fewer than 1% of the EBI Metagenomics datasets (Donovan et al., 2018).

2.2 TYPES OF NUCLEIC ACIDS IN FUNGI

2.2.1 Deoxyribonucleic Acid

Fungi have deoxyribonucleic acid (DNA) that serves as the organism's growth, development, and operating manual, just like every other eukaryote species. The genetic material known as DNA holds the genetic information for the organism and is found in the nucleus of all eukaryotic cells, including those of fungi. DNA contains the instructions for building and maintaining the organisms, and it is passed from one organism to the next. DNA in fungi, as in all eukaryotes, consists of a double-stranded helix. It carries genetic information that determines the characteristics of the fungus. This information includes instructions for the synthesis of proteins, which are essential for the formation of the cell. In fungal cells, DNA is primarily located within the nucleus. The nucleus is a membrane-bound organelle that houses the genetic material, protecting it and regulating its access. The DNA in fungi is organized into structures called chromosomes. The number of chromosomes can vary among different fungal species. Chromosomes contain genes, which are specific segments of DNA that code for protein and functional RNA molecules. DNA in fungi, as in all organisms, undergoes replication during the cell cycle, and each cell formed through cell division receives a complete and accurate copy of genetic information. Gene expression, which is essential to how fungi function and react to their environment, occurs when fungi use the information encoded in DNA to make RNA, which is subsequently translated into proteins.

DNA plays a critical role in generating genetic diversity within fungal populations. Variations in characteristics can result from DNA mutations, and genetic variety is crucial for both the long-term evolution of new fungus species and adaptation to changes in environmental conditions. DNA determines the mating types of many fungi, which influences how they reproduce. Some fungi have complex life cycles that involve sexual reproduction; this interaction between different mating types is controlled by specific genes (Roth et al., 2023).

2.2.2 Ribonucleic Acid

Ribonucleic acid (RNA) is a fundamental molecule in fungi, as it is in all living things. RNA is involved in a number of crucial biological activities in fungi. Transcription, or the conversion of DNA into RNA and RNA into protein, is a crucial step in gene expression. In fungi, there are various kinds of RNA.

2.2.2.1 Messenger RNA

Through messenger RNA (mRNA), genetic information is transmitted from the ribosomes in the cytoplasm to the DNA in the nucleus. During translation, it serves as a template for the creation of proteins. The amino acid sequence of the resulting protein is determined by the nucleotide sequence of the mRNA.

2.2.2.2 Transfer RNA

Transfer RNA (tRNA) is necessary for translating the genetic code from mRNA into a specific sequence of amino acids during protein synthesis. Each type of transfer RNA is associated with a certain amino acid. The tRNA molecules' anticodon regions pair with mRNA's complementary codons.

2.2.2.3 Ribosomal RNA

The key structural and functional element of ribosomes, the cellular structure where protein synthesis takes place, is ribosomal RNA (rRNA). Based on instructions from mRNA, ribosomes, which are made up of both protein and rRNA, catalyze the creation of polypeptide chains from amino acids.

2.2.2.4 Small Nuclear RNA

Small nuclear RNA (snRNA) is involved in the process of RNA splicing. It produces mature mRNA that can be translated into a protein by splicing out introns (non coding regions) from mRNA molecules and joining together the exon coding regions.

2.2.2.5 Small Nucleolar RNA

Small nucleolar RNAs (snoRNAs) are mainly involved in modifying rRNA molecules. They guide the chemical modifications that occur on specific nucleotides in rRNA, which are essential for the proper functioning of ribosomes.

2.2.2.6 Small Interfering RNA

Small interfering RNAs (siRNAs), in some fungi, are involved in post-transcriptional gene silencing. Through the targeting and degradation of complementary mRNA molecules, they can block the expression of particular genes.

2.2.2.7 Micro RNA

Another type of short RNA, called micro RNA (miRNA), can control gene expression by attaching to particular mRNA molecules and either encouraging or inhibiting the translation of these mRNA molecules (Henras et al., 2015).

2.3 STRUCTURE AND ARCHITECTURE OF THE FUNGAL GENOME

The structure and architecture of the fungal genome vary among different fungal species. The architecture of a fungal genome refers to the structural organization and arrangement of genetic material within the genome of the fungal species. Understanding the genetics, evolution, and functional capacities of fungi depends heavily on architecture. Fungal genomes can be as small as 10 Mbp or as large as hundreds of Mbp. In Ascomycota, the average genome size is about 37 Mbp; in Basidiomycota, 47 Mbp; and in Oomycota, 75 Mbp. The smallest fungal genomes of *Malassezia restricta*, *Wallemia mellicola*, and *Wallemia ichthyophaga* are similar in size and number to bacterial genomes. The first fully sequenced eukaryotic genome was that of the extensively studied yeast *Saccharomyces cerevisiae*, which has a genome size of about 12 Mbp. Fungal genomes are significant study models because of their small size, which makes them easier to sequence than most other eukaryotic genomes (Mohanta and Bae, 2015). Fungi are among the most diverse group of organisms; they exhibit various genetic characteristics that distinguish them from other fungi. The arrangement and makeup of the DNA present in fungus cells is referred to as the fungal genome structure. It includes various aspects of the genome, such as the arrangement of the genes, the presence of repetitive elements, and the number and size of chromosomes.

2.3.1 Chromosomes

Unlike fungi, which have a special region of the genome called the mating-type locus that regulates the establishment of cell type identification and variations in DNA sequence between cells of various mating types, other organisms lack this feature; vertebrates have dimorphic sex chromosomes to determine sexual identity. The nuclei of fungal cells normally include linear chromosomal arrangements of the fungal genome. Fungal species have varying numbers of chromosomes. A few chromosomes or even hundreds of chromosomes may be present in some fungi (Fraser and Heitman, 2004).

2.3.2 Haploid and Diploid Phases

The haploid phase, which is the predominant stage in the life cycle of fungi and in which they only have one set of chromosomes (n), is common. When two haploid cells unite during sexual reproduction, a diploid phase (2n) may result. This diploid phase normally lasts for a short time, and meiosis produces haploid spores (Sharp et al., 2018).

2.3.3 Gene Content

Microbial communities interact with their gene content via functional features that can have an impact on the environment or act as a response. To comprehend connections between the makeup and function of microbial communities, it is essential to examine how distinct ecological traits and strategies interact with the environment.

The direct evaluation of the characteristics of microbes colonizing natural substrates is extremely difficult; however, indirect methods based on gene expression and content offer new perspectives. Fungi have regulatory elements, non-coding RNAs, and genes that encode proteins. The complexity of the organism, its ecological niche, and its lifestyle can all have a significant impact on how many genes are present in a fungal genome (Barbi et al., 2020).

2.3.4 Introns and Exons

Many fungal genes contain introns, which are non-coding regions within the gene sequence. A gene's exon–intron structure is an essential component. A wide range of exon and intron lengths occurs, contributing to the complexity and diversity of fungal genomes. In unicellular eukaryotes, the intron length is equivalent to the exon length. Introns are transcribed into precursor mRNA (pre mRNA) but are usually removed through a process called splicing before the mature mRNA is translated, while exons are the coding regions of the genes (Ivashchenko et al., 2009).

2.3.5 Gene Clusters

The intimate interaction of two or more genes participating in a single metabolic or developmental process is a general description of a fungal gene cluster. Additionally, certain fungi have gene clusters that may be in charge of producing secondary metabolites like mycotoxins or antibiotics. On the same chromosome, the gene clusters may be close to one another (Lind et al., 2017).

2.3.6 Telomeres

Telomeres, which are repeated DNA sequences found at the ends of fungal chromosomes, aid in preserving chromosome stability during replication and guard against destruction. Telomeres, which make up the ends of linear chromosomes, are often protein complexes that bind to short, repeating sequence motifs added to the 3′ ends of DNA by the telomerase. One of the main functions of telomeres is to address the "end-replication problem," which, if left unaddressed, can cause the progressive, irreversible attrition of sequences from the ends of chromosomes and ultimately lead to viability loss. The telomere binding proteins also shield the chromosome from 5′ to 3′ exonuclease action and hide the chromosomal ends from the machinery that repairs double-strand breaks (Rahnama et al., 2021).

2.3.7 Mitochondrial DNA

A distinct mitochondrial genome is frequently present in fungus genomes. The circular mitochondrial DNA (mtDNA) of fungi carries the genes required for generating energy and other mitochondrial processes (Williamson, 2002). Depending on the type of fungus species, their ecological niche, and their evolutionary history, different aspects of the fungal genome structure may exist. The complex architecture of fungal genomes continues to be better understood due to developments in genomics.

2.4 EVOLUTION OF THE FUNGAL GENOME

The kingdom of fungi comprises species that inhabit all ecosystems; nearly all environments are suitable for species of fungi. Fungi are multicellular and unicellular organisms with a variety of morphologies that are both free-living and symbiotic. Invading plant and animal cells, cycling nutrients in terrestrial and aquatic ecosystems, and thriving in a variety of settings are all made possible by the genes included in fungal genomes. Individual and massive efforts have been used to create the ever-growing databases of fungal genome sequences (Stajich et al., 2009). It is possible to use genomic databases to learn more about the ecological niches and lifestyle adaptations of fungi. Large fungal genomic and transcriptome datasets have allowed for the use of fresh techniques, revolutionizing the study of fungus evolution from a molecular sequence perspective (Hibbett et al., 2013). Understanding phylogenetic relationships and the relative evolutionary divergence of organisms is necessary for studies of fungal evolution. The earliest methods of classifying fungi into related groups were based on morphological traits. Based on identifiable physical properties such as spore shape and asexual and sexual structures, these strategies were successful in building a comprehensive framework for the arrangement of fungal species for taxonomic classification (James et al., 2006). However, the microscopic nature of many fungi, particularly yeast-forming fungi, and the frequent occurrence of evolution that results in identical traits across a tree have made the taxonomic classification of fungi very difficult or readily misleading. The development of primers that amplify fungi's ribosomal RNA (rDNA), as well as DNA sequencing and polymerase chain reaction (PCR), has opened a new age of molecular phylogenetic research (Moncalvo et al., 2000). The ability to determine the evolutionary relationships among fungi and other species has improved due to technological advancements such as high-performance computing, cost-effective high-throughput genome sequencing, and improved phylogenetic software. Phylogenomic methods have been adopted for fungi, one of the first eukaryotes to profit from these discoveries. Phylogenomics is the process of assembling concatenated datasets or individual gene trees produced from several orthologous genes selected from the genomes of various species to create a comprehensive phylogenetic tree. Fungal genome evolution has been influenced by a number of variables, including environmental stresses, interactions with other organisms, and the inherent genetic pathways that promote adaptability and diversification (Nagy et al., 2014).

2.4.1 Diversification of Fungal Lineages

Our knowledge of the evolution of land plants has greatly improved as a result of the molecular revolution in plant systematics, particularly the understanding of the evolutionary processes underlying changes from autotrophy to full mycoheterotrophy. The lifestyle, ecological niches, and reproductive methods of fungi are remarkably diverse. Due to this diversity, various fungal lineages with unique genomic adaptations have developed (Merckx, 2012).

2.4.2 Adaptation to Different Environments

The ability to compare gene content and sequence alterations based on genome sequencing has made it possible to start predicting the likely molecular bases for characteristics and adaptations. The fungal kingdom exhibits a wide range of interactions between fungi and other creatures, both cooperative and parasitic. In addition to coexisting with plants and animals as commensal companions, such as asymptomatic skin fungi or plant endophytes, fungi can also spread deadly diseases to both plants and animals through a variety of invasive techniques. Some pathogens are opportunistic in nature and only cause diseases to hosts having weakened immune systems. To harm host cells or impair the host's immune system, several pathogenic fungi produce toxins or effector proteins. Overall, fungi have developed to be successful in a variety of settings, including soil, water, plant surfaces, and animal hosts. These adaptations frequently alter gene expression, the creation of certain metabolic pathways, and the emergence of distinctive cell wall components (Stajich, 2017).

2.4.3 Gene Duplication Functional Diversification

In order for organisms to acquire additional genes and develop unique genetic combinations, gene duplication is a critical process. The discovery of multiple unique gene functions as a result of this mechanism has greatly aided in the advancement of developmental programs in a variety of animals. Retroposition, chromosomal or genome duplication, and unequal crossing over are all possible causes of gene duplication. Gene duplication events have played a significant role in fungal genome evolution. These duplicated genes can undergo divergence, leading to the evolution of new functions. This process is necessary for the acquisition of traits that enhance fitness in a specific environment (Magadum et al., 2013).

2.4.4 Rapid Evolution of Pathogens

Genetic diversity among plant infections and diseases has been found to vary significantly, not just between closely related species but occasionally even within the same species (Möller and Stukenbrock, 2017). Numerous fungal diseases with highly dynamic genome architectures have been discovered through comparative genomic investigations within and between species; several of these infections may be related to host specialization or quick adaptability. There is evidence that many of these fast-changing genes have an impact on pathogen pathogenicity. Numerous pathogenic fungi have quickly evolved as a result of interactions with their hosts and the desire to get past the host immune system. As a result, a variety of virulence factors, host manipulation mechanisms, and adaptive responses to host defense mechanisms have all been developed (De Vries et al., 2020).

2.4.5 Symbiotic Relationships

In fungal phylogenies, parasitic and symbiotic species coexist with saprophytes. There are many examples where the relative benefit and harm to the host plant depend heavily on the circumstances. They must infect another organism in order to multiply, consume nutrients from the host tissues, get past the host's defenses, and eventually create propagules that infect more than one new host. Fungi are involved in various symbiotic relationships; mycorrhizal associations with plants and lichen formations with algae are only a few examples of the many symbiotic relationships that fungi participate in. As they need particular genetic adaptations to initiate and maintain symbiosis, these interactions have had an impact on the evolution of fungal genomes (Aguileta et al., 2009).

2.5 TECHNIQUES FOR FUNGAL GENOME SEQUENCING

The success of plants depends on their tight relationships with fungal pathogens, symbionts, and endophytes. Therefore, it is crucial to have a thorough grasp of these interactions in order to establish the ideal conditions for plant growth. In addition to revealing the genomes of plants and fungi linked with plants, genomics also provides a useful method for tracking the shifting patterns of these connections through gene expression and other genome-driven studies. Advanced sequencing technologies generate enormous amounts of data, providing the means for an unparalleled depth of investigation into the complex plant–fungus connections found in nature (Kuo et al., 2014). Genomics uses bioinformatics and DNA sequencing to assemble, annotate, and analyze genomes. Since Tom Roderick first coined the term "genomics" in 1986, it has widened to include a wide range of sequence-based methods, including transcriptomics and epigenomics. In 1975, a method employing DNA polymerase with radio-labeled nucleotides was developed and later used to sequence the 5 kbp genome of a bacteriophage, marking the first fully sequenced genome. The Sanger sequencing method was developed as a result of advancements in DNA sequencing techniques. Following the 1992sequencing of the brewer's yeast *Saccharomyces cerevisiae* chromosome III, a collaborative effort by scientists from various laboratories led to the revelation of its full genome in 1996. This achievement marked the beginning of the field of fungal genomics, as it represented the sequencing of the first eukaryotic genome (Yadav, 2007). There are currently many DNA sequencing platforms, and they differ more in terms of their features. The development of sequencing technologies has made it simpler and more efficient to sequence the fungal genome. The choice of sequencing technique for a fungal genome is influenced by several elements, including size, complexity, cost, and research goals (Schuster, 2008). There are some techniques that are used for fungal genome sequencing.

2.5.1 Whole-Genome Shotgun Sequencing (WGS)

Some pathogenic fungus can infect people, animals, or plants, resulting in the development of disease. As high-throughput sequencing technologies have matured, the cost of genome sequencing has significantly decreased, making it possible to quickly and affordably read fungal genomes. Pathogenic fungi, genes involved in the interaction between pathogens and hosts, and the evolutionary links between closely related species can all be found using fungus whole-genome sequencing. It is the most common approach for fungal genome sequencing. In this sequencing method, the genome is fragmented into very fine pieces, and these fragments are sequenced randomly. The resulting shotgun reads are then assembled using specialized software to reconstruct the whole genome. A complete representation of the genome, including both coding and non-coding sections, is provided through whole-genome sequencing (WGS) (Keepers et al., 2019).

2.5.2 Next-Generation Sequencing

High-throughput sequencing (HTS), also known as next-generation sequencing (NGS), is a technique that can reveal novel variations and mutations in the genome. NGS is performed using a reversible stop-codon determination procedure with PCR and gene chip technologies (Nowrousian, 2010).

2.5.3 Illumina Sequencing

Fungal genome sequencing makes considerable use of Illumina technology for short-read sequencing. Illumina sequencing has been used to evaluate several genomes, making it possible to juxtapose DNA sequences to gain deeper insights into health and diseases. Illumina sequencing is significantly faster and more affordable than alternative sequencing techniques because it generates millions of reads with astonishing precision. With this method, the DNA is divided into smaller pieces for easier processing, typically spanning 200 to 600 base pairs. It generates short reads of good quality, making it appropriate for assembling genomes of relatively modest complexity (Abbas et al., 2014).

2.5.4 PacBio Sequencing (Single Molecule Real Time, SMRT)

Real-time single-molecule DNA sequencing is made possible by Pacific Biosciences' sequencing of a single molecule in real time (SMRT) technology. Long reads produced by this approach are highly valuable for tackling complex genomic regions, repetitive sequences, and structural alterations. The method depends on a polymerase molecule sequentially integrating fluorescently labeled nucleotides during unconstrained DNA synthesis. Zero-mode waveguide nanostructure arrays on newly formed DNA strands enable optical observation volume confinement as well as the simultaneous evaluation of many single-molecule sequencing operations. SMRT can improve genome assembly quality and aid in the identification of large-scale genomic features (Jain et al., 2021).

2.5.5 Oxford Nanopore Technologies (ONT) Sequencing

Long reads are provided via ONT sequencing, which is quite similar to PacBio sequencing. This technology is also used to sequence fungal genomes. Although it may have higher error rates than short-read technologies, it is particularly useful for quickly getting long-range information. By analyzing the current distortion caused by DNA flowing through a nanopore, Oxford Nanopore Technologies (ONT) creatively exploited nanopore technology to identify base sequences. ONT sequencing has developed since its launch into one of the most potent sequencing technologies. It performs quickly and efficiently while offering the benefits of whole-genome sequencing and disease diagnostics (Lin et al., 2021).

2.6 PREDICTION AND ANNOTATION OF GENES IN FUNGI

2.6.1 Prediction

Gene prediction is a basic step in the annotation process. Eukaryotic genomes exhibit complex gene structures and architecture, to navigate this complexity, gene prediction has aided by expressed sequence tags (ESTs), repeat-masking and the identification of homologus from other fungi using various prediction methodologies. Additionally, multiple lines of evidence are employed to validate projected gene models. The phrase "gene annotation" refers to comprehensive data provided about a particular gene and its translated protein products. An uncharacterized gene or protein is often annotated using reference-based or homology-based sequence similarity techniques. These are the most widely used methods for gene annotation and prediction (Taher et al., 2003).

2.6.2 Annotation

Annotating fungal genomes is the first step in genome content analysis. This approach often uses a variety of techniques to identify different parts of a genome assembly, such as pseudogenes, repetitive and transposable elements, and protein-coding and non-coding genes (Stanke et al., 2006). Transcriptome sequences closely resemble genome sequences, and using them for gene annotation has been shown to be quite successful. These sequences clearly define the intron–exon boundaries, which makes them a perfect source of trustworthy data for figuring out the architecture of the genes inside the same organism whose genome is being sequenced. These databases contain a variety of transcript sequences, such as expressed sequence tags (ESTs), which are produced by single-pass Sanger sequencing reads that are positioned at the 5′ or 3′ termini of cDNA clones. Other examples of these transcript sequences include full-length cDNAs (FL-cDNAs) and, more recently, cDNA sequences derived from next-generation short-read transcriptome sequencing (referred to as RNA-Seq). The best substrate for gene structure annotation is FL-cDNA sequences, which ideally encode the transcriptional start site, all of the exons of a fully spliced transcript, and the polyadenylation cleavage site at the 3′ end. All elements of the gene structure, including the open reading frame (ORF) and terminal untranslated regions (UTRs) of exons, are frequently disclosed with precise spliced alignments of FL-cDNAs in the genome (Haas et al., 2002). To predict and annotate genes in fungi, coding sequences within the genome must be located, and functional annotations must be provided to comprehend their probable functions. Understanding the biology, evolution, and pathogenicity of fungi depends on this process. There are some aspects that are involved in the prediction and annotation of genes in fungi.

2.6.3 Gene Sequencing

Getting the DNA sequence of the fungal genome is the first step. Most often, this is accomplished using high-throughput sequencing techniques, which produce millions of short DNA reads that must be assembled into longer contiguous sequences. The prediction and annotation of genes in fungi are both highly dependent on genome sequencing. Because entire genome sequences are now available, it is possible to identify coding sequences and comprehend how the functional components of the genome work (Ter-Hovhannisyan et al., 2008).

2.6.4 Functional Annotation

The prediction and annotation of genes in fungus require a critical step called functional annotation. This entails determining potential roles and functions for projected genes based on a variety of evidence. After a fungal genome's gene structural annotation phase is finished, attention turns to the gene set's functional annotation. The goal is to give gene products names that are mostly based on the functional characterization of the genes in silico (Finn et al., 2010).

2.6.5 Comparative Genomics

Pathogens have developed a variety of lifestyles and taken on crucial new functions in both human habitats and natural ecosystems. Different fungal species' annotated genes can be compared to learn more about evolutionary links, gene gain or loss, and adaptive features. The molecular processes behind their adjustment to new lifestyles are unknown, however. It was decided to use comparative genomics to identify the differential virulence methods of plant ascomycete fungal pathogens with various lifestyles (Wang et al., 2022).

2.6.6 Experimental Validation

Detailed analyses of molecular pathogen–host interactions using reverse genetics have been made possible by the expanding number of fungal and oomycete plant pathogen genomes that have been sequenced (Dean et al., 2012). Extracellular protein identification may be complicated or excluded by incomplete or erroneous gene annotations. The existence and unique labeling of genes' 5′ exons, which carry N-terminal signal peptides, are crucial for the prediction of secretion. A key area of research in plant disease is the exact computer prediction of secreted proteins from full genome sequences. Despite the difficulties brought on by these technical complications, this aims to quickly identify secreted proteins for future effector prediction (Raffaele et al., 2010). Various techniques are used for experimental validation of the prediction and annotation of genes in fungi, and genes can be experimentally validated using techniques like reverse transcription polymerase chain reaction (RT-PCR), functional assays, and quantitative PCR, also called qPCR (Urban et al., 2015).

2.6.7 Data Integration

About 200 genomes have been entirely sequenced or are currently being analyzed, and the number of fungal genomes that have been studied is constantly growing. Only a small portion of these genomes has been thoroughly examined, for instance utilizing functional genomics methods. Comparative analysis has been shown to be a useful method for improving our comprehension of evolutionary biology and the complexity of less-studied genomes. This strategy has improved our understanding of these topics (Hedeler et al., 2007). Systematic comparison studies can be difficult to conduct since the necessary data is frequently dispersed throughout a wide range of different sources. The meticulous fusion of created datasets, which classify genes or creatures in a way that makes it easier to formulate queries illuminating similarities and differences between species, is another advantage of good comparative investigations. The maintenance of genomic data in multi-species databases, as opposed to exclusively depending on model genome databases centered on individual genomes, is essential for advancing the study of comparative genomics. Transcriptomics, proteomics, and metabolomics are just a few of the data types that can be integrated to provide a thorough understanding of gene function (Foissac et al., 2008).

2.6.8 Public Databases

More and more fully sequenced fungal genomes are becoming available, and many more sequencing projects are in the pipeline. Our understanding of several main and secondary metabolisms of fungi, including how they produce advantageous compounds like antibiotics or detrimental ones like mycotoxins, should be improved by this research. Since many fungal genomes' functional annotation is not perfect, especially for genes producing enzymes, we require specialized methods to thoroughly examine their metabolic pathways. Annotated genomes are frequently made available to the scientific community by being deposited in open databases like GenBank, Ensemble Fungi, or JGI MycoCosm (Grossetête et al., 2010). There is no doubt that the accuracy of gene prediction and annotation directly depends on the quality of the genome sequence, the experimental data that is available, and the sophistication of the data that is used in annotation.

2.7 GENES INVOLVED IN FUNGAL PATHOGENESIS IN PLANTS

Different methods are used by fungi to infect their host plants. As more fungi become amenable to molecular genetic research, these tactics are being considerably well understood. For example, methods like labeled mutagenesis have made it possible to identify fungal genes that are essential for disease development. Before these methods were widely used, pathogenicity genes or factors were discovered by comparing naturally occurring pathogenic and non-pathogenic isolates. "Pathogenicity" genes are a hotly debated idea (Idnurm and Howlett, 2001), and the efficiency of transformation technologies and gene mutation has improved, making it simpler to functionally analyze pathogenicity genes presently. As a result, too many fungal pathogenicity genes have been discovered today to count. They are listed in many databases, some of which fungi, are dedicated to specific organisms and relationships while others, are meant to find gene families involved in pathogenicity (Winnenburg et al., 2007). Pathogenic genes are of interest because they could serve as targets for disease prevention in addition to helping us learn more about disease processes in general. Over the past ten years, reviews of fungus pathogenicity processes and pathogenicity genes have been conducted. Numerous new genes that cause disease have been discovered over the past few years. Similar to the rice blast fungus known as *Magnaporthe grisea*, the majority of these genes have been discovered in easily controlled fungi. Depending on how a particular fungus infects its host, different genes are important for developing disease (Aslam et al., 2019). While some fungi use the plant's cuticle and cell wall as entry points, others penetrate the epidermis by building specialized structures like an appressorium. Some fungi, however, use wounds or other naturally occurring openings to penetrate their host (Dufresne and Osbourn, 2001). Fungal infections can cause a variety of plant diseases, which can result in large losses in the agricultural sector. Numerous genes, many of which are linked to the virulence and infection mechanisms of fungi, are involved in fungal disease in plants.

2.7.1 Effector Genes

Effector genes play a critical role in fungal pathogenesis in plants by enabling the fungus to manipulate the host plant's immune response. These genes produce effector proteins, which the fungus secretes into the plant's cells to infect it. The main sources of the complex molecular interactions between the pathogen and the host are effector proteins. Effector proteins are secreted into plant cells by fungal infections in order to influence the host's immune response and spread infection. These effector genes are essential for reducing plant defenses and are frequently specific to particular diseases. *Phytophthora* species, with particular emphasis on *P. cinnamomi*, are associated with specific molecular factors that contribute to pathogenicity despite having limited saprophyte capability, which means they display weak growth and competitiveness in the presence of other soil and plant microorganisms. Notably, NPP1 (*P. cinnamomi* necrosis-inducing protein 1) causes plant necrosis, which kills the plant by causing it to lose its leaves and roots. Glucanases are crucial for the adhesion, penetration, and colonization of host tissues, whereas GIP (glucanase inhibitor protein) dampens the host's defense response. These molecular elements have attracted a lot of interest in figuring out how *P. cinnamomi* causes disease (Winnenburg et al., 2007).

2.7.2 Pathogen-Associated Molecular Patterns (PAMPs)

Every multicellular organism is known for its capacity to recognize unknowns and afterwards launch a protective reaction against the invading disease. When the host species' pattern recognition receptors (PRRs) detect microbe- or pathogen-associated molecular patterns (MAMPs/PAMPs), this reaction is triggered. These patterns are invariant surface features that are crucial to the microbe but do not present in the host, enabling the host to identify them as foreign invaders and ward them off. PAMPs are present in various pathogens and can trigger plant immunological reactions. Examples include chitin, which is detected by plant chitin receptors and is a component of fungal cell walls (Postel and Kemmerling, 2009).

2.7.3 Toxin Production Genes

As low-weight secondary metabolites, host selective toxins (HSTs) typically induce a wide variety of structural changes. In particular, interactions between plants and pathogens serve as effectors by controlling pathogenicity or virulence. There are some fungi that produces toxins that damage plant cells and invade to form disease symptoms. For instance, some isolates of the fungus *Pyrenophora tritici-repentis* carry the "ToxA" gene, which results in tan spot disease in wheat. Seven diseases caused by *Alternaria alternata* have been recognized, and fungal pathogenesis in these diseases is mediated by HSTs (Tsuge et al., 2013).

2.7.4 Signal Transduction Pathway Genes

To infect their hosts, plant-pathogenic fungi produce an infection cell type known as an appressorium. The host plant's particular physical or chemical signals stimulate the creation of the appressorium. Components of signal transduction pathways have been linked to morphogenesis, virulence, and pathogenicity in infections. Signal transduction pathway components' roles in influencing virulence and pathogenicity, as well as in the morphogenesis of infections, have been established. To avoid host defenses, fungi can modify plant signaling pathways, for example, by interfering with plant hormones like salicylic acid (SA) and jasmonicacid (JA), which are essential for triggering defense reactions (Deising et al., 2000).

2.7.5 Secondary Metabolism Genes

By affecting numerous aspects of the host–pathogen relationship, secondary metabolites have a substantial impact on the development of fungal pathogenesis in plants. Secondary metabolite genes are involved in fungal pathogenesis by various means, including the production of toxins, suppression of plant defense responses, manipulation of host cell physiology, formation of infection structures, and so on. Secondary metabolites produced by fungi, such as toxins and other bioactive substances, might influence how plants and pathogens interact. These substances may aid in virulence or host defense suppression (Fox and Howlett, 2008).

2.8 ENGINEERING GENOMES IN FUNGI

The biological characteristics of natural fungi are very useful in manufacturing, agriculture, and medicine. The tremendous diversity of fungal taxa is highlighted by recent metagenomic studies, and the related specialized metabolic diversity promises a valuable and mostly unexplored avenue for natural product discovery. Despite having a remarkable range of chemical compositions and biological functions, fungal natural products have a number of fundamental traits in common with one another. First, biosynthetic gene clusters (BGCs) are groups of genes on a chromosome that encode the various phases of a biosynthetic pathway. Second, these BGCs frequently have discrete distributions among genetically related species and isolates and originate in specific areas of the genome. Third, numerous distinct chemicals are produced by the same enzyme (super) families. Fourth, genes that share a pathway are frequently co-regulated and co-expressed in a variety of growth circumstances (Skellam, 2019). The importance of synthetic biology will increase with the discovery of novel substances and the reengineering of their pathways. New natural products and the biosynthetic gene clusters (BGCs) that are connected to them will both experience an acceleration of discovery made possible by developments in computation and experimentation (Mattern et al., 2015). It is significant to note that the synthetic biology-based refactoring of BGCs (Shao et al., 2013; Smanski et al., 2014) allows for access to fungal strains that are difficult to grow and may allow for the characterization of chemicals made from clusters that are now only detected through metagenomic assays. A number of heterologous expression mechanisms have previously been developed, and they have successfully produced a number of chemicals outside of their original host (Chiang et al., 2013). Such techniques enable the efficient separation of BGCs from metabolically complex animals that code for cryptic pathways in order to study them in a neutral environment (Rutledge and Challis, 2015).

2.8.1 Manipulating Global Transcription

The transcription of these silent BGCs may be increased by inhibiting the histone deacetylase (HDAC) enzyme because many of them are found in transcriptionally reserved portions of the genome in fungi (Asai et al., 2013). Two new cyclic peptides and two new meroterpenoids were identified, and their biological functions were still unclear when the HDAC was deleted from 42 of 68 BGCs. *Calcarisporium arbuscula* usually exclusively produces the ATPase suppressors aurovertin B and D (Mao et al., 2015). In contrast, an unexpected result emerged from the HDAC rpdA knockdown in *Aspergillus nidulans* during metabolomics analysis: 61 molecules simultaneously showed a 100-fold increase in relative expression, while 47 compounds had a 100-fold decrease. This emphasizes how unpredictable using pleiotropic methods is by nature (Albright et al., 2015). Eight of the most highly expressed biosynthetic gene clusters (BGCs) were removed from a genetic replication strain. In order to prevent the re-identification of known compounds in *A. nidulans*, these BGCs were responsible for the removal of 244 kbp from the genome. This method allowed for the initial isolation of aspercryptin, which was followed by investigations that identified the regulatory gene mcrA as being in charge of several gene clusters. The antibiotic felinone A and two new forms of the phytotoxin cichorine were discovered as a result of the deletion of mcrA. Over 10 BGCs produced more natural compounds as a result of this deletion-induced change. Numerous fungi have been reported to include mcrA homologs, and deletions of the related genes in *Aspergillus terreus* and *Penicillium canescens* similarly altered their Natural products (NP) profiles, although the changes have not yet been fully understood (Oakley et al., 2017).

2.8.2 Pathway Engineering

By replacing native promoters with inducible or constitutive promoters or by overexpressing the pathway transcription factor, individual BGCs can be activated, which may stimulate dormant BGCs or increase the concentration of a particular natural product (NP). Pox, a transcription factor with properties similar to Gal4, was over expressed in *Penicillium oxalicum*, which led to the synthesis of oxaleimides, some of which were cytotoxic. A novel type of biosynthetic enzyme relationship was discovered after new research, revealing new engineering options for NP production (Sato et al., 2017). Because of the overexpression of the pathway transcription factor fsqA, fumisoquins were found in *Aspergillus fumigatus* (Baccile et al., 2016). Surprisingly, other fungal species share identical fumisoquin BGCs, and their production is comparable to that of plant-derived isoquinolines (such as codeine, morphine, and berberine). Cordyceptin and its protective molecule pentostatin are two adenosine analogs that are encoded by the *Cordyceps militaris* BGC. The titers of cordyceptin were raised by the selective overexpression of cns1/cns2 fusion, whereas the titers of pentostatin were increased by the overexpression of cns3. Similar to this, the kojic acid pathway genes kojA and kojT were over expressed in *Aspergillus oryzae* utilizing self-excising Cre/loxP marker recycling to increase the production of a cosmetic ingredient (Zhang et al., 2017)

A technique for prioritizing the setup of quiet BGCs is called resistance-guided genome mining, and the discovery of the function of a self-resistant gene, which is also present in the BGC, is the basis of this idea. Consider the gene inpE, a possible proteasome subunit, as an example. Since it was discovered in *A. nidulans*'s buried inp cluster, it is possible that this cluster is where a proteasome inhibitor is encoded. The BGC activates its own expression as the genes in the inp cluster are gradually replaced with a controlled promoter, which makes it possible to identify the proteasome inhibitor fellutamide B (Yeh et al., 2016). It is now possible to find uncharacterized intermediate metabolites by using targeted gene inactivation. This is accomplished by either stopping the production of the pathway at a certain junction or by correlating a biosynthetic gene cluster (BGC) with a recognized natural product (NP). Due to fungi's propensity for frequent non-homologous end joining (NHEJ), targeted gene disruption in these organisms has historically been difficult. However, a number of methods have been developed to increase the effectiveness of gene-targeting techniques. The first fungal ribosomally synthesized and post-translationally modified peptide (RiPP) BGC was discovered by the use of the Cre/loxP marker recycling system and heterologous gene expression. In *Ustilaginoidea virens*, this cluster, which expresses ustiloxin B, was found (Ye et al., 2016). More biosynthetic pathways have been comprehensively characterized using gene disruption approaches. The pathways utilized by the mycotoxin cercosporin, the halogenated mycotoxin asirochlorine, and the protein phosphatase-2 inhibitor rubratoxin A are among them. Numerous fungi have been reported using CRISPR-Cas9 genome editing, which can be used to construct pathways by inactivating genes. The unknown fungus *Talaromyces atroroseus* produces ZG-1494a, an inhibitor of PAF acetyl transferase. Since there were no genetic tools available, the BGC encoding ZG-1494 was found using CRISPR-Cas 9. By deleting a single-nucleotide insertion, CRISPR-Cas9 was also employed to restore the manufacturing of the anti-protozoal NP trypacidin in an *A. fumigatus* strain. This nucleotide insertion caused a premature stop codon and frameshift, which prevented the polyketide synthases (PKSs) that make trypacidin function properly. The pace of finding, researching, and engineering fungal NPs has quickened with the emergence of heterologous expression platforms. Transformation techniques and host selection are influenced by selection markers and technology for multiple gene expression, even though fungal NPs are now usually produced in a variety of expression hosts. Furthermore, fungi have been produced to serve as a platform for the heterologous production of essential pharmaceuticals generated from plant resources (Shi et al., 2017).

2.8.3 Combinatorial Biosynthesis

This is a technique for creating unique BGCs by adding genes from various biosynthetic pathways. By combining unique or previously undetected components, cooperative enzyme activity makes it possible to synthesize novel non-natural nanoparticles, sometimes known as "unnatural" nanoparticles. For instance, when IDT cyclases from several fungi and enzymes from the indole diterpene (IDT) pathway worked together, previously unreported nanoparticles were produced. Similar to this, 20 glycosylated polyketides were produced by the co-expression of several fungal PKSs and a gene pair encoding glycosylated transferase and methyl transferase from *Beauveria bassiana*. The polyketides showed increased cytotoxicity and better solubility (Xie et al., 2018). By introducing the compacting pathway of *Penicillium chrysogenum* DS50662, the common cholesterol-lowering drug pravastatin was generated by introducing an engineered P450 from the bacterium *Amycolatopsis orientalis* in 6 g/l of pravastatin. A PKS, a phenyl transferase, and a halogenase from the fungus *Stachybotrys bisbyi*, as well as a daurischromenic acid synthase from the plant *Rhododendron dauricum*, were used to create an artificial pathway that produces (+)-5-chlorodaurichromenic acid. The combined NP becomes more effective against bacteria than the parts of it individually (Okada et al., 2017).

2.9 UNDERSTANDING METAGENOMICS

Major participants in the ecosystems of Earth are microbial communities. In addition to ecology and evolution, the study of their genomes may be crucial for the identification of novel, possibly beneficial enzymes and metabolites. Metagenomics is the study of the complete genetic content of microorganisms living in a given environment using methods and processes independent of culture (Daniel, 2005). Without the need for clean cultures, metagenomics is a molecular technique that analyzes DNA from ambient materials to examine the population of microorganisms present. Functional metagenomics enables the high-resolution genomic investigation of non-cultivable bacteria and correlates the genomes with specific environmental roles. Recently, the subfields of metagenomics known as metatranscriptomics and metaproteomics emerged to permit more in-depth functional analyses of microbial populations. Metaproteomics is a complement to metagenomics, which aims to characterize microbiota proteins. Using an integrative omics method, detailed information about community genes, RNA, proteins, and metabolites may be obtained (Ghosh et al., 2019). Metagenomics investigations have grown in popularity as a result of next-generation sequencing techniques. Metagenomics, which was initially used to describe microbiomes, is now being offered as a diagnostic tool in clinical microbiology; however, its use is still only being studied. NGS can be used for metagenomics analysis in two different ways: focused metagenomics, which focuses on an amplified region of interest (such as the 16S region), or shotgun metagenomics, which amplifies every sequence in a sample without making any assumptions about what it contains (Lavezzo et al., 2016). There are three generations of DNA sequencing methods that are being employed by the scientific community. The use of the chain termination method and read lengths of 500–1000 base pairs (bp) are the tenets of first-generation methodologies. Pyrosequencing, sequencing by synthesis, and sequencing by ligation are examples of second-generation approaches, commonly referred to as next-generation sequencing (NGS), with read lengths between 50 and 500 bp. The read length of third-generation sequencing techniques, sometimes referred to as third-generation technologies (TGS), is measured in tens of kilo bases (kb). These techniques sequence single molecules. The first-generation sequencer was Sanger; the second-generation sequencers were Roche, Illumina, and SOLID; and the third-generation sequencers are SMRT and Oxford Nanopore Technologies (Wilson et al., 2014).

2.10 APPLICATIONS OF METAGENOMICS IN FUNGI

Fungi are important representatives of the eukaryotic domain of life. Some species have significant economic value because they produce antibiotics, ferment foods like bread and beer, and break down cellulose. In recent years, the emphasis of research has steadily switched from the study of isolated species to the investigation of how those species relate to one another in settings more like those that are present in their natural niches. This tendency is reflected in the growing usage of nucleic acid sequencing from environmental samples without any previous understanding of the species present. While the terms "microbiome" and "microbiota" relate to the collection of bacteria present in any given environment, "metagenomics" is defined as the study of the genetic material contained in the microbiota. Metagenome and microbiotae are concepts that are frequently used interchangeably (Ursell et al., 2012).

2.10.1 Detection of Pathogens

Metagenomics has been used for the detection of pathogens in samples. In the majority of studies, comparisons of metagenomics to the conventional culture approach have been studied. Metagenomic analysis has made diagnoses that could not have been made by culture or conventional molecular methods. Two methods are sequence based and are frequently used to classify the fungi that make up a mycobiome. The most common method used involves sequencing the internal transcribed spacer (ITS) segments within rRNA operons after PCR amplification, particularly focusing on ITS2 which is positioned between the 5.8S and 28S genes. In fungal research, they have been chosen them as the industry standard barcode for identifying their species. Plutof, Clotu, PIPITS, and CloVR-ITS are just a few of the pipelines that have been created to identify different fungal species and determine how common each species is. Using these tools, researchers can analyze fungal communities more efficiently, gaining insights into biodiversity and ecological dynamics. Additionally, ITS data, Mothur, and Qiime, amplicon readings for fungi and also for bacteria can be utilized to further enhance the understanding of microbial communities and their interactions in various environments (Caporaso et al., 2010). The second method uses shotgun metagenomes to identify species. Most solutions combine exclusive databases with search methods like BLAST, USEARCH and UBLAST, GhostX, and DIAMOND. With the help of software, database sequences can be found that resemble metagenomes. An approach is to use algorithms similar to KAIJU and Kraken to give readings to a lowest common ancestor (LCA) (Wood and Salzberg, 2014). While KAIJU interprets readings and aligns them with a reference protein database, Kraken aligns nucleotide queries to a nucleotide database. For the use of exact k-mer matches relatively more time-consuming alignment-based techniques are often required for other methods.

2.10.2 Identification of Fungi in Metagenomics Datasets

Using the EBI Metagenomics database, 57 metagenomics datasets from the "Host-associated Mammals" collection and 13 additional datasets from the MG-RAST database were selected and subjected to the Find Fungi v0.23 method. There were 2.5 billion readings among the 70 datasets, and 39 datasets contained 77 fungal species that Find Fungi predicted could be present. All 77 species readings in the sample were calculated and aligned to the NCBI nt/nr database with BLAST to see whether they included any false positive predictions (Botelho et al., 2012). These genome assemblies most likely contained contaminants that are excluded from the prediction, including *Puccinia arachidis*, *Amauroascus mutatus*, *Allomyces macrogynus*, *Chryosporium queenslandicum*, *Byssoonygena ceratinophila* and *Amauroascus niger*.

2.10.3 Identification of Potentially Pathogenic Fungi

To identify potentially pathogenic fungi, a variety of clinical, laboratory, and diagnostic methods are used to ascertain whether a fungal infection is present as well as the particular fungus that is causing it. Fungi identified in 16 datasets were associated with human fungal infection, notably *Candida* species. Only one dataset contained an unidentified *Candida albicans*, which is the main common *Candida* species in human fungal infections. Pathogenic fungi that effect plants have also been found in other environments; *Aspergillus niger*, responsible for the black mold on fruits and vegetables, has been identified in the mouse microbiome. A barley-infecting fungus named *Ustilago hordei* was projected to be the source of 122 readings from a sample of cow faces. The linked vertebrate microbiomes of a mouse and an unidentified vertebrate were discovered to have the associated to grain pathogens *Ustilago esculenta* and *Ustilago maydis*. *Verticillium tricorpus* (opportunistic plant pathogen), *Colletotrichum gloeosporioides*, *Phialocephala subalpine*, and *Rhizoctonia solani* were among the other plant pathogens that were discovered (Schlegel et al., 2016).

2.11 ECOLOGICAL NICHES OF FUNGI

Fungal niches, as they currently exist, are larger than we previously believed and are the outcome of evolution. This is predicated on fungi's evolutionary propensity to change ecological niches. While some individuals or animal species might stay in both the old and the new niches over time, others may not enter the expanded niche or may only partially transition to it. For instance, *Metarhizium* was discovered to be both an insect parasite and a plant endophyte (Behie et al., 2015). The Clavicipitaceae family repeated evolutionary changes from animal to plant hosts (Spatafora et al., 2007). While certain clavicipitaceous species are firmly connected with either animals or plants, some species are "lost in translation" and colonize hosts from both kingdoms. Numerous earlier examples, particularly those provided, are pertinent in the context of the transition to a mycorrhizal habit, a scenario that has frequently occurred during the history of fungi (Kohler et al., 2015). Two possible paths are recommended: first, it is possible to directly develop from saprotrophic soil ancestors by going backward in time and losing many of the genes required for a protrophic free life. The second possibility is that the mycorrhizal habit originated from ancestors that can already colonize living host plant tissues as endophytes. The "waiting room" theory has been used to describe the latter route (Selosse et al., 2009; Van der Heijden et al., 2015). According to this theory, root endophytes serve as a symbiotic "waiting room" in which biotrophic coexistence encourages evolution toward a closer mutualism with a more intricate cooperative mycorrhizal morphogenesis. The latter approach appears relevant for mycorrhizal fungi within ericoid hosts and orchids. *Sebacinales*, which evolved from endophytic forebears to develop ectomycorrhizal behavior and took on ericoid and orchid mycorrhizal tendencies, shows continuity on the fungal side of the tree of life. The idea of a "waiting room" could provide insights into the explanation of this phenomena, suggesting that certain ectomycorrhizal species, such as *Tricholoma*, preserve their native habitat. The existence of saprotrophic *Mycena* species as mycorrhizal fungi in orchids or *Ericaceae* plants, which are known for their endophytic character, may also be explained by this idea. The possibility of the opposite tendency, in which mycorrhizal properties promote endophytic colonization in non-mycorrhizal hosts, cannot be completely ruled out. Both situations would result in the surveillance of two niches as a natural course of evolution. Among fungi that colonize the aerial parts of plants, most *Ascomycetes* species, which are both litter decomposers and endophytes, show a noticeable convergence in the transition from saprotrophs to endophytes or vice versa. The prevalence of roots with facultative endophytes, which also serve as soil saprotrophs, like those of *Microthlaspi* (Brassicaceae) or *Populus*, may be explained by repeated transitions from saprotrophs to endophytes (Bonito et al., 2016). It is noteworthy that the transition from neutral or harmful endophytes to mutualistic endophytes is particularly important for plant physiology among endophytes of aerial tissues. Plants provide a potent selective pressure—vertical transmission—when considering the evolution of mutualism in entophytic interactions. Seeds can be colonized by mother fungus during development in other plants, infecting seeds, fruit envelopes, or even the embryo itself, in contrast to animals, where newborns are axenic from birth (Tobias et al., 2017). Vertical transmission typically favors mutualism, since any partner that causes harm to its symbiotic partner will result in offspring engaging with inferior partners, whereas any partner that is more altruistic will increase the quality of companions produced for its own offspring (Bebber and Chaloner, 2022). For instance, several parasitic fungi cause damage to seeds by generating spores for horizontal transmission, and protective endophytes with many favorable effects emerge as a result of vertical transmission in grasses. From such pathogenic predecessors, the mutualistic *Neotyphodium* emerged from the loss of sporulation and autonomous proliferation by means of vertical transmission (Selosse and Schardl, 2007). *Fusarium graminearum* is observed to be vertically transmitted with native North American grasses, which can account for the lack of mycotoxin production and pathogenicity in the hosts. On the other hand, the change to new hosts (the introduced grasses) has led to harmful interactions because there had not been any prior co-evolution or effective vertical transmission. Depending on the mechanism of transmission, a given species may have a distinct effect on the host, and an evolutionary trajectory that involves an increase in vertical transmission can favor mutualism.

According to recent studies on fungal genomics, three Leotiomycetes (Ascomycota) species, *Meliniomyces bicolor*, *M. variabilis*, and *Rhizoscyphus ericae*, have another ecological niche. They form a specific type of mycorrhiza with some *Ericaceae* plants called an ericoid mycorrhiza. The analysis of these genomes supports patterns found in *Oidiodendron maius*, another sequenced ericoid mycorrhizal leotiomycete. Specifically, this ericoid mycorrhizal fungus possesses several enzyme-coding genes involved in the breakdown of polysaccharides, proteins, and lipids which are similar to those found in related saprotrophic species. This may account for their long-term saprotrophic survival as well as the amazing degradation capacities that have been noted during in vitro cultivation. There are a few traits that they share with mycorrhizal fungi. These include the increase in the number of short secreted proteins (SSPs), which are essential for biotrophic interactions, as well as the gene family responsible for nutrient intake and exchange additionally. According to gene expression studies, between 10 and 20% of these SSPs are activated during mycorrhizal interaction, a percentage that is comparable to that of fungus belonging to other mycorrhizal types. Since the majority of phosphorus (P) and nitrogen (N) in soils with gradual organic matter turnover exist as organic compounds, the capacity of ericoid mycorrhizal fungi to breakdown organic matter becomes crucial for promoting host plant adaptation.

2.12 TECHNIQUES FOR STUDYING FUNGAL COMMUNITIES

In the evaluation of fungal microbial diversity, a molecular method has been developed for complex communities. DNA-based techniques are able to highlight the variety of fungi present in ecosystems and have the potential to provide accurate and quick detection (Saad et al., 2004). Internal transcribed spacer (ITS) regions and variable ribosomal DNA (rDNA) are used for the species-level molecular identification of fungi. The non-coding ITS region, which is composed of the nuclear rDNAs ITS1, 5.8S, and ITS2, is widely distributed in the fungal genome. Due to its large copy number, this characteristic produces a very sensitive assay when used as the amplification target. These areas experience fast evolutionary changes, in contrast to the rRNA genes' conserved coding portions, leading to significant sequence difference among closely related species. Therefore, compared to those from coding areas, DNA sequences from the ITS region typically offer a higher level of taxonomic resolution (Anderson et al., 2016; Lord et al., 2002). Furthermore, the ITS region's DNA sequences are very varied and may act as taxonomically distant group indicators. Molecular techniques began to be used for the examination of identification and classification issues as a result of the limitations of conventional methods. In order to address different aspects of fungal diagnosis, numerous molecular methods have evolved into significant tools. These techniques consist of immunological techniques, PCR technology, and nucleic acid-based probe technologies. The latter approach, which relies on genotypic characteristics, delivers outcomes more quickly, efficiently, and potentially effectively than the other approaches, which are based on phenotypic traits. Contrary to standard methods, culture is not required for the isolation of organisms.

Enzyme-linked immunosorbent assay, an immunological technique, is used for the quantification and detection of fungi when morphological characteristics are not evident. This immunological technique uses the antigen–antibody response in conjunction with a fluorescent dye or enzyme to get results that are both quick and easy (Bouasria et al., 2012). The development of antibodies has generally been successful against viruses, but it has been less successful against complex organisms like fungi. However, making highly specific antibodies is expensive and complicated. These techniques only consistently function at the genus level, among other restrictions. In diagnostic applications, specificity and accuracy are crucial, which may limit the use of immunological techniques. Southern hybridization, in situ hybridization, fluorescent in situ hybridization, microarrays, and macroarrays are all molecular techniques that use DNA/RNA probe technology (Aslam et al., 2022). The field of isothermal amplification technology is exemplified by the methods known as loop-mediated isothermal amplification (LAMP), rolling circle amplification (RCA), and nucleic acid sequence-based amplification (NASBA). Reverse transcriptase (RT)-PCR, multiplex PCR, nested PCR, real-time PCR, and DNA barcoding are examples of PCR technology (Borman et al., 2008).

2.13 FUNGAL COMPARATIVE GENOMICS

The ability to induce infection in plants is a characteristic that many members of the fungus kingdom share. However, during the evolution of fungi, plant-pathogenic lineages often emerge independently. As a result, different plant pathogen lineages presumably gained the capacity to transmit disease in various ways. Comparative genomics tools are extensively used throughout the fungal kingdom to provide information on the mechanisms regulating the evolution of pathogenicity. Creating high-quality genome assemblies is one of the most important first stages in comparative genomic analysis (Gibriel et al., 2016). A complete genome assembly, specifically an assembly without gaps stretching from one telomere to the next, is necessary for accurate genome annotations that reflect genome gene content and for identifying certain breakpoints across homologous chromosomes of closely related species. For the purpose of identifying evolutionary originality, like recently evolved effector genes, accurate gene predictions are especially essential. Biotrophic and endophytic fungi require living plant cells to survive and receive nutrients, while saprophytic and necrotrophic fungi feed on the remains of dead plant tissues. As a result, several fungus species have developed unique gene sets to successfully carry out their mode of infection. Comparative genomics is a potent approach to identify specific gene functions associated with fungal lifestyles. A surprising finding was made when 89 different fungus species with different lifestyles had their genetic compositions examined: saprotrophs, necrotrophs, and hemibiotrophs had considerably higher concentrations of the enzymes that broke down plant cell walls. On the other hand, genomes encoding secreted proteins were remarkably plentiful in obligate biotrophs and their symbionts. Compared to hemibiotrophic and necrotrophic fungi, biotrophic fungi exhibit a larger percentage of species-specific predicted effectors (Kim et al., 2016). Comparative genetic analysis was performed on two ascomycetes and 10 basidiomycetes. *Melampsora laricipopulina* and *Puccinia graminis* f. sp. *tritici* received the most attention. These two rust species, which cause poplar leaf rust and wheat/barley stem rust, respectively, are obligate biotrophs. Additionally, biotrophic fungi frequently have fewer carbohydrate active enzymes (Zhao et al., 2014). Among the biggest fungal genomes are those of these obligate pathogens. Transposable elements, as well as large gene families encoding copper/zinc superoxide dismutase, oligopeptide and amino acid transporters, helicases, kinases, and transcription factors, all contribute to the expansion of genomes. A lineage-dependent mechanism governs the expansion of gene sets responsible for creating compact secreted proteins. These putative effectors, which show significant plant regulation, might contribute to host specialization. Due to diminished polysaccharide-degrading enzymes and impairments in nitrate and sulfate absorption routes, both of the rust species under investigation developed an obligate biotrophic lifestyle. It can be difficult to classify different fungus species into a clear infection lifestyle group. For instance, *Zymoseptoria tritici* was once thought to have a hemibiotrophic (coexisting with necrotrophic) existence. According to recent theories, *Z. tritici* mostly functions as a necrotrophic fungus with an early latency period. Similar to this, *Ramularia collo-cygni* was formerly thought to be an asymptomatic endophyte of barley. However, current research has revealed that the fungus is capable of causing leaf senescence and necrotic lesions (Walters et al., 2008). Genomes of fungi that cause plant diseases can be compared to learn more about lineage-specific innovations and mechanisms for pathogenicity. In contrast to non-pathogens, the research discovered that plant-pathogenic species do not share any special gene sets. This shows that variables with universal pathogenicity might be uncommon. However, the genomes of pathogens are broadened to include protein families that may participate in pathogenic activities like plant cell wall disintegration, toxin production, the generation of reactive oxygen species, and detoxification (Soanes et al., 2007).

Numerous genes participate in host–pathogen interactions under ongoing selection pressure, which leaves traces in the genomes of both the pathogen and the host (Aguileta et al., 2009). The assessment of the ratio between the rates of synonymous and non-synonymous substitutions (dN/dS) is commonly used to detect positive selection. The direction and strength of selection are both indicated by the dN/dS ratio. Genes implicated in virulence are expected to show signs of positive selection as a result of the "arms race" that plant-pathogenic fungi engage in with their host plants. In *Puccinia graminis*, for example, selection analysis shows that pathogen-associated gene families are highly polymorphic and almost separate families under positive selection (Upadhyaya et al., 2015). It is possible to pinpoint the genes in *Z. tritici* that are involved in adaptation to wheat by contrasting the genome to two closely related species that infect wild grasses. There were found to be 27 positively selected genes in all, but none of them has a known function (Poppe et al., 2015). The trademarks of recent, strong selection are frequently missed by analyses of selection among various species. Different approaches are needed for intra-specific assessments of selection, which are covered herein. Recent research has demonstrated that hybridization results in the appearance of novel plant pathogens. For instance, *Blumeria graminis* produces powdery mildew in triticale. There were 46 isolates from four different formae speciales that were studied, and their genomes were sequenced by Menardo et al., in (2016). *B. graminis* f. sp. *triticale* adapted to triticale is produced by a hybridization event between the two formae speciales *B. graminis* f. sp. *tritici* and *B. graminis* f. sp. *secalis*. Genome-wide analyses for nucleotide diversity in *B. graminis* f. sp. *triticale* isolates revealed that large chromosomal segments are either nearly similar or very varied. However, in isolates of other formae speciales, nucleotide diversity is lower. The study by Menardo et al. (2016) provides an exceptional example of "mirrored" development in both the host, which evolved from the cross breeding of wheat and rye, and the pathogen, which evolved from the cross breeding of *B. graminis* f. sp. *tritici* and *B. graminis* f. sp. *secalis*. Additionally, the hybridization of two distinct *Z. tritici* lineages led to the emergence of the species *Z. pseudotritici* (Stukenbrock et al., 2012). About 380 sexual generations have passed since the hybridization event. These illustrations show how hybridization occurrences can contribute to the emergence of diseases.

2.14 CHALLENGES IN FUNGAL GENOMICS AND METAGENOMICS

Sequences are the primary source of data for investigations of molecular evolution, and we now have practically unlimited access to them. Recently, the roots of the fungal tree of life (fTOL) have evolved, but only a small portion of its lineages have been sampled thus far, and our knowledge of the earliest branching patterns is still incomplete. We may anticipate that we will have representation for the main fTOL lineages sooner rather than later because it is likely that the solution to the missing lineage problem will just require patience and persistence. But it is also possible that this will not be enough on its own to get rid of every problem with the fTOL. There are many issues with phylogenomic analysis that cannot be resolved by simply collecting more data (Jeffroy et al., 2006), and fixing specific fTOL issues will undoubtedly be difficult for years to come. Understanding how crucial ecological and morphological changes in mushrooms, such their transition to terrestrial habitats and the development of multicellularity, have been impacted by evolutionary changes in their genomes over time is a difficulty related to understanding the evolutionary history of fungi (Kiss et al., 2019). We are able to analyze evolutionary links at increasing resolutions as the number of genomes that have been sequenced rises. Resolution does not always imply solution, such as when enlarging a fractal image. For instance, the idea of a pan-genome for fungal organism becomes increasingly relevant as additional genomes for a particular species become accessible allowing researchers to explore the full-genetic diversity and functional potential withing that species (Naranjo-Ortiz and Gabaldón, 2019). Furthermore, there is still disagreement about the best method for classifying species, and access to more genomes has shown complex patterns of species linkages where introgression and hybridization seem to be frequent (Gabaldón, 2020). Therefore, among the urgent problems we face is improving our understanding of the processes of genome evolution, adaptation, and speciation, as well as developing a new framework for conceptualizing the idea of fungal species.

2.14.1 The Ecology of Fungi

Fungi are an essential part of microbial communities that exist in a variety of habitats, including on or within plants and animals that are not isolated entities (Peay et al., 2016). The study of microbial ecosystems has been transformed by metagenomics, a branch of genomics that focuses on analyzing nucleic acids isolated from complex habitats. Its applications in agriculture, environmental studies, human and animal health, and industry are diverse. But current approaches still primarily concentrate on the bacterial component of the microbiome, ignoring the pervasive but challenging-to-research mycobiome, which is composed of fungi. Compared to bacteria, fungi are often present in smaller quantities, but it has been demonstrated that they play crucial roles in many microbial ecosystems. Our understanding of the microbial environment is severely constrained by its limited proportions, the presence of cell walls, and the lack of universal markers and reference databases, which presents a tremendous obstacle to research. For metagenomics to reach its full potential, these obstacles must be removed. In the upcoming years, exploring the as yet unexplored territory of fungal functions in many ecosystems and their relationships with other creatures promises to provide insightful information.

2.14.2 Functional Genomics and Other "Omics"

Depending on how much we can connect a sequence to functional data, we can extract knowledge from it. Assigning roles to the increasing number of genes identified by genome sequencing that have not yet been characterized is one of biology's biggest difficulties in this area. A rising variety of in silico methods enables making more educated assumptions about what a gene's potential function might be. Numerous developments in high-throughput technologies to screen for gene function and symptoms in more complex contexts are also revolutionizing the experimental characterization of genes (Sharon et al., 2018). The use of several large-scale data formats in conjunction with genomics is expanding. These include "omics" techniques such as transcriptomics, proteomics, metabolomics, epigenomics, phenomics, and others. These data categories stand out because of their depth and focus on numbers. As a result, there is an increasing need to integrate these various data formats. Traditionally, this has been done by using the genome as a central focus and creating links that make sense between various data kinds and certain loci. Despite advancements, there is still a pronounced preference for a small number of fungal species models that do not adequately capture the broad range of fungal diversity.

2.15 CHALLENGES

Although there have been notable advancements in fungal genomics and metagenomics, a number of challenges still exist. The vast array of fungi, many of which remain unidentified, is a significant obstacle to thorough research on fungal genetics. Comparative genomic investigations are further complicated by the lack of reference genomes for many fungal species. The interpretation of data becomes more challenging due to factors such as structural complexity, genomic heterogeneity, and fungal genetic diversity. Furthermore, in order to account for the unique characteristics of fungal cells, extraction and sequencing technologies must be continuously improved. Determining the functional significance of different species within complex fungal communities and analyzing the environmental influences impacting fungal genetics are still challenging tasks. As fungal genomics develops, ethical and legal concerns about data exchange and biotechnology uses must also be addressed. In order to fully utilize genomics and metagenomics in understanding the workings of the fungal world, several obstacles must be overcome.

2.16 FUTURE PROSPECTS OF GENOMICS AND METAGENOMICS IN FUNGI

Fungi has a bright future in "omics" science. The metabolic fluxome, interactome, and metabolome will provide new approaches to the study of fungi and their use in biotechnology. For instance, we will have a greater understanding of the years-long evolution of fungi and how it has affected both the environment and us. Scientists can influence evolution by predicting how it will proceed in the coming years. Fungi will be used as bioreactors in medical mycology to convert drugs or produce vaccines. Agriculture, the food industry, and biotechnology for the manufacture of biofuels will all be transformed by using fungi as a vector for the creation of enzymes. We will have a better understanding of how fungi interact with organisms as either enemies or helpers. Understanding how fungus interacts with plants and modifying their genetic makeup can increase crop output. Finally, in order to reduce the cost of producing numerous essential compounds for use in medicine, we will use fungi as models for the interactions of various proteins in biotechnological applications. So, in light of that, fungi have been utilized as sources for antibiotics and enzymes as well as significant plant pathogens and devastating human diseases. Furthermore, several kinds of fungi serve as model organisms to study the regulation of gene expression as well as the expression of therapeutic proteins. Since the development of fungal molecular biology in the early 1960s, there have been more opportunities to investigate fungi as biotechnological agents. In order to accomplish this, fungal researchers use DNA modification to attain concepts from genetics and biochemistry, resulting in the manipulation of metabolic pathways and the creation of novel compounds for human use. Fungal genomics is a field of study that focuses on identifying and characterizing every sequence in the complete genome. The discovery of the genome sequence is only the start of genomics.

Moreover, it is necessary to examine gene functions (functional genomics), compare genes to those of other organisms (comparative genomics), study expression profiles (transcriptomics), and finally examine the proteins' "expression profiles" (proteomics), as well as creating references for the comprehensive study of low-molecular-weight compounds in the fungal cell.

REFERENCES

Abbas, M.M., Q.M. Malluhi and P. Balakrishnan. 2014. Assessment of de novo assemblers for draft genomes: A case study with fungal genomes. *BMC Genom.* 15:1–12.

Aguileta, G., M.E. Hood, G. Refrégier and T. Giraud. 2009. Chapter 3 genome evolution in plant pathogenic and symbiotic fungi. *Adv. Bot. Res.* 49:151–193.

Albright, J.C., M.T. Henke, A.A. Soukup, R.A. McClure, R.J. Thomson, N.P. Keller and N.L. Kelleher. 2015. Large-scale metabolomics reveals a complex response of *Aspergillus nidulans* to epigenetic perturbation. *ACS Chem. Biol.* 10:1535–1541.

Anastassiou, D. 2001. Genome Signal Processing. *IEEE Signal Processing*. 18(4):8–20.

Anderson, J.P., J.K. Hane, T. Stoll, N. Pain, M.L. Hastie and P. Kaur. 2016. Proteomic analysis of *Rhizoctonia solani* identifies infection-specific, redox associated proteins and insight into adaptation to different plant hosts. *Mol. Cell. Proteomics*. 154:1188–1203.

Asai, T., T. Yamamoto, N. Shirata, T. Taniguchi, K. Monde, I. Fujii and Y. Oshima. 2013. Structurally diverse chaetophenol productions induced by chemically mediated epigenetic manipulation of fungal gene expression. *Org. Lett.* 1513:3346–3349.

Aslam, H.M.U., M.L. Gleason, A. Abbas and L. Amrao. 2019. Molecular characterization of *Magnaporthe oryzae* in Punjab, Pakistan and it's *in vitro* suppression by fungicides and botanicals. *Int. J. Agric. Biol.* 1459–1466.

Aslam, H.M.U., H. Riaz, N. Killiny, X.-G.Zhou, L.S. Thomashow, N.T. Peters and A.K. Chanda. 2022. Microarray technology for detection of plant diseases. *Trends Plant Dis. Assess.* 203–223.

Baccile, J.A., J.E. Spraker, H.H. Le, E. Brandenburger, C. Gomez, J.W. Bok and F.C. Schroeder. 2016. Plant-like biosynthesis of isoquinoline alkaloids in *Aspergillus fumigatus*. *Nat. Chem. Biol.* 12(6):419–424.

Barbi, F., A. Kohler, K. Barry, P. Baskaran, C. Daum, L. Fauchery, K. Ihrmark, A. Kuo, K. LaButti, A. Lipzen, E. Morin, I.V. Grigoriev, B. Henrissat, B.D. Lindahl and F. Martin. 2020. Fungal ecological strategies reflected in gene transcription—A case study of two litter decomposers. *Environ. Microbiol.* 22:1089–1103.

Bebber, D.P. and T.M. Chaloner. 2022. Specialists, generalists and the shape of the ecological niche in fungi. *New Phytol.* 234:345.

Begerow, D., H. Nilsson, M. Unterseher and W. Maier. 2010. Current state and perspectives of fungal DNA barcoding and rapid identification procedures. *Appl. Microbiol. Biotechnol.* 87:99–108.

Behie, S.W., S.J. Jones and M.J. Bidochka. 2015. Plant tissue localization of the endophytic insect pathogenic fungi *Metarhizium* and *Beauveria*. *Fungal Ecol.* 13:112–119.

Borman, A.M., C.J. Linton, S.-J. Miles and E.M. Johnson. 2008. Molecular identification of pathogenic fungi. *J. Antimicrob. Chemother*. 61:i7–i12.

Botelho, N.S., S.B. de Paula, L.A. Panagio, P. Pinge-Filho, L.M. Yamauchi and S.F. Yamada-Ogatta. 2012. Candida species isolated from urban bats of Londrina-Parana, Brazil and their potential virulence. *Zoonoses Public Health* 59(1):16–22.

Bouasria, A., T. Mustafa, F. De Bello, L. Zinger, G. Lemperiere, R.A. Geremia and P. Choler. 2012. Changes in root-associated microbial communities are determined by species-specific plant growth responses to stress and disturbance. *Eur. J. Soil Biol.* 52:59–66.

Buchfink, B., C. Xie and D.H. Huson. 2014. Fast and sensitive protein alignment using DIAMOND. *Nat. Methods* 12:59–60.

Caporaso, J.G., J. Kuczynski, J. Stombaugh, K. Bittinger, F.D. Bushman and E.K. Costello. 2010. QIIME allows analysis of high-throughput community sequencing data. *Nat. Meth.* 7(5):335–336.

Chiang, Y.M., C.E. Oakley, M. Ahuja, R. Entwistle, A. Schultz, S.L. Chang, C.T. Sung, C.C.C. Wang and B.R. Oakley. 2013. An efficient system for heterologous expression of secondary metabolite genes in *Aspergillus nidulans*. *J. Am. Chem. Soc.* 135:7720–7731.

Cuomo, C.A. and B.W. Birren. 2010. The fungal genome initiative and lessons learned from genome sequencing. *Meth. Enzymol.* 470:833–855.

Daniel, R. 2005. The metagenomics of soil. *Nat. Rev. Microbiol.* 3(6):470–478.

De Vries, S., E.H. Stukenbrock and L.E. Rose. 2020. Rapid evolution in plant–microbe interactions—An evolutionary genomics perspective. *New Phytol.* 226:1256–1262.

Dean, R., J.A. Van Kan, Z.A. Pretorius, K.E. Hammond-Kosack, A. Di Pietro, P.D. Spanu and G.D. Foster. 2012. The Top 10 fungal pathogens in molecular plant pathology. *Mol. Plant. Pathol.* 13(4):414–430.

Deising, H.B., S. Werner and M. Wernitz. 2000. The role of fungal appressoriain plant infection. *Microb. Infect.* 2(13):1631–1641.

Donovan, P.D., G. Gonzalez, D.G. Higgins, G. Butler and K. Ito. 2018. Identification of fungi in shotgun metagenomics datasets. *PLoS One* 13:1–16.

Dufresne, M. and A.E. Osbourn. 2001. Definition of tissue-specific and general requirements for plant infection in a phytopathogenic fungus. *Mol. Plant Microbe Interact.* 14:300–307.

Finn, R.D., J. Mistry, J. Tate, P. Coggill, A. Heger, J.E. Pollington and A. Bateman. 2010. The Pfam protein family's database. *Nucl. Acids Res.* 38: D211-D222.

Foissac, S., J. Gouzy, S. Rombauts, C. Mathé, J. Amselem, L. Sterck and T. Schiex. 2008. Genome annotation in plants and fungi: EuGene as a model platform. *Curr. Bioinform.* 3(2):87–97.

Fox, E.M. and B.J. Howlett. 2008. Secondary metabolism: Regulation and role in fungal biology. *Curr. Op. Microbiol.* 11(6):481–487.

Fraser, J.A. and J. Heitman. 2004. Evolution of fungal sex chromosomes. *Mol. Microbiol.* 51:299–306.

Gabaldón, T. 2020. Hybridization and the origin of new yeast lineages. *FEMS Yeast Res.* 20:foaa040.

Ghosh, A., A. Mehta and A.M. Khan. 2019. Metagenomic analysis and its applications. *Encycl. Bioinforma. Comput. Biol.* 184–193.

Gibriel, H.A.Y., B.P.H.J. Thomma and M.F. Seidl. 2016. The age of effectors: Genome-based discovery and applications. *Phytopathology* 106:1206–1212.

Grossetête, S., B. Labedan and O. Lespinet. 2010. FUNGIpath: A tool to assess fungal metabolic pathways predicted by orthology. *BMC Genome*11:1–15.

Haas, B.J., N. Volfovsky, C.D. Town, M. Troukhan, N. Alexandrov, K.A. Feldmann and S.L. Salzberg. 2002. Full-length messenger RNA sequences greatly improve genome annotation. *Genome Biol.* 3:1–12.

Hedeler, C., H.M, Wong, M.J. Cornell, I. Alam, D.M. Soanes, M. Rattray and N.W. Paton. 2007. e-Fungi: A data resource for comparative analysis of fungal genomes. *BMC Genom.* 8(1):1–15.

Henras, A.K., C. Plisson-Chastang, M.F. O'Donohue, A. Chakraborty and P.E. Gleizes. 2015. An overview of pre-ribosomal RNA processing in eukaryotes. *Wiley Interdiscip. Rev. RNA* 6:225–242.

Hibbett, D.S., J.E. Stajich and J.W. Spatafora. 2013. Toward genome-enabled mycology. *Mycologia* 105:1339–1349.

Hunter, S., M. Corbett, H. Denise, M. Fraser, A. Gonzalez-Beltran, C. Hunter, P. Jones, R. Leinonen, C. McAnulla, E. Maguire, J. Maslen, A. Mitchell, G. Nuka, A. Oisel, S. Pesseat, R. Radhakrishnan, P. Rocca-Serra, M. Scheremetjew, P. Sterk, D. Vaughan, G. Cochrane, D. Field and S.A. Sansone. 2014. EBI metagenomics—A new resource for the analysis and archiving of metagenomic data. *Nucl. Acids Res.* 42:1–7.

Idnurm, A and B.J. Howlett. 2001. Pathogenicity genes of phytopathogenic fungi. *Mol. Plant. Pathol.* 2(4):241–255.

Ivashchenko, A.T., M.I. Tauasarova and S.A. Atambayeva. 2009. Exon-intron structure of genes in complete fungal genomes. *Mol. Biol.* 43:24–31.

Jain, S., M. Nehra, R. Kumar, N. Dilbaghi, TonyY. Hu, S. Kumar, A. Kaushik and C. Li. 2021. Internet of medical things (IoMT)-integrated biosensors for point-of-care testing of infectious diseases. *Biosens. Bioelectron.* 179:113074.

James, T.Y., F. Kauff, C.L. Schoch, P.B. Matheny, V. Hofstetter, C.J. Cox, G. Celio, C. Gueidan, E. Fraker and R. Vilgalys. 2006. Reconstructing the early evolution of Fungi using a six-gene phylogeny. *Nature* 443:818–822.

Jeffroy, O., H. Brinkmann, F. Delsuc and H. Philippe. 2006. Phylogenomics: The beginning of incongruence? *Trends Genet.* 22:225–231.

Keepers, K.G., C.S. Pogoda, K.H. White, C.R. Anderson Stewart, J.R. Hoffman, A.M. Ruiz, C.M. McCain, J.C. Lendemer, N.C. Kane and E.A. Tripp. 2019. Whole genome shotgun sequencing detects greater lichen fungal diversity than amplicon-based methods in environmental samples. *Front. Ecol. Evol.* 7:1–14.

Kim, K.-T., J. Jeon, J. Choi, K. Cheong, H. Song, G. Choi, S. Kang and Y.-H. Lee. 2016. Kingdom-wide analysis of fungal small secreted proteins (SSPs) reveals their potential role in host association. *Front. Plant Sci.* 7.

Kiss, E., B. Hegedüs, M. Virágh, T. Varga, Z. Merényi and T. Kószó. 2019. Comparative genomics reveals the origin of fungal hyphae and multicellularity. *Nat. Commun.* 10:4080. doi: 10.1038/s41467-019-12085.

Kohler, A., A. Kuo, L.G. Nagy, E. Morin, K.W. Barry, F. Buscot and F. Martin. 2015. Convergent losses of decay mechanisms and rapid turnover of symbiosis genes in mycorrhizal mutualists. *Nat. Genet.* 47:410–415.

Kuo, A., B. Bushnell and I.V. Grigoriev. 2014. *Fungal genomics: Sequencing and annotation*, 1st Ed. Elsevier Ltd.

Lavezzo, E., L. Barzon, S. Toppo and G. Palu. 2016. Third generation sequencing technologies applied to diagnostic microbiology: Benefits and challenges in applications and data analysis. *Exp. Rev. Mol. Diagnos.* 16(9):1011–1023.

Lin, B., J. Hui and H. Mao. 2021. Nanopore technology and its applications in gene sequencing. *Biosensors* 11.

Lind, A.L., J.H. Wisecaver, C. Lameiras, P. Wiemann, J.M. Palmer, N.P. Keller, F. Rodrigues, G.H. Goldman and A. Rokas. 2017. Drivers of genetic diversity in secondary metabolic gene clusters within a fungal species. *PLoS Biol.* 15:1–26.

Lockhart, D.J. and E.A. Winzeler. 2000. Genomics, gene expression and DNA arrays. *Nature* 405:827–836.

Lord, N.S., C.W. Kaplan, P. Shank, C.L. Kitts and S.L. Elrod. 2002. Assessment of fungal diversity using terminal restriction fragment (TRF) pattern analysis: Comparison of 18S and ITS ribosomal regions. *FEMS Microbiol. Ecol.* 42:327–337.

Magadum, S., U. Banerjee, P. Murugan, D. Gangapur and R. Ravikesavan. 2013. Gene duplication as a major force in evolution. *J. Genet.* 92:155–161.

Mao, X.M., W. Xu, D. Li, W.B. Yin, Y.H. Chooi, Y.Q. Li and Y. Hu. 2015. Epigenetic genome mining of an endophytic fungus leads to the pleiotropic biosynthesis of natural products. *Angewandte Chemie* 127(26):7702–7706.

Martin, F., A. Aerts, D. Ahrén, A. Brun, E.G.J. Danchin, F. Duchaussoy, J. Gibon, A. Kohler, E. Lindquist, V. Pereda and I.V. Grigoriev. 2008. The genome of *Laccaria bicolor* provides insights into mycorrhizal symbiosis. *Nature* 452:88–92.

Mattern, D.J., V. Valiante, S.E. Unkles and A.A. Brakhage. 2015. Synthetic biology of fungal natural products. *Front. Microbiol.* 6:775.

Menardo, F., C.R. Praz, S. Wyder, R. Ben-David, S. Bourras and H. Matsumae. 2016. Hybridization of powdery mildew strains gives rise to pathogens on novel agricultural crop species. *Nat. Genet.* 48:201–205.

Merckx, V.S.F.T. 2012. *Mycoheterotrophy: The biology of plants living on fungi.* Springer.

Mohanta, T.K. and H. Bae. 2015. The diversity of fungal genome. *Biol. Proced.* 17:1–9.

Möller, M. and E.H. Stukenbrock. 2017. Evolution and genome architecture in fungal plant pathogens. *Nat. Rev. Microbiol.* 15:756–771.

Moncalvo, J.M., F.M. Lutzoni, S.A. Rehner, J. Johnson and R. Vilgalys. 2000. Phylogenetic relationships of agaric fungi based on nuclear large subunit ribosomal DNA sequences. *Syst. Biol.* 49:278–305.

Nagy, L.G., R.A. Ohm, G.M. Kovács, D. Floudas, R. Riley, A. Gácser, M. Sipiczki, J.M. Davis, S.L. Doty, G.S. De Hoog, B.F. Lang, J.W. Spatafora, F.M. Martin, I.V. Grigoriev and D.S. Hibbett. 2014. Latent homology and convergent regulatory evolution underlies the repeated emergence of yeasts. *Nat. Commun.* 5.

Naranjo-Ortiz, M.A and T. Gabaldón. 2019. Fungal evolution: Diversity, taxonomy and phylogeny of the Fungi. *Biol. Rev.* 94:2101–2137.

Nowrousian, M. 2010. Next-generation sequencing techniques for eukaryotic microorganisms: Sequencing-based solutions to biological problems. *Eukaryot. Cell* 9:1300–1310.

Oakley, C.E., M. Ahuja, W.W. Sun, R. Entwistle, T. Akashi, J. Yaegashi and B.R. Oakley. 2017. Discovery of McrA, a master regulator of *Aspergillus* secondary metabolism. *Mol. Microbiol.* 103(2):347–365.

Okada, M., K. Saito, C.P. Wong, C. Li, D. Wang, M. Iijima and I. Abe. 2017. Combinatorial biosynthesis of (+)-daurichromenic acid and its halogenated analogue. *Org. Lett.* 19(12):3183–3186.

Onito G., K. Hameed, R. Ventura, J. Krishnan, C.W. Schadt and R. Vilgalys. 2016. Isolating a functionally relevant guild of fungi from the root microbiome of Populus. *Fung. Ecol.* 22:35–42.

Peay, K.G., P.G. Kennedy and J.M. Talbot. 2016. Dimensions of biodiversity in the Earth mycobiome. *Nat. Rev. Microbiol.* 14:434–447.

Poppe, S., L. Dorsheimer, P. Happel and E.H. Stukenbrock. 2015. Rapidly evolving genes are key players in host specialization and virulence of the fungal wheat pathogen *Zymoseptoria tritici* (*Mycosphaerella graminicola*). *PLoS Pathog.* 11:e1005055.

Postel, S. and B. Kemmerling. 2009. Plant systems for recognition of pathogen-associated molecular patterns. *Semin. Cell Dev. Biol.* 20:1025–1031.

Raffaele, S., J. Win, L.M. Cano and S. Kamoun. 2010. Analyses of genome architecture and gene expression reveal novel candidate virulence factors in the secretome of *Phytophthora infestans*. *BMC Genom.* 11:1–18.

Rahnama, M., B. Wang, J. Dostart, O. Novikova, D. Yackzan, A. Yackzan, H. Bruss, M. Baker, H. Jacob, X. Zhang, A. Lamb, A. Stewart, M. Heist, J. Hoover, P. Calie, L. Chen, J. Liu and M.L. Farman. 2021. Telomere roles in fungal genome evolution and adaptation. *Front. Genet.* 12:1–22.

Roth, M.G., N.M. Westrick and T.T. Baldwin. 2023. Fungal biotechnology: From yesterday to tomorrow. *Front. Fungal Biol.* 4:1–11.

Rutledge, P.J and G.L. Challis. 2015. Discovery of microbial natural products by activation of silent biosynthetic gene clusters. *Nat. Rev. Microbiol.* 13:509–523.

Saad, D.S., G.C. Kinsey, S. Kim and C.C. Gaylarde. 2004. Extraction of genomic DNA from filamentous fungi in biofilms on water-based paint coatings. *Int. Biodeterior. Biodegrad.* 54:99–103.

Sato, M., J.E. Dander, C. Sato, Y.S. Hung, S.S. Gao, M.C. Tang and Y. Tang. 2017. Collaborative biosynthesis of maleimide-and succinimide-containing natural products by fungal polyketide megasynthases. *J. Am. Chem. Soc.* 139(15):5317–5320.

Schlegel, M., M. Munsterkotter, U. Guldener, R. Bruggmann, A. Duo and M. Hainaut. 2016. Globally distributed root endophyte *Phialocephala subalpina* links pathogenic and saprophytic lifestyles. *BMC Genom.* 17(1):1015.

Schoch, C.L., K.A. Seifert, S. Huhndorf, V. Robert, J.L. Spouge, C.A. Levesque, W. Chen, E. Bolchacova, K. Voigt, P.W. Crous, N. Zhang, W.Y. Zhuang and D. Schindel. 2012. Nuclear ribosomal internal transcribed spacer (ITS) region as a universal DNA barcode marker for Fungi. *Proc. Natl. Acad. Sci. U.S.A.* 109:6241–6246.

Schuster, S.C. 2008. Next-generation sequencing transforms today's biology. *Nat. Methods* 5:16–18.

Selosse, M.A. and C.L. Schardl. 2007. Fungal endophytes of grasses: Hybrids rescued byvertical transmission? An evolutionary perspective. *New Phytol.* 17:452–458.

Selosse, M.A., M.P. Dubois and N. Alvarez. 2009. Do Sebacinales commonly associate with plant roots as endophytes? *Mycol. Res.* 113:1062–1069.

Shao, Z., G. Rao, C. Li, Z. Abil, Y. Luo and H. Zhao. 2013. Refactoring the silent spectinabilin gene cluster using a plug-and-play scaffold. *ACS Synth. Biol.* 2:662–669.

Sharon, E., S.A.A. Chen, N.M. Khosla, J.D. Smith, J.K. Pritchard and H.B. Fraser. 2018. Functional genetic variants revealed by massively parallel precise genome editing. *Cell* 175:544–557.e16.

Sharp, N.P., L. Sandell, C.G. James and S.P. Otto. 2018. The genome-wide rate and spectrum of spontaneous mutations differ between haploid and diploid yeast. *Proc. Natl. Acad. Sci. U.S.A.* 115:E5046–E5055.

Shi, T.Q., G.N. Liu, R.Y. Ji, K. Shi, P. Song, L.J. Ren and X.J. Ji. 2017. CRISPR/Cas9-based genome editing of the filamentous fungi: The state of the art. *Appl. Microbiol. Biotechnol.* 101:7435–7443.

Skellam, E. 2019. Strategies for engineering natural product biosynthesis in fungi. *Trends Biotechnol.* 37(4):416–427.

Smanski, M.J., S. Bhatia, D. Zhao, Y. Park, L. Woodruff, G. Giannoukos, D. Ciulla, M. Busby, J. Calderon, R. Nicol, D.B. Gordon, D. Densmore and C.A. Voigt. 2014. Functional optimization of gene clusters by combinatorial design and assembly. *Nat. Biotechnol.* 32:1241–1249.

Soanes, D.M., T.A. Richards and N.J. Talbot. 2007. Insights from sequencing fungal and oomycete genomes: What can we learn about plant disease and the evolution of pathogenicity? *Plant Cell* 19:3318–3326.

Soanes, D.M., W. Skinner, J. Keon, J. Hargreaves and N.J. Talbot. 2002. Genomics of phytopathogenic fungi and the development of bioinformatic resources. *Mol. Plant. Microbe Interact.* 15:421–427.

Spatafora, J.W., G-H.Sung, J-M.Sung, N.L. Hywel-Jones and J.F. White. 2007. Phylogenetic evidence for an animal pathogen origin of ergot and the grassendophytes. *Mol. Ecol.* 16:1701–1711.

Stajich, J.E. 2017. Fungal genomes and insights into the evolution of the kingdom. *Microbiol. Spectrum* 5. doi: 10.1128/microbiolspec.funk-0055–2016.

Stajich, J.E., M.L. Berbee, M. Blackwell, D.S. Hibbett, T.Y. James, J.W. Spatafora and J.W. Taylor. 2009. The fungi. *Curr. Biol.* 19:840–845. https://doi/10.1016/j.cub.2009.07.004

Stanke, M., O. Schöffmann, B. Morgenstern and S. Waack. 2006. Gene prediction in eukaryotes with a generalized hidden Markov model that uses hints from external sources. *BMC Bioinform.* 7:62.

Stukenbrock, E.H., F.B. Christiansen, T.T. Hansen, J.Y. Dutheil and M.H. Schierup. 2012. Fusion of two divergent fungal individuals led to the recent emergence of a unique widespread pathogen species. *Proc. Natl. Acad. Sci. U.S.A.* 109:10954–10959.

Taher, L., O. Rinner, S. Garg, A. Sczyrba, M. Brudno, S. Batzoglou and B. Morgenstern. 2003. AGenDA: Homology-based gene prediction. *Bioinformatics* 19(12):1575–1577.

Taylor, J.W. 2011. One Fungus = One Name: DNA and fungal nomenclature twenty years after PCR. *IMA Fungus* 2:113–120.

Tedersoo, L., I. Liiv, P.A. Kivistik, S. Anslan, U. Kõljalg and M. Bahram. 2016. Genomics and metagenomics technologies to recover ribosomal DNA and single-copy genes from old fruitbody and ectomycorrhiza specimens. *MycoKeys* 13:1–20.

Ter-Hovhannisyan, V., A. Lomsadze, Y.O. Chernoff and M. Borodovsky. 2008. Gene prediction in novel fungal genomes using an ab initio algorithm with unsupervised training. *Genome Res.* 18(12):1979–1990.

Tobias, T.B., E.C. Farrer, A. Rosales, R.L. Sinsabaugh, K.N. Suding and A. Porras-Alfaro. 2017. Seed-associated fungi in the alpine tundra: Both mutualists and pathogenscould impact plant recruitment. *Fungal Ecol.* 30:10–18.

Tsuge, T., Y. Harimoto, K. Akimitsu, K. Ohtani, M. Kodama, Y. Akagi and H. Otani. 2013. Host-selective toxins produced by the plant pathogenic fungus *Alternaria alternata*. *FEMS Microbiol. Rev.* 37(1):44–66.

Upadhyaya, N.M., D.P. Garnica, H. Karaoglu, J. Sperschneider, A. Nemri, B. Xu, R. Mago, C.A. Cuomo, J.P. Rathjen, R.F. Park, J.G. Ellis and P.N. Dodds. 2015. Comparative genomics of Australian isolates of the wheat stem rust pathogen *Puccinia graminis* f. sp. tritici reveals extensive polymorphism in candidate effector genes. *Front. Plant Sci.* 5.

Urban, M., R. Pant, A. Raghunath, A.G. Irvine, H. Pedro and K.E. Hammond-Kosack. 2015. The Pathogen-Host Interactions database (PHI-base): Additions and future developments. *Nucl. Acids Res.* 43:D645–D655.

Ursell, L.K., J.L. Metcalf, L.W. Parfrey and R. Knight. 2012. Defining the human microbiome. *Nutr. Rev.* 70.

Van der Heijden, M.G.A., F.M. Martin, M.-A. Selosse and I.R. Sanders. 2015. Mycorrhizal ecology and evolution: The past, the present, and the future. *New Phytol.* 205:1406–1423.

Walters, D.R., N.D. Havis and S.J.P. Oxley. 2008. Ramularia collo-cygni: The biology of an emerging pathogen of barley. *FEMS Microbiol. Lett.* 279:1–7.

Wang, Y., J. Wu, J. Yan, M. Guo, L. Xu, L. Hou and Q. Zou. 2022. Comparative genome analysis of plant ascomycete fungal pathogens with different lifestyles reveals distinctive virulence strategies. *BMC Genom.* 23(1):34.

Williamson, D. 2002. The curious history of yeast mitochondrial DNA. *Nat. Rev. Genet.* 3:475–481.

Wilson, M.R., S.N. Naccache, E. Samayoa, M. Biagtan, H. Bashir and G. Yu. 2014. Actionable diagnosis of neuroleptospirosis by next-generation sequencing. *N. Engl. J. Med.* 370(25):2408–2417.

Winnenburg, R., M. Urban, A. Beacham, T.K. Baldwin, S. Holland, M. Lindeberg and J. Köhler. 2007. PHI-base update: Additions to the pathogen–host interaction database. *Nucl. Acids Res.* 36:D572–D576.

Wood, D. and S.S. Kraken. 2014. Ultrafast metagenomic sequence classification using exact alignments. *Genome Biol.* 15:R46.

Xie, L., L. Zhang, C. Wang, X. Wang, Y.M. Xu, H. Yu and Y. Xu. 2018. Methylglucosylation of aromatic amino and phenolic moieties of drug-like biosynthons by combinatorial biosynthesis. *Proc. Natl. Acad. Sci. U.S.A.* 115(22):E4980–E4989.

Yadav, S.P. 2007. The wholeness in suffix—omics,—omes, and the word om. *J. Biomol. Tech.* 18:277.

Ye, Y., A. Minami, Y. Igarashi, M. Izumikawa, M. Umemura, N. Nagano and H. Oikawa. 2016. Unveiling the biosynthetic pathway of the ribosomally synthesized and post-translationally modified peptide ustiloxin B in filamentous fungi. *Angewandte Chemie* 128(28):8204–8207.

Yeh, H.H., M. Ahuja, Y.M. Chiang, C.E. Oakley, S. Moore, O. Yoon and B.R. Oakley. 2016. Resistance gene-guided genome mining: Serial promoter exchanges in *Aspergillus nidulans* reveal the biosynthetic pathway for fellutamide B, a proteasome inhibitor. *ACS Chem. Biol.* 11(8):2275–2284.

Zhang, S., A. Ban, N. Ebara, O. Mizutani, M. Tanaka, T. Shintani and K. Gomi. 2017. Self-excising Cre/mutant lox marker recycling system for multiple gene integrations and consecutive gene deletions in *Aspergillus oryzae*. *J. Biosci. Bioeng.* 123(4):403–411.

Zhao, Z., H. Liu, C. Wang and J.R. Xu. 2014. Correction: Comparative analysis of fungal genomes reveals different plant cell wall degrading capacity in fungi. *BMC Genom.* 15:6.

3 Fungal Genomics
From the Perspective of Successful Fungal Parasitism in Plants and Toxin Production

Amjad Abbas, Maham Sarwar, Saman Arif, Muhammad Huzaifa Tanveer, Effat Naz, Sobeeqa Khalid, Laraib Zulfiqar, Muhammad Jabran, and Muhammad Amjad Ali

3.1 INTRODUCTION

Fungi are very important organisms and are well known for their ecological and commercial importance in agriculture. Fungi are diverse, ranging from beneficial to pathogenic, and ubiquitous. Many pathogenic fungi have evolved over the last few decades for survival in different ecological niches, and they cause huge yield losses in agricultural production. For their successful attack at the host sites, they have developed different pathogenic tools depending on their lifestyles and the way they take up nutrients from the host plants. Talking about the feeding habits of fungi, various categories are seen, as some of them are biotrophic and take nutrients from the living host. In contrast, the others are necrotrophic in nature, and such fungi feed only on the non-living tissues of the host, or they may trigger cell death in the host for nutrient uptake.

Some fungi are in between the aforementioned categories and are known as hemibiotrophic, spending the initial phase of their life as biotrophic, and then taking up nutrients from the dead tissues of plants. The most basic tool for fungal entrance into a host is the formation of an appressorium, which is hyphal penetration into the host for the purpose of the uptake of nutrition, and it also allows for parasitism by fungi. These structures are compound or single-celled appressoria formed by a combination of different cells and collectively forming infection cushions (Armentrout & Downer, 1987).

The formation of these infection structures is controlled by complex regulatory pathways, which are further accompanied by regulated gene expression (Kahmann & Basse, 2001). These appressorium cells operate as physical entities and can exert physical force to rupture the cuticle of the plant cell wall for its entrance into the host

DOI: 10.1201/9781003452362-3

cell. Appressorium development is mainly achieved by the fungus by altering the structures of the actin cytoskeleton, moderated by septin GTPases and abrupt cell wall differentiation (Ryder & Talbot, 2015).

Transcription factors play a key role in the infection process of filamentous fungi. Filamentous fungi encode putative transcription factors belonging to an important gene family. During plant invasion caused by the fungal appressorium, transcription factors control cytoskeletal dynamics and infection-related autophagy (Galhano et al., 2017).

Fungi are complex pathogens of plants, and they start the inoculation or infection procedure by the production of many enzymes for the breakdown of plant cell wall components like cellulose, hemicellulose, cutin, lignin, pectin, etc. Enzymes like cellulases, hemicellulases, pectinases, xylanases, kinases, etc., are involved in the degradation of host cell wall components. Depending on the host and site of infection, hydrolytic enzymes are released by fungi for their successful entrance to the host cell. Many fungi like *Aspergillus fumigatus* and *Trichoderma* spp. are highly known for their ability to release such degrading enzymes.

Many fungi also release virulence factors or molecules to overcome plant cell resistance known as effectors. Typically, effectors have following molecular features: approximately 50–300 residues or amino acids in length, further containing an N-terminal signal peptide of a highly specific sequence, absence of a transmembrane structural domain, no subcellular localized signaling for mitochondria or other organelles, no attachment sites for glycosylphosphatidylinositol (GPI), and an abundance of cysteine residues (Stergiopoulos & de Wit, 2009; Jones & Dangl, 2006). Effectors can be secreted intracellularly as well between the cells. Effectors mostly comprise low-molecular-weight proteins, sRNA (Chanclud & Morel, 2016), toxins (Ciuffetti et al., 2010), and phytohormones (Weiberg et al., 2013). LysM is a protein effector released by *Trichoderma atroviride* for successful mycoparasitism and hyphal growth (Romero-Contreras et al., 2019).

Among approximately 500 candidate secreted effector proteins (CSEPs), CSEP0081 and CSEP0254 are the only ones known as virulence factors that increase powdery mildew infection in barley (Ahmed et al., 2016). In smut disease of maize, *Ustilago maydis* releases the effector protein Pep1 for successful invasion (Doehlemann et al., 2009).

Plants have continuously battled pathogens, and they have evolved strong defense responses, PAMP (pathogen associated molecular pattern)-triggered immunity (PTI) and effector triggered immunity (ETI), over the years to combat the virulence factors of pathogens. PAMP-triggered immunity, or PTI, is the activation of resources of the plant against pathogens, whereas effector-triggered immunity is the most updated version of immunity working in plants. These mechanisms are the basis for monogenic resistance in modern crop plant varieties. These resistance mechanisms are activated in plants in response to pathogens, partial or broken organelles, and chemicals produced by the pathogens.

Pathogens, mainly *Phytophthora infestans*, use many strategies to overcome PTI. Using hundreds of effectors, they can control host defense responses and avoid detection by the host by altering or suppressing their elicitors. These effectors interfere with physiological and host defense processes by altering calcium and MAPK signaling, RNA interference, alternative splicing, vesicle trafficking, proteolysis, cell-to-cell trafficking, and phytohormone signaling pathways, among other pathways. Pathogen movement into the host cell has been discovered from the fact that oomycete pathogens also have a fungus-like mode of infection, but they are members of the Stramenopile clade that releases effector proteins with an N-terminal RXLR (Arg-X-Leu-Arg) host-targeting signal (Morgan & Kamoun, 2007). *P. infestans* effector Avr3's entrance into the host cell is mediated by this signal, along with the downstream dEER (Asp-Glu-Glu-Arg) motif (Whisson et al., 2007). To cause endocytosis, some oomycete effector RXLR motifs may bind to phosphatidyl inositol phosphates (PIPs) on the surface of the plant plasma membrane (Kale et al., 2010).

Talking about the hijacking of the plant immune system, plant pathogens use an advanced virulence mechanism involving their sRNA for the silencing of host cell immune genes. Pathogens use a strategy to suppress the RNA silencing of plants for successful infection, although research on this aspect is quite limited, with no characterization of these fungi (Yin et al., 2019). Recent studies have shown that small RNAs of *Botrytis cinerea* can manipulate the immune genes of host plants like Arabidopsis and tomato. This pathogen transfers its highly virulent factors into host tissue, which takes control of the immune genes of the host, silences them, and causes successful infection (Weiberg et al., 2013).

When a pathogen attacks the plant, then the machinery of the plant is activated, and responses like systemic acquired resistance or hypersensitive responses are activated because of the production of different chemicals like salicylic acid, jasmonic acid, and ethylene. In plants, there is a phenomenon of reactive oxygen species (ROS) production because of environmental stress and pathogen attack. These reactive oxygen species help in the defense of plants against pathogens. Pathogens counteract or limit the production of these ROS for successful infection. For instance, when *Magnaporthe oryzae*, an important pathogen of rice, attacks a rice plant to cause blast disease, many defense responses become active, including the production of ROS. This pathogen contains the gene MoHYR1, which helps in improving these defense reactions and contributes to hyphal expansion and the successful colonization of host tissues (Huang et al., 2011). It is informative to know about the fungal tools and secretions that are involved in causing successful parasitism in plants and how fungi use various strategies to monitor the defensive response of plants.

3.2 FUNGAL GENOME AND UNDERSTANDING FUNGAL TOXINS

The advent of DNA sequencing technologies in the late 20th century marked a new era in fungal understanding and genomics. The first fungal genome to be fully sequenced was that of *Saccharomyces cerevisiae* in 1996 as part of the Saccharomyces Genome Project. This milestone was important, and it provided a template for genome studies in other complex organisms and led to the genomics revolution in the mycology field. Fungal genomes are diverse, and they vary greatly in size, ranging from as small as 8 Mb in some yeast species to over 100 Mb in certain filamentous fungi. Despite this variability, fungal genomes share several common features.

3.3 HOW HAS FUNGAL GENOME RESEARCH CHANGED OUR UNDERSTANDING?

Genome sequencing is an important technique for understanding the biology, survival, and pathogenicity of fungal organisms. Like other higher organisms, fungi also utilize different enzymes, which are proteins, to perform multiple functions inside the body and to communicate with the micro-environment. The sequence of genes after genome sequencing is helpful to make a layout of the genes involved in a particular function. Such information will lead us to understand the type of enzymes produced by the fungi and regulators that are important in forming these enzymes. Important traits related to the survival of fungi like cell cycles, the communication of microbes and plants with the fungi, and development and differentiation could be assessed with the sequencing of genes responsible for these important functions (Sieber et al., 2013).

The available information on related fungi with similar genes is also important for predicting the function of a fungal gene or an entire gene family. Knowledge of the genome of a particular organism is helpful to make an inference about the close relationship between two species, and the possibility of sharing a common ancestor is also considered. Such ancestral information helps us to understand the function of a particular gene or a cluster of genes. Other sophisticated tools also depict the sequence of enzymes or proteins based on the genome sequencing information (Werner et al., 2007).

It is not the norm to use fungal genomes for the similarity in ancestry; they are equally important for monitoring the differences as well. The dynamic nature of fungal genomes is also clear in sequencing data; for example, there is more than 68% amino acid sequence identity between *Aspergillus fumigatus* and *A. oryzae craze* (Wortman et al., 2006). The data also inform us about genome duplications and the evolution of similar plant pathogenic fungi (Margis et al., 2006).

Plants and fungi interact closely in the soil, air, rhizosphere, or anywhere in the field. The host plant is either resistant or susceptible. A resistant host plant carries resistance mechanisms to overcome the invasion of pathogenic fungi. This plant's resistance against the established fungal pathogen is not the first step but is the climax reached after a series of events between the plant and pathogen. The main theme of any disease resistance breeding program in crop plants relies on the understanding of interactions between plants and pathogens, which are decisive and lead to resistance. The background of these interactions is also solid information on virulence genes of the fungi and resistant genes or R genes of the plant (Kawakubo et al., 1990).

The presence of R genes in the plant provides the grounds for vertical resistance against a particular plant pathogenic fungus (Zahid et al., 2023). This could be applied to any of the model plant pathovar systems. The best example may be rust disease in wheat. Proteins encoded by virulence genes present in the leaf rust fungus *Puccinia striiformis* interact with R proteins of the wheat plant. The genome data of the rust fungus helps researchers make better decisions and guide the pyramid of more specific resistance genes in wheat varieties (de Vries et al., 2011).

Similarly, the fungus's uptake of nutrition from the host is also a vital event of life. Appressoria are very specific structures a fungus uses to penetrate the host cells and absorb nutrition. This is one of the critical steps in the life events of fungi and is controlled by a series of regulating genes and pathways. For example, the MAP-kinase pathway is crucial for appressorium formation, along with chitin synthase, which is a key component of fungi. Similarly, PBS2 and HOG1 genes are involved in the creation of osmotic pressure, which is a major component during the penetration of the appressorium into the plant cell. This depends on the sequence of genes and genome information, which is best to understand and make valid decisions (Liu et al., 2005).

3.4 MECHANISMS USED BY FUNGI FOR PARASITISM

Fungi have developed multiple strategies for successful penetration into host cells by breaking down or digesting physical structures of the plant. The main penetration method is the formation of appressoria, which generate and exert turgor pressure to break the cell wall of the plant by a physical force (Ryder & Talbot, 2015). Many carbohydrate-active enzymes like glycoside hydrolases, carbohydrate esterase, and polysaccharide lyases are mainly used to degrade cell walls. These enzyme-coding genes are present in higher amounts in plant pathogenic species than in any other organism like animals or saprophytes. Fungi also can inhibit host defense responses by using different strategies. A high amount of reactive oxygen species (ROS) is very dangerous for the host cell, and an oxidative degradation of cells occurs that helps in fungal infection (Levine et al., 1994; Wu et al., 1997). On the other hand, a normal amount of ROS enhances host defense response. Many fungi like *Sclerotinia sclerotiorum* induce a pH shift and release oxalic acid at the host surface, which degrades host tissues (Bolton et al., 2006). In addition, many fungi have mechanisms like the inhibition of host cell signaling and protease secretion that aid in successful fungal penetration and infection (Rodriguez-Moreno et al., 2018).

3.5 FUNGAL ENTRANCE INTO PLANT SYSTEMS BY USING CELL WALL-DEGRADING ENZYMES

Plant cells comprised cell walls, cell membranes, Golgi bodies, and various other organelles. Plants' outer surfaces consist of epidermal cells covered with a cuticle, which are further layered with wax. These slippery structures are very important and keep plants safe from invading microbes. The first hindrance that gets in the way of a pathogen attack is the cell wall. If the cell wall is hard, then it will keep the pathogen away from penetration into the plant system. The cell wall is diverse and is made up of components that include cellulose, cutin polymers, cuticular wax, hemicellulose, lignin, pectin, etc. These are the structures that pathogens attack for their food, shelter, and reproduction. Pathogens attack plant cells successfully because they have developed substances that can break plant cell hindrances. Many pathogens like nematodes and fungi can invade host cells by mechanical and chemical means. Here is a brief description of fungal enzymes used for the degradation of cell wall components.

Cellulose is the most abundant and is a long, straight chain of β-linked D-glucose units. It is an enormous polysaccharide with an annual production of millions of tons in plant bodies. However, its production on Earth does not increase due to fungal and bacterial degradation (Aro et al., 2005).

Many members of Ascomycota and Basidiomycota are involved in the degradation of cellulose, while the most effective and best degradation is done by members of the phylum Basidiomycota. Basidiomycetes are important fungi and have been studied for decades for their involvement in the degradation of fungi (Reese & Levinson, 1952). However, the idea of cellulose degradation has been altered during the last few years. The identification of endoglucanases in brown rot has an important role in understanding the degradation process (Cohen et al., 2005; Yoon et al., 2007) and the formulation of the role of oxidative systems in cellulose breakdown, whereas the attempt to quantify their significance is more interesting to understand the scientific bases (Suzuki et al., 2006). The release of the first genome sequence of a wood-rotting Basidiomycota named *Phanerochaete chrysosporium* (Martinez et al., 2004) was helpful to understand the plant pathogen relation and genomics, which increased the power of proteomic and computational methods for the identification of individual components of its model cellulolytic system (Wymelenberg et al., 2005, 2006; Kersten & Cullen, 2007; Sato et al., 2007). Natural cleaners of the environment like litter decomposers and wood-rotting basidiomycetes play an important role in the biodegradation of cellulose for successful entrance into host cells.

Many enzymes work synergistically to break cellulose into simple forms of carbon. Fungi take up these simple carbon-containing compounds as a nutrient source. Three classes of enzymes are produced by fungi for the degradation of cellulose: β-glucosidase (BGL), β-1,4-endoglucanases (EGL), and cellobiohydrolases/exoglucanases, which are further divided into eight GH families (de Vries et al., 2011). CBHs cleave the cellulose units from the long polysaccharide chain. Endoglucanases cut the chain internally and provide more sites for CBHs to act upon (Teeri, 1997). And, finally, BGL cleaves the polysaccharide into monomers as a carbon source for fungi (Beguin, 1990).

Many genes have been isolated and characterized; for example, genes in the fungus *T. reesei* that encode for the degradation of cellulose (Aro et al., 2005) are helpful for the decay of plants and debris. These include two cellobiohydrolase-encoding genes (Shoemaker et al., 1983; Teeri et al., 1983, 1987; Chen et al., 1987), seven genes encoding β-glucosidases (Barnett et al., 1991; Mach et al., 1995; Takashima et al., 1999; Foreman et al., 2003), and eight endoglucanase genes (Penttilä et al., 1986; Saloheimo et al., 1988, 1994; Ward et al., 1993; Foreman et al., 2003). They work synergistically and act on the reducing and non-reducing ends of the cellulose chain, respectively (Nutt et al., 1998). The breakdown of crystalline parts of the polysaccharide chain is also done by CBHs. The breakdown of this polysaccharide chain results in the successful entrance of the pathogen into the host cell.

Hemicellulose is an important polysaccharide present in plants and is the second most abundant polysaccharide, having diverse compositions. In this class, xylan is the most significant hemicellulose and is a part of cereals and hardwood, comprising β-1,4 linked D-xylose and β-1,4 linked D-mannose in the long polymer backbone (Aro et al., 2005). Xylan is the main part of hemicellulose, and its degradation is a complex process that is accompanied by many hydrolytic enzymes. Basically, two enzymes, endo-1,4-β-xylanase and xylan 1,4-β-xylosidase, are involved, which are generally produced by many filamentous fungi and other microbial communities. Endo-1,4-β-xylosidase breaks down xylan and generates oligosaccharides, while xylan 1,4-β-xylosidase cleaves xylan oligosaccharides and produces xylose (De Vries & Visser, 2001). Some auxiliary enzymes are also needed for xylan degradation, like ferulic and p-coumaric esterases, xylan esterases, and α-4-O-methyl glucuronidase working synergistically (Kirk & Cullen, 1998).

Acetyl esterases eliminate acetyl groups of polymers, while α-galactosidases remove residues of galactose. Finally, oligomer β-1,4 bonds are generated by the breakdown of Endoglucanases in the presence of β-mannosidase and β-glycosidase (Pérez et al., 2002). Multiple types of endoxylanases are produced by the white-rot fungus *Phanerochaete chrysosporium* (Kirk & Cullen, 1998).

Even though hemicellulose has a very complicated structure, its enzymatic degradation is well understood and a known factor. The combined effect of various enzymes, co-enzymes, and other molecules helps in the breakdown by using water. This process converts bigger molecules into smaller and simpler ones in plants. Endo-enzymes cleave the main chain internally, exo-enzymes convert the polymers into monomers, and ancillary enzymes have a role in the hydrolysis of long side chains into simple mono- or disaccharides (Aro et al., 2005). Through this process, the highly complex and large structure of hemicellulose is degraded slowly by fungal enzymes for successful attack. The degradation of hemicellulose is done by many microbes like filamentous fungi and bacteria as well.

Pectin is a complex polysaccharide in living organisms and reacts as a backbone of α-1,4 linked to D-galacturonic acid. Pectin is a natural chemical in plants and is very different from the other natural polysaccharides (Pérez et al., 2000; Ridley et al., 2001). In higher plants like dicotyledons, pectin is an integral component of the middle lamella of primary cell walls (Williamson et al., 1998; Van der Vlugt-Bergmans et al., 2000)

The dissolution of the bond that is present between galacturonic acid and rhamnose is catalyzed by the enzyme rhamnogalacturonan I (RGI). This enzyme has been characterized and isolated from many species of fungi, mainly from *Aspergillus aculeatus* (Mutter, 1997). It may further have different action patterns like the random dissolution of endo- or terminal pectin bonds (Silva et al., 2016).

Up to now, 23 different enzymes have been characterized that specifically attack on the pectin RGI backbone (Lombard et al., 2014). RGI endo-hydrolases, RGHs, mainly attack at the RGI backbone in an endo-pattern and cleave the bond. "The mechanism involves an inversion of the anomeric C1 configuration in galacturonic acid releasing oligosaccharides with β-D-Gal*p*A at the reducing end and L-Rha*p* at the non-reducing end as products" (Azadi et al., 1995; Pitson et al., 1996).

RGI exo-hydrolases consist of RGI galacturonohydrolases, RGI rhamnohydrolases, RGGH, URGH, and RGRH (Ochiai et al., 2007). All these enzymes act on the non-reducing sides of the RGI backbone (Mutter, 1997; Mutter et al., 1998; Ochiai et al., 2006; Zandleven et al., 2005).

Experimental results have confirmed that about 126 proteins are secreted by *Botrytis cinerea* that have degrading effects on the pectin and the sucrose components of plant cells grown in broth media (Shah et al., 2009). Many enzymes are used by *Botrytis cinerea* for pectin dissolution, including pectin methyl esterases

(PMEs), endopolygalacturonases (endo PGs), exopolygalacturonases (exo PGs), pectin and pectate lyases (PLs), and rhamnogalacturonan hydrolase and lyase (Geraeds et al., 1990; Wubben et al., 2000). After dissolving the plant cell wall component pectin, fungi can easily spread hyphae inside the cell by developing haustoria.

3.6 LIGNINS

These important polymers are primarily deposited in thickened cells, making them hard to break and impermeable. In addition, lignins are classified into the two major classes: 1. lignin-oxidizing enzymes and 2. lignin-degrading auxiliary enzymes (Levasseur et al., 2008; Kameshwar & Qin, 2016a). Formation is also triggered due to biotic and abiotic stresses such as pathogen attack, wound, drought, heat stress, etc. (Caño-Delgado et al., 2003; Tronchet et al., 2010). Lignin is formed by the combination of three compounds, namely, coniferyl alcohol, p-coumaryl alcohol, and sinapyl alcohol, which are synthesized by the phenylpropanoid pathway (Boerjan et al., 2003). Many oxidative and degrading enzymes are secreted by fungi for lignin degradation. Lignin-degrading enzymes are very important.

The degradation of lignin by fungal enzymes involves the depolymerization and dissolution of the aromatic ring (Datta et al., 2017). A sizable group of enzymes that code for cytochrome P450 monooxygenases, iron reductase, ferredoxin, alcohol dehydrogenases (Fenton's reagents), aromatic ring and epoxide hydroxylases, and intra- and extra-dioxygenases, along with other large sets of enzymes, is involved in the decomposition of lignin (Kameshwar & Qin, 2017). Lignin degradation occurs by many species of soft-rot fungi, brown-rot fungi, and deuteromycetes. It was shown that the brown-rot decay of lignin occurs in a nonenzymatic way in the initial stages due to the production of highly degrading hydroxyl radicles (Goodell et al., 1997; Arantes & Milagres, 2009, 2006; Jensen et al., 2002; Hammel et al., 2002; Korripally et al., 2013; Martinez et al., 2009). Different phenolic compounds like 2,5-dimethoxy-1,4-bezoquinone and 4,5-dimethoxy-1,2-bezoquinone are produced by these fungi, which aid in the further advancement of Fenton reactions by ferric iron chelators and electrons for iron reduction (Goodell et al., 1997; Enoki et al., 1997; Kerem et al., 1999; Paszczynski et al., 1999).

The Fenton system is activated and releases reactive species to act on the cellulose or lignin. Reactive species like ·OH are released at a distance from fungal spores or hyphae but close to cellulose or lignin so that degradation can occur more precisely. Furthermore, oxalic acid is also released and accumulated by brown-rot fungi in the lumen of the host plant (Takao, 1965; Dutton et al., 1993). This oxalic acid reacts with the iron present in the plant tissue in Fe oxide form (Goodell et al., 1997) and helps in lignin degradation. Although brown-rot fungi participate little in the degradation of wood, they are of great significance. White-rot fungi (basidiomycetes) also colonize the cellulosic and lignin parts of the wood and cause excellent decay of plant cell wall components.

Soft-rot fungi mainly belong to ascomycetes. They cause wood degeneration, mainly of lignin, slowly in the absence of extracellular nutrients (Worrall et al.,

1991). Lignin degeneration occurs more prominently in woody plants than in herbaceous plants (Fernandez-Fueyo et al., 2012; Kuhad et al., 1997).

In Deuteromycetes, two strains of *Paecilomyces inflatus* were isolated and found to show the degradative effects of lignin by the enzyme laccase. The mineralization of 14Cβ-labeled lignin into CO_2 occurs. The optimum temperature and pH for laccase production are 30°C and 4.5–5.5, respectively (Kluczek-Turpeinen et al., 2003).

3.7 CUTIN AND SUBERIN

A hydrophobic lipid layer that covers the epidermis of fruits, leaves, flowers, and newly emerging stems is termed as a cuticle (Samuels et al., 2008). This hydrophobic layer comprises various wax crystals that are embedded in a cutin layer, a non-soluble polyester of lipid (Kolattukudy, 2001). Cutin makes up 40 to 80% by weight of the cuticle, and chemically, it is a polymeric network of ester cross-linked oxygenated C16 and C18 fatty acids (Walton, 1990; Osman et al., 1995). The breakdown of the ester bonds occurs either by alkaline hydrolysis, transesterification, or any other method (Heredia, 2003).

Analytical modeling of cutin molecules has revealed that cavities with a spacing of 0.4–0.5 nm are present between cutin oligomer chains (Matas & Heredia, 1999), while X-ray diffraction has given values of 0.45 and 1.0 nm (Luque et al., 1995). These cavities are further occupied by other cuticle components. Cutin is a major barrier between host cells and pathogen interaction, and many enzymes are released by fungal pathogens for their degradation.

Cutinase is produced by *Fusarium* spp. fungal pathogens, which can degrade cutin polyester (Baker & Bateman, 1978). It has also been suggested that *Fusarium* PCL degraders are the true cutinases. Non-mutant strains of *Fusarium* spp. had the ability to degrade both PCL and cutin, while mutant strains could not break down both cutin and PCL (Murphy et al., 1996). Cutinase is the enzyme that gives virulence to the fungi for their entrance into host cells. Cutinases are produced by different species of fungi like *Fusarium solani* (Alisch-Mark et al., 2006; O'Neill et al., 2007), *Pichia pastoris* (Munari, 2019), *Aspergillus oryzae*, *Penicillium citrinum*, and *Humicola insolens*, which were identified by Liebminger et al. in 2007.

Suberin is a macromolecule formed in terrestrial plants when a barrier toward the surrounding cells is needed. A polyaliphatic domain closely linked with the polyaromatic domain produced from ferulic acid makes up suberin (Kolattukudy, 2001; Steinbuchel, 2001; Bernards, 2002). A few soil fungal pathogens were observed to be involved in suberin degradation, yet its degradation had yet to be examined extensively (Kolattukudy, 1981). The breakdown of the suberin aromatic domain when *Aspergillus nidulans* was cultured on suberized cell walls was confirmed by the presence of some monomers while the aliphatic domain remained intact (Martins et al., 2014). It was demonstrated that *Rosellinia desmazieresii* produced a lot of cell wall-degrading enzymes, which included esterase enzymes that could breakdown suberin (Ofong & Pearce, 1994).

3.8 PLANT DISEASE RESISTANCE AND FUNGAL TOXINS

Mycotoxins are secondary metabolites produced by fungi that are toxic to humans, animals, microbes, and plants. Most mycotoxins are produced by three fungal genera: *Aspergillus*, *Penicillium*, and *Fusarium*. Seed quality, germination, viability, seedling vigor, root growth, and coleoptile integrity are especially influenced by these metabolites. Biological effects on plant physiology occur due to mycotoxins, even without disease symptoms.

The term "mycotoxin" first came into use in the 1960s. Bennett defined mycotoxins as "natural products produced by fungi that evoke a toxic response" (Bennett, 1987).

> Bennett and Bentley define mycotoxins as secondary metabolites that are metabolic products, found as a differentiation product in restricted taxonomic groups, not essential to growth and life of the producing organism, and biosynthesized from one or more general metabolites by a wider variety of pathways than is available in general metabolism.
>
> **(Bennett & Klich, 2003)**

The two categories of toxic fungi that damage seed crops are known as field pathogens (e.g., *Cladosporium*, *Fusarium*, *Alternaria* spp.) and storage fungi, which access seeds while the seeds are being stored, like *Aspergillus* and *Penicillium* spp. (Martín et al., 2022).

Members of the *Penicillium* genus include mycotoxigenic species such as *P. crustosum*, *P. hirsutum*, *P. commune*, *P. cyclopim*, *P. madriti*, and *P. palitans*. These fungi produce several toxins such as citrinin (CTN), patulin (PAT), penicillic acid (PA), penitrem A (PNT), palitantin (PAL), gliotoxin (GT), and rubratoxin B (RB) (Frisvad & Filtenborg, 1983; Frisvad et al., 1996; Müller & Amend, 1997; Frisvad & Samson, 2004; Yamaji et al., 1999, 2005; Ismaiel & Papenbrock, 2014). In many regions of the world, these fungi and their mycotoxin components infect seeds and cause losses in agricultural products (Keromnes & Thouvenot, 1985; Koteswara et al., 2014).

Many *Fusarium* members also produce different phytotoxic compounds, such as fusaric acid (FA), fumonisins (fumonisin B1, FB1), enniatin (ENN), moniliformin (MON), and trichothecenes. These toxins have biological activities and cause morphological, physiological, and metabolic changes in plants like necrosis, chlorosis, wilting, and growth inhibition (Van Asch et al., 1992; Nelson et al., 1994; McLean, 1996a; Desjardins & Hohn, 1997).

3.9 MYCOTOXINS OF *ASPERGILLUS* AND *PENICILLIUM* SPP.

3.9.1 Aflatoxins and Toxicity Levels

The discovery of aflatoxin is unique, and it has opened the gates for research on common molds because of its toxicity and carcinogenic effects on food. Aflatoxins are synthesized by more than 16 species (Frisvad et al., 2019), especially strains of *Aspergillus flavus* and *A. parasiticides*, which are very common in maize grains and cobs. Aflatoxins are diverse and can be classified as B1, B2, G1, G2, M1, M2, AFL, and Q1 types. *A. flavus* alone produces 18 different types of aflatoxins, from which aflatoxin B1 is considered the major threat, and it is extremely toxic (Richard et al., 2003; Ismaiel & Tharwat, 2014). Aflatoxins produced by pathogens in plants have multilayer effects, such as changes in the photosynthesis ability of plants and the interruption of chlorophyll and carotenoid synthesis, which results in albinism in affected plants (Reiss, 1978).

Aflatoxin interferes with protein interactions and metabolic enzyme inhibition and affects RNA polymerase. Aflatoxin binding occurs in plasma albumin. AFB1 is oxidized by cytochrome P450s, which produces two epoxides, which subsequently engage in an interaction with the lysine amino group to produce adducts of aflatoxin B1 with albumin (Haschek et al., 2001; Scholl & Groopman, 2008; Wu et al., 2005). Aflatoxins employ phagocytosis and chemotaxis, as well as reduced immunological responses mediated by the cell Awuchi et al., 2021. The metabolism of carbohydrates is influenced by AFB1 (Domijan & Fumonisin, 2012; Jenkins et al., 2000).

Plants treated with a high concentration of aflatoxin have reduced tillers, and a non-aflatoxin strain of this mycotoxin has the capacity to colonize plant hosts such as cotton bolls (Cotty, 2007). *Aspergillus flavus* is a soil-dwelling organism that grows there. Aflatoxin B1 temporarily lives in soil for approximately 5 days at 28°C; however, when a substrate is available for fungal growth, such as maize wastes, it is continually produced.

Aflatoxins produced by *Aspergillus* spp. can be accumulated in roots, enter the different organs degraded there, or return to the medium by diffusing. The hyperparasitism of *A. flavus* and *A. terreus* was also investigated regarding *Fusarium oxysporum* f. sp. *melonis*. As a result of climatic effects, if the environment is good for the growth of molds, there is an increase in aflatoxin contamination in crops like maize (Battilani et al., 2016). Temperatures between 13°C and 40°C are considered the best to produce aflatoxins, but still, it is unknown when or under which conditions the fungus decides to produce the toxins, though it is possible only when favorable conditions prevail to produce aflatoxins.

3.10 CITRININ TOXINS AND THE FUNGAL GENOME

Hetherington and Raistrick first isolated citrinin from cultures of *Penicillium citrinum* Thom (Hetherington & Raistrick, 1931). Other fungi from three different genera, *Penicillium* (*P. expansum*, *P. verrucosum*), *Aspergillus* (*A. terreus*), and *Monascus* (*M. ruber*) also generate citrinin (Ciegler et al., 1977; Bragulat et al., 2008). CTN infects maize (Nelson et al., 1985), wheat, rye, barley, oats (Scott et al., 1972), and rice (Tanaka et al., 2007). CTN also has antibiotic effects against gram-positive bacteria.

A CTN screening test showed phytotoxic effects in many trials (White & Truelove, 1972; Damodaran et al., 1975; Betina, 1989; Maćias et al., 2000). Bean plants exhibiting wilting due to CTN (*Dolichos lablab*, *Phaseolus vulgaris*, and *Phaseolus mungo* Co-1), as did two varieties of cotton (*Gossypium arboreum* K6 and 0320–1) and sorghum (*Sorghum vulgare* Co-20) (Damodaran et al., 1975). Damodaran et al. (1975) explained that CTN binds with the cellular components of the leaf and vascular system and, in this way, affects the osmotic balance and transfer of minerals. CTN also affects seedlings and the respiration of germinating seeds. CTN is correlated with yellow rice disease in Japan (Bennett & Klich, 2003).

3.11 FUNGAL PATULIN AND ITS TOXICITY LEVEL

The polyketide mycotoxin patulin is primarily produced by *Penicillium* species and less so by *Aspergillus*. Chain et al. (1942) isolated it from *Penicillium claviforme*; that is why it is also called claviforme. Blue mold rot mostly infects fresh fruits, juices, and jams. Fungus, which produces PAT, mostly infested the fruits at 0°C and increased in the rotten parts of the fruits when they were kept at room temperature (20°C) after storage (Sanderson & Spotts, 1995). The phytotoxicity of PAT has been noticed in isolated tissues of maize, wheat, and germinating seeds. Restraining the formation of proteins, RNA, and DNA is the primary effect of PAT. RNA polymerase is also inhibited by patulin (Tripathi & Misra, 1981; Tashiro et al., 1979). The application of PAT on cultivated plants reduces seed number, seed weight, and flowers (Ismaiel & Papenbrock, 2014; Ellis & McCalla, 1973).

3.12 OCHRATOXIN AS A MAJOR FUNGAL TOXIN

Ochratoxin is produced by *Aspergillus* spp. and *Penicillium* spp. present in grapes, nuts, and dried fruits. In 1965, ochratoxin was found in *A. ochraceus* as a new mycotoxin (Van Der Merwe, 1965). *P. verrucosum* can also produce ochratoxins, which are major types of toxins. OTA is mostly extracted from different cheese and cereal samples and is produced by *P. nordicum* and *P. verrucosum* (Bogs et al., 2006). Advanced investigations on OTA indicated that cell death occurs in plants because when the suspension of OTA is infiltered into the leaves of *Arabidopsis thaliana,* macroscopic lesions are observed within 1–2 days after application (Wang et al., 2012). By starting an oxidative burst, ochratoxin causes damaging lesions on *A. thaliana* leaves and slows the growth of the roots and plantlets (Peng et al., 2010). Many experiments showed the effects of OTA toxin on *Zea mays* embryos (McLean, 1996b). Researchers also studied the impacts on the growth of the roots and the shoots. Ochratoxin A suppresses the production of protein by competing with phenylalanine-tRNA ligase (Foroud & Eudes, 2009; Pestka, 2008). The metabolism of carbohydrates is influenced by OTA (Domijan & Fumonisin, 2012; Jenkins et al., 2000).

3.13 SECONDARY FUNGAL METABOLITES AS TOXINS

Penicillium tuberculum Bainier produces the mycotoxin penicillic acid as a secondary metabolite. During experimental work, fungi were identified and isolated to show a relation between the incidence of pellagra and mold decay of maize (Alsberg & Black, 1913). *P. verrucosum* (Gedek et al., 1981), *P. aurantiovirens* (Lund & Frisvad, 1994), and *P. aurantiogriseum* also produce this toxin. PA could reduce seedling respiration in plants (Bastin & van Roey, 1954). The primary modification PA causes in seedlings is an important change in the root system of lettuce plants (Yamaji et al., 2005). Penicillic acid prevents or halts the formation of DNA, RNA, proteins, and certain enzymes. Ambiguity exists regarding the precise mode of action (Wilson, 1976).

3.14 MYCOTOXINS OF *FUSARIUM* AND OTHER SPECIES

3.14.1 Trichothecenes

Mycotoxin-manufacturing *Fusarium* species are important pathogens in cereals such as wheat, oats, barley, and maize (Nganje et al., 2004). More than 120 trichothecenes are known. In 1948, Freeman and Morrison first (Freeman & Morrison, 1948) isolated trichothecenes from *Trichotheceum roseum*. Trichothecenes are classified as the strongest mycotoxins because they cause toxicity in plants and animals (Proctor et al., 2009). *Fusarium* spp. release trichothecene toxins in plants, which causes wilt, necrosis, and mortality in plants because many diseases occur such as stalk rot, root decline, and leaf crumbling in many major crops and ornamental plants (Ueno, 1983; Cutler, 1988; Cheeke, 1998; CAST, 2003; Harris et al., 1999; Wang et al., 2006). Examples of this kind of wilt include kudzu (Abbas et al., 2002) and duckweed (Abbas et al., 1998). Many studies showed that trichothecenes also inhibit protein formation in eukaryotes using different mechanisms (Cundliffe et al., 1974; Cundliffe & Davies, 1977). They interact with eukaryotic ribosomes and can permeate into cells, where they can impede translation (McCormick et al., 2011; Pestka, 2008). Trichothecene prevents mitochondrial activities, lipid peroxidation, apoptosis, and the formation of DNA, RNA, and proteins (Mostrom & Raisbeck, 2012; Wan et al., 2015). Trichothecenes break the polysome, which is why they are categorized into two types. The inhibition of protein translation is done by type I (T type), and the inhibition of elongation and termination is done by type II (ET type) (McLaughlin et al., 1977). The effect of trichothecenes is highly dependent on the host plant species. For example, *Arabidopsis* and wheat are mainly affected by T-2 toxin and 15-acetyldeoxynivalenol, respectively (Perincherry et al., 2019). After *Arabidopsis* was treated with T-2 toxin, many morphological changes were observed in this plant, like the curling of dark green leaves and shrunken cells (Masuda et al., 2007).

3.15 FUSARIUM TOXINS IN CEREALS

Such toxins were first recognized by Cole et al. (1973). *Fusarium proliferatum*, *F. nygamai*, *F. subglutinans*, *F. thapsinum*, *F. avenaceum*, *F. tricinctum*, and *F. fujikuroi* are some of the strains that produce them (Desjardins, 2006). T-2 toxin is more toxic than moniliformin, but the capability of photosynthetic pigments is highly affected by moniliformin toxin and inhibits plant growth and development (Vesonder et al., 1992). The inclusion of pyruvate into the tricarboxylic acid cycle is inhibited by MONs. Additionally, MON prevents the oxidation of TCA intermediates (Thiel, 1978). Gathercole et al. (1986) observed that MON decreased the activity of the enzyme pyruvate dehydrogenase and suggested that due to its structural resemblance, MON competes with pyruvate for complex formation with thiamin pyrophosphate (co-factor) and, consequently, the binding site of the enzyme. Significant impeded effects of this toxin on wheat leaf decreased the fresh mass of wheat plantlets (Wakulinski, 1989).

3.15.1 Fumonisin B1

Many *Fusarium* spp. produce the toxin fumonisin or fusaric acid, like the important fungi *F. verticillioides* and *F. nygamai* (Thiel et al., 1991; Rheeder et al., 2002). *F.*

moniliforme MRC 826 was the first *Fusarium* spp. from which fumonisin was isolated by Gelderblom et al. (1988). FB1 causes many toxic effects on plants in general and in different economically vital crops especially (Van Asch et al., 1992; Abbas et al., 1998, 1991; Danielsen & Jensen, 1998; Lamprecht et al., 1994). For example, soybeans were devastatingly affected by FB1 (Abbas & Boyette, 1992). These toxins of *Fusarium* attack the membranes of the cells during their interactions with plants and animals (Abbas et al., 1993). Fumonisin is involved in protein and mitochondrial interactions. Fumonisin B1 has a higher toxicity level and can change the levels of mRNA (Liu et al., 2005; Hanumegowda et al., 2002). This toxin impeded the synthesis of the enzyme ceramide, and consequently, cell signaling and function were stopped (Merrill et al., 2001). When maize seeds are soaked in fumonisin toxin, 75% radical prolongation occurs (Doehlert et al., 1994). When plants are infected by fumonisin, many physiological changes occur, such as necrosis and wilting (Abbas & Ocamb, 1995). The starch content of the endosperm is substantially degraded in grains when infected by fumonisin (Jackowiak et al., 2002).

3.15.2 Fusaric Acid (FA)

Some *Fusarium* spp. and entomopathogenic fungi produce fusarins (Krasnoff et al., 2006). Fusarin A, B, C, and D were recognized in 1981 by Wiebe and Bjeldanes in Berkeley, California (Wiebe et al., 1981). Fusarium species that are more distantly related to one another, such as *F. heterosporum*, *F. napiforme*, *F. oxysporum*, and *F. solani*, as well as members of the *Gibberella fujikuroi* species complex (GFC), such as *F. fujikuroi*, *F. proliferatum*, and *F. circinatum* produce different kinds of fusarins. In some species like *F. mangiferae* and *F. verticillio*, high toxins levels are detected during fungal infection (Bacon et al., 1996; Sieber et al., 2013; Brown et al., 2012; Abbas et al., 1989). As a breathing inhibitor, fusaric acid induces similar symptoms to tuberculosis in humans, and these toxins can alter the roots of maize as well (Arias, 1985; Telles-Pupulin et al., 1996). Fusaric acid inhibits electron flow between succinate dehydrogenase and coenzyme Q, as well as the activities of ATPase/ATP-synthase and, possibly, ketoglutarate dehydrogenase in the isolated mitochondrial metabolism of corn roots (Telles-Pupulin et al., 1996). Fusaric acid weakened the leaf cell defense mechanism of watermelon seedlings by accelerating lipid peroxidase activity in the leaves (Wu et al., 2008). Plants exposed to FA experience harmful consequences such as altered membrane permeability (modification of cell membrane potential), decreased mitochondrial activity and oxygen uptake, the suppression of ATP production, and a reduction in root growth (Shilina et al., 1973; D'Alton & Etherton, 1984; Marré et al., 1993; Bouizgarne et al., 2006). When fusaric acid is applied to tomato plants, chlorophyll pigmentation decreases, and proteolytic enzyme activity increases, which ultimately lessens the cellular metabolic process and photosynthesis rate, affects the cell structure, and causes wilt (Singh et al., 2017).

3.15.3 Zearalenone (ZEA)

The leading toxin of cereals and grains is the mycotoxin ZEA, also known as F-2 toxin (Cheeke & Shull, 1985). Moreover, this toxin has also been witnessed in liquids like malt and beer. Fungi of the genus *Fusarium* contaminate cereals both in the field and in storage. Usually, this toxin is produced in the field, but due to mismanagement, this toxin is also found in storehouses (CCFAC, 2000). Zearalenone (ZEA) is mainly generated by *F. graminearum* and *F. culmorum* and predominantly occurs in maize and other grain crops (Al-Jaal et al., 2019; Caglayan et al., 2020). Many genes are individually or together responsible for the production of toxins in these species of fungi. ZEB1, ZEB2, PKS4, and PKS13 are four neighboring genes that make up a gene cluster necessary to produce ZEN (Kim et al., 2018). Various *Fusarium* fungi produce ZEA, such as *F. graminearum*, *F. cerealis*, *F. equiseti*, and *F. semitectum*; these are topsoil fungi and infect global crops as per the reports of Bennett and Klich (2003).

Zearalenone is less toxic but reduces the length of roots and shoots and fresh mass storage in growing embryos of *Z. mays* (McLean, 1996b). ATPase enzyme activity is also decreased in maize coleoptiles by this toxin. The penetrability of the plasmalemma and tonoplast is also altered according to the results of Vianello and Marci (Vianello & Macri, 1978). According to this study, up to 50% of the mycotoxin was found to bind to starch, hemicellulose, and especially lignin fractions, indicating that ZEA metabolites bind to the elements of cell walls.

3.16 TOXINS OF *FUSARIUM* IN FOOD ITEMS

Fusarium species produce enniatins (ENs) and beauvericins (BEA) that are frequently found in many food items (Urbaniak et al., 2019). The pathogen *Beauverina bassiana* is the first fungus, from which beauvericin toxin was isolated for the first time (Hamill et al., 1969). *Fusarium subglutinans* was the first species discovered to synthesize BEA (Gupta et al., 1991). It inhibits enzymes and induces oxidative stress at the cellular level (Juan et al., 2014).

Fusarium orthoceras, later termed *Fusarium oxysporum*, produces enniatins (ENs) (Plattner & Nager, 1947). Seven of the 29 known naturally occurring enniatin analogs (ENs A, A1, B, B1, B2, B3, and B4) have been discovered in cereals (Sy-Cordero et al., 2012). According to a recent study, EN B1 can destabilize the membrane proteins that relate to lysosomes, which causes the lysosomes to become alkaline and portions of their membranes to become permeable (Oliveira et al., 2020). Damaging mitochondrial pathways triggers DNA fragmentation and apoptosis (Mallebrera et al., 2018). Paciolla et al. described that phytotoxicity in tomato protoplasts was mainly mediated by BEA and T-2 toxin (Paciolla et al., 2004).

3.17 *ASPERGILLUS*, *PENICILLIUM*, AND *FUSARIUM* INFECTIVITY IN PLANTS

Fungi have been known for centuries as foremost plant pathogens. They target different parts of the plant for nutrition. Fungi not only parasitize plants for nutrition but also cause typical symptoms. These fungal pathogens also produce toxins or secondary metabolites that are not safe for human consumption.

3.18 SUPPRESSION OF RESISTANCE IN PLANTS BY FUNGAL TOOLS

Pathogens and plants have engaged in a ferocious but quiet tug-of-war from the beginning of plant–pathogen interaction (De Wit et al., 2009). Pathogenic microorganisms take nutrients from plants for their existence and reproduction; in return, plants activate many defense mechanisms to impede the survival of pathogens (Ngou et al., 2022). Pathogens release effectors that suppress the immunity of plants and cause physiological changes. Effectors are proteins that are excreted by pathogens on plant surfaces or in intracellular spaces (Duplessis et al., 2011; Saitoh et al., 2012; Gan et al., 2013; Giraldo et al., 2013); these proteins either trigger immunity in plants or cause diseases in plants. Harold Flor demonstrated that comparable gene pairs determine how plants inherit their resistance to diseases and the pathogens' capacity to produce disease (Flor, 1942). The plant R gene and pathogen avr gene interact with each other, and this interaction is conferred by a resistant response (Kawakubo et al., 1990). Pathogens suppress the immunity of the host during infection. By growing specialized infection hyphae inside the plant cells from which nutrients are absorbed, biotrophic fungi form a strong bond with their host (Mendgen & Hahn, 2002). However, necrotrophic fungi release enzymes and toxins in plant cells that kill the cells and tissues and absorb nutrients from dead tissue (Horbach et al., 2011). Both mechanisms are combined in hemibiotrophic fungus (Horbach et al., 2011). Plant diseases release many effector proteins into the plant apoplast or cytoplasm to aid in infection. The majority of our knowledge of plant pathogen effectors comes from research on prokaryotic pathogens, such the bacterium *Pseudomonas syringae*, which secretes roughly 30 effectors into the plant cell via a type three secretion system (T3SS) (Chang et al., 2005; Schechter et al., 2006). It is not clear whether filamentous pathogens such as fungi and oomycetes release effectors in plant cytoplasm or not to inhibit the defensive responses and to change the plant metabolism (Ellis et al., 2009; Kamoun, 2009; Panstruga & Dodds, 2009).

A successful pathogen must first protect itself from deteriorating enzymes or harmful substances like H_2O_2. Then, it must prevent the production of signals that could trigger or even amp up defense responses. There are three ways for pathogens to deal with this difficulty: masking, detoxification/sequestering, and suppression (Doehlemann & Hemetsberger, 2013). Many pathogen-associated molecular patterns (PAMPs) are present in the invading oomycetes and filamentous fungi in certain plant diseases. These PAMPs are identified by associated PRRs of the plant, which, among other things, cause the development of cell wall-degrading enzymes (CWDEs), or pathogenesis-related (PR) proteins, the activation of which released additional defense elicitors. Recognition between the pathogen's elicitors and the plant's receptors during interaction is crucial for a resistant or susceptible response. Microbes try to alter the composition and structure of their cell walls to prevent or, at the very least, reduce recognition by plants to break out of this vicious cycle. Chitin, mannans, and β-glucans make up the majority of the components of the fungal cell wall, with β-glucan cross-linked to chitin comprising the essential elements (Latge, 2007). Since chitin and β-glucans are the two main

PAMPs found in fungal cell walls, many alterations are made to lessen their perception. For example, the rust fungi *Uromyces fabae* and *Puccinia graminis* f. sp. *tritici* can alter chitin to chitosan, which is present in the infection structures of many plant diseases.

In addition to decreasing PAMP release, this change also defends the cell wall against plant chitinases (El Gueddari et al., 2002; Werner et al., 2007). *Magnaporthe oryzae*, a rice blast fungus found in plants, modulates the cellular surface by α-glucan apposition during pathogenic growth (Fujikawa et al., 2009). Recent reports show that surface α-(1,3)-glucan, which plants cannot digest, enhances "silent infection" by hiding the pathogen from being recognized by plant host defense (Fujikawa et al., 2012). This observation led to the development of hybrid rice plants exhibiting a bacterial α-(1–3)-glucanase that were highly resistant to fungal infection (Fujikawa et al., 2012). These processes depend on the interaction between host cell surface proteins and pathogen effector proteins.

For example, HBGAs (Lewis b) glycosylated the mucin MUC5AC, which led to *Helicobacter pylori* infection, but GlcNAc-capped glycans glycosylated the mucin MUC6, which impeded *H. pylori* infection (Kawakubo et al., 1990). Host protein glycosylation has a significant impact on pathogen interaction and penetration (Hooper & Gordon, 2001; Bhat et al., 2019).

Pathogens use their glycosylated proteins as tools to attack their hosts (Lin et al., 2020). The primary glycoproteins in pathogens that can facilitate infection are cell surface proteins and secretory proteins (Lin et al., 2020). Increased secreted proteins, also known as effectors, boost pathogen virulence (Li et al., 2013; Chen et al., 2017). For example, the pathogenic fungus *Magnaporthe oryzae* secreted the LysM protein1 (Slp1). Rice (*Oryza sativa*) chitin elicitor binding protein (CEBiP) detects PAMPs of *M. oryzae*. However, Slp1 is a major virulence component of *M. oryzae* that can bond with the PAMP chitin to avoid detection, allowing rice blast fungus to resist host innate immune response successfully (Mentlak et al., 2001). N-glycosylation is necessary for the Slp1 chitin-binding process to be active and stable. N-glycosylation is a vital component for pathogen virulence, confirmed by many other effectors such as BAS4 (from *M. oryzae*) (Chen et al., 2014), CBH1 (from *Trichoderma reesei*) (Qi et al., 2014), and PCIPGII (from *Phytophthora capsici*) (Sun et al., 2009).

According to the zigzag model, *Phytophthora* species have different proteins called effectors to interact with plant proteins to alter PTI and cause effector-triggered susceptibility (ETS) (Jones & Dangl, 2006). According to their location in the plant cell, effectors from *Phytophthora* spp. can be categorized either as cytoplasmic effectors or apoplastic actuators. Cytoplasmic effectors are transmitted into cells, and apoplastic actuators are released in the apoplast of the host cells (Mcgowan & Fitzpatrick, 2017). Blocking the action of positive immune regulatory bodies and supporting the activity of negative immunological regulators are techniques used by phytophthora pathogens to restrain the immune system. While cytoplasmic effectors and RXLR effectors have been found to target both the positive and negative regulators of defense responsible for a variety of cellular activities at multiple subcellular sites, to date, all apoplastic actuators have been found to target the positive immune regulatory bodies (Naveed et al., 2020).

3.19 FUNGI EXPLOIT RNA SILENCING FOR PARASITISM

RNA silencing is a mechanism that regulates genes with specific nucleotide sequences. It controls development and defends several eukaryotes against infection-causing organisms. It is common in antiviral immunity. RNA silencing has remained a topic of interest in recent years (Molnar et al., 2011). The first paper on RNA silencing was published in 1928. In this paper, Wingard explained the recovery of tobacco plants from the tobacco ringspot virus by silencing virus RNA (Wingard, 1928). In plants, there are a minimum of three pathways of RNA silencing for the silencing of a particular gene. The first is cytoplasmic RNA silencing, the second is miRNA silencing of the endogenous messenger RNA, and the methylation of DNA and inhibition of transcription are linked with the third pathway of RNA silencing. Green plants can silence using these three methods (Baulcombe, 2004).

The RNA silencing mechanism is mediated by a protein present in plants, fungi, and animals named Argonaute proteins (Baulcombe, 2004). For example, *Arabidopsis thaliana* is a model plant with four Dicer-like and ten Argonaute paralogs. These have a specific role in RNA silencing (Margis et al., 2006; Hutvagner & Simard, 2008).

In plants, Dicer obtains single-stranded RNAs from the dsRNA, integrates them into Argonaute, and uses the base pairing of Watson–Crick to guide the complex of AGO to point out the nucleic acid. This process helps in the transcriptional or post-transcriptional regulation of selected sequences. This is the mechanism involved in inhibiting translation (Ghildiyal & Zamore, 2009).

Recent research on RNA silencing indicates that silencing movement is limited to a range of about 10 to 15 cells. This is distinct from transmission, which is long distance. This shows that silencing signals may depend upon more than one silencing pathway (Molnar et al., 2011).

When a pathogen attacks, it introduces its nucleic acid to plant cells. As a result, dsRNA is produced and processed in sRNA, which specifies the sequence of observed silencing. The length of these sRNA species is 21 to 24 nucleotides. These are also called small interfering RNAs (Eamens et al., 2008).

Aspergillus, *Fusarium*, and *Penicillium* cause major damage to food due to the production of mycotoxins and other secondary metabolites (Majumdar et al., 2017). Mycotoxins have a bad impact on the economy. In Canada and the USA, losses are almost US 0.5–1.5 billion dollars/year (Wu, 2006). In recent years, it has been estimated that the USA could lose 52.1 million to 1.68 billion dollars annually from the corn industry due to aflatoxin contamination (Mitchell et al., 2016). The host shows a response to this act in different ways, one of which is RNA interference-based host-induced gene silencing. This targets the specific genes that help in proper growth and is very efficient, helping the host to combat mycotoxigenic fungi that cause loss (Majumdar et al., 2017).

In a previous report, it was reported that host infection can be prevented by producing Dicer-like (dcl) mutant and change in Argonaute (AGO) proteins. *Colletotrichum higginsianum* causes anthracnose disease, which particularly infects species of Brassicaceae. In *Colletotrichum higginsianum* the dcl1. dcl1dcl2 double mutant and modified ago1 strain produce irregularity in conidiation and conidia morphology

(Campo Sánchez et al., 2016). This minimizes the expression of dsRNA mycovirus (ChNRV1) and causes the development of defective conidia (Majumdar et al., 2017).

Fungi attack plants mainly for food. To fulfill their requirements, they must overcome the host defense system. One of these mechanisms is RNA silencing, which prevents plants from pathogen attacks. Pathogen sRNA silences the genes of the host to suppress its immunity and for infection. In the case of *Botrytis cinerea*, its sRNA takes over the RNA interference machinery of a plant by binding to an Argonaute protein. This whole process causes the silencing of the host gene. In the same way, *Verticillium dahlia* works to cause infection and pathogenicity (Ellendorff et al., 2009).

Small RNAs (sRNAs) in the fungal pathogen *Botrytis cinerea* can use the host RNA silencing machinery to suppress resistance (Weiberg et al., 2013). Through genomic analysis, scientists have discovered that rust fungi have hundreds of genes expressed in haustorial cells. These genes are predicted to encode small proteins secreted into living host plant cells, and while some have been shown to inhibit plant defense, their precise molecular functions are mostly unknown (Petre et al., 2014; Ramachandran et al., 2017). One possible way that these proteins might achieve their inhibitory effects is by interfering with RNA silencing.

Researchers have discovered a new protein called PgtSR1, produced by the rust-causing fungi *P. graminis*. PgtSR1 could suppress RNA silencing in plants, and its expression in wheat and Arabidopsis affects the abundance of small RNAs important for plant immunity. When PgtSR1 is expressed, it makes the plant more susceptible to various pathogens and suppresses the cell death response triggered by R proteins. This suggests that PgtSR1 is capable of suppressing both basal defenses and effector-triggered immunity. The discovery of PgtSR1 represents the first evidence that *P. graminis* has evolved to produce fungal suppressors of RNA silencing, and it also highlights the common strategy employed by many plant pathogens to evade host defense responses (Yin et al., 2019).

3.20 FUNGI OVERCOME ROS PATHWAYS FOR COMPATIBLE INTERACTIONS

Fungi cause up to 60% losses in plants, encouraging us to study fungal–plant interactions in more detail. During this research, we can learn about morphological and physiological changes that occur in plants. The establishment of disease fungi overcomes the ROS pathway of plants. Reactive oxygen species (ROS) are produced by NADPH-dependent oxidase enzyme complexes in many multicellular organisms and were first identified in mammalian phagocytes. NADPH Oxidase (NOX) complexes also exist in plants, which act against pathogen attacks (Segal & Wilson, 2018). ROS is a component of the innate immune response. ROS production sites include chloroplasts, cell walls, the endoplasmic reticulum, apoplasts, peroxisomes, plasma membranes, and mitochondria in plant cells (Sharma et al., 2012). ROS include singlet oxygen, superoxide, hydroxyl, and hydrogen peroxide. They can interact with biomolecules and cause the inactivation and the degeneration of enzymes, DNA damage, and the peroxidation of lipids (Beckman & Ames, 1998). ROS work with other molecules like nitric acid and calcium ions to fight against fungal pathogens and act as a hindrance against their spread. *Verticillium dahliae* is a fatal fungal pathogen that causes wilting in many species of plants. During the study of their interaction with different hosts, it was concluded that ROS-mediated defenses help us select transgenic and resistant genotypes (Ali et al., 2018).

Mannitol is defined as a six-carbon acyclic polyol that is present in plants as well as fungi (Meena et al., 2015). But here, we will discuss the role of mannitol in fungi against plant defense. When pathogenic fungi invade a host such as a plant, ROS generation will be activated by NADPH oxidase embedded in the plasmalemma (Bolwell & Wojtaszek, 1997). ROS will act as signals for the activation of resistant components. These include hypersensitive responses and systemic acquired resistance. Some fungi that cause disease in plants use mannitol, which restrains plant defense mediated by reactive oxygen species. For example, *Alternaria alternata*, which is phytopathogenic fungi, releases mannitol during its attack on plants (Jennings et al., 1998).

Pathogens show different behaviors in overcoming high levels of reactive oxygen species. Yapl, which is known as yeast-activated protein 1, is a common example of a particular mechanism used by eukaryotic cells for sensing ROS and regulating genes. Overall, Yapl homologs and the peroxiredoxins that activate transcriptional factors (Yapl) by H_2O_2 are important regulators of antioxidant activity. This is mainly prominent in filamentous fungi (Segal & Wilson, 2018).

3.21 THE ROLE OF YAP1 HOMOLOGS IN FUNGI AGAINST ROS DETOXIFICATION

Botrytis cinerea is a plant pathogen that infects many plant species, especially grapevines. Plants produce reactive oxygen species (ROS) in response to this fungus, which is an early plant response. But instead of avoiding the ROS environment, the fungus also promotes the production of ROS in plants. The AP-1 transcription factor Bapl has been identified as a key regulator of ROS detoxification *in vitro*. However, Bapl is not essential for the fungus's pathogenicity. This suggests that the fungus does not experience H_2O_2-induced oxidative stress *in planta*, challenging traditional views on the role of the oxidative burst in the infection process (Temme & Tudzynski, 2009). *Ustilago maydis* is biotrophic in nature. The transcriptional regulator of this fungus for the detoxification of host plant ROS is UmYapl (Molina & Kahmann, 2007). *Magnaporthe oryzae* is a hemibiotroph with MoApl, which detoxifies reactive oxygen species of rice (Guo et al., 2011).

Stripe rust fungus, also known as *Puccinia striiformis* f. sp. *tritici* (PST), is a plant pathogen that releases effectors to harmonize with the defense systems of wheat. During early infection, the glycine–serine-rich effector gene PstGSRE1 is produced by the pathogen in excess amounts. PstGSRE1 mainly focuses on the transcriptional factor TaLOL2, which is correlated with reactive oxygen species. Then, it alters the amino acid sequences of TaLOL2, which points out proteins for transportation into the nucleus of a cell, and controls the ROS-mediated death of the cell, which is prompted by TaLOL2. By using this strategy, stripe rust fungi overcome the ROS defense pathway of the wheat plant (Qi et al., 2019). Fungi adapted themselves in such a way that they combat the ROS of plants and cause infection. Every pathogen causes infection in the host according to specific patterns.

REFERENCES

Abbas, H. K., & Boyette, C. D. (1992). Phytotoxicity of fumonisin B1 on weed and crop species. *Weed Technology*, 6, 548–552.

Abbas, H. K., Boyette, C. D., Hoagland, R. E., & Vesonder, R. E. (1991). Bioherbicidal potential of Fusarium moniliforme and its phytotoxin, fumonisin B1. *Weed Science*, 139, 673–677.

Abbas, H. K., Duke, S. O., Merrill, A. H. Jr., Wang, E., & Shier, W. T. (1998). Phytotoxicity of australifungin, AAL-toxins and fumonisin B 1 to Lemna pausicostata. *Phytochemistry*, 47, 1509–1514.

Abbas, H. K., Duke, S. O., & Tanaka, T. (1993). Phytotoxicity of Fumonisins and Related Compounds. *Journal of Toxicology: Toxin Reviews*, 12, 225–251.

Abbas, H. K., Johnson, B. B., Shier, W. T., Tak, H., Jarvis, B. B., & Boyette, C. D. (2002). Phytotoxicity and mammalian cytotoxicity of macrocyclic trichothecene mycotocins from Myrothecium verrucaria. *Phytochemistry*, 59, 309–313.

Abbas, H. K., Mirocha, C. J., Kommedahl, T., Vesonder, R. F., & Golinski, P. (1989). Production of trichothecene and non-trichothecene mycotoxins by Fusarium species isolated from maize in Minnesota. *Mycopathologia*, 108, 55–58.

Abbas, H. K., & Ocamb, C. M. (1995). First report of production of fumonisin B1 by Fusarium polyphialidicum collected from seeds of Pinus strobus. *Plant Disease*, 79, 642.

Ahmed, A. A., Pedersen, C., & Thordal-Christensen, H. (2016). The barley powdery mildew effector candidates CSEP0081 and CSEP0254 promote fungal infection success. *PLoS One*, 11(6), e0157586.

Ali, M., Cheng, Z., Ahmad, H., & Hayat, S. (2018). Reactive oxygen species (ROS) as defenses against a broad range of plant fungal infections and case study on ROS employed by crops against Verticillium dahliae wilts. *Journal of Plant Interactions*, 13(1), 353–363.

Alisch-Mark, M., Herrmann, A., & Zimmermann, W. (2006). Increase of the hydrophilicity of polyethylene terephthalate fibres by hydrolases from Thermomonospora fusca and Fusarium solani f. sp. pisi. *Biotechnology Letters*, 28(10), 681–685.

Al-Jaal, B. A., Jaganjac, M., Barcaru, A., Horvatovich, P., & Latiff, A. (2019). Aflatoxin, fumonisin, ochratoxin, zearalenone and deoxynivalenol biomarkers in human biological fluids: A systematic literature review, 2001–2018. *Food Chemical Toxicology*, 129, 211–228.

Alsberg, G. L., & Black, O. F. (1913). Contributions to the study of maize deterioration: Biochemical and toxicological investigations of Penicillium puberulum and P. stoloniferum. *U.S. Department of Agriculture Bureau Plant Indian Bulletin*, 270, 1–47.

Arantes, V., & Milagres, A. M. (2009). Relevância de compostos de baixa massa molar produzidos por fungos e envolvidos na biodegradação da madeira. *Química Nova*, 32, 1586–1595.

Arantes, V., & Milagres, A. M. F. (2006). Degradation of cellulosic and hemicellulosic substrates using a chelator-mediated Fenton reaction. *Journal of Chemical Technology & Biotechnology: International Research in Process, Environmental & Clean Technology*, 81(3), 413–419.

Arias, J. A. (1985). Secretory organelle and mitochondrial alterations induced by fusaric in root of Zea mays. *Physiological Plant Pathology*, 27, 149–158.

Armentrout, V. N., & Downer, A. J. (1987). Infection cushion development by Rhizoctonia solani on cotton. *Phytopathology*, 77(4), 619–623.

Aro, N., Pakula, T., & Penttilä, M. (2005). Transcriptional regulation of plant cell wall degradation by filamentous fungi. *FEMS Microbiology Reviews*, 29(4), 719–739.

Awuchi, C. G., Ondari, E. N., Ogbonna, C. U., Upadhyay, A. K., Baran, K., Okpala, C. O. R., et al. (2021). Mycotoxins affecting animals, foods, humans, and plants: Types, occurrence, toxicities, action mechanisms, prevention, and detoxification strategies—A revisit. *Foods*, 10(6), 1279.

Azadi, P., O'Neill, M. A., Bergmann, C., Darvill, A. G., & Albersheim, P. (1995). The backbone of the pectic polysaccharide rhamnogalacturonan I is cleaved by an endo hydrolase and an endo lyase. *Glycobiology*, 5(8), 783–789.

Bacon, C. W., Porter, J. K., Norred, W. P., & Leslie, J. F. (1996). Production of fusaric acid by Fusarium species. *Applied and Environmental Microbiology*, 62, 4039–4043.

Baker, C. J., & Bateman, D. F. (1978). Cutin degradation by plant pathogenic fungi. *Phytopathology*, 68, 1577–1584.

Barnett, C. C., Berka, R. M., & Fowler, T. (1991). Cloning and amplification of the gene encoding an extracellular beta-glucosidase from Trichoderma reesei: Evidence for improved rates of saccharification of cellulosic substrates. *Biotechnology*, 9, 562–567.

Bastin, R., & van Roey, G. (1954). Action of some antibiotics and inhibiting substances on cellular stretching. P. *Pharmacy in Belgium*, 9, 112–127.

Battilani, P., Toscano, P., Van der Fels-Klerx, H. J., Moretti, A., Camardo Leggieri, M., Brera, C., Rortais, A., Goumperis, T., & Robinson, T. (2016). Aflatoxin B1 contamination in maize in Europe increases due to climate change. *Scientific Reports*, 6(1), 24328. doi: 10.1038/srep24328.

Baulcombe, D. (2004). RNA silencing in plants. *Nature*, 431(7006), 356–363.

Beckman, K. B., & Ames, B. N. (1998). The free radical theory of aging matures. *Physiological Reviews,* 78, 547–581.

Beguin, P. (1990). Molecular biology of cellulose degradation. *Annual Review of Micriobiology*, 44, 219–248.

Bennett, J. W. (1987). Mycotoxins, mycotoxicoses, mycotoxicology and mycopathologia. *Mycopathologia*, 100, 3–5.

Bennett, J. W., & Klich, M. (2003). Mycotoxins. *Clinical Microbiology Reviews*, 16, 497–516.

Bernards, M. A. (2002). Demystifying suberin. *Canadian Journal of Botany*, 80(3), 227–240.

Betina, V. (1989). *Mycotoxins: Chemical, Biological and Environmental Aspects*, Elsevier: New York, NY, USA.

Bhat, A. H., Maity, S., Giri, K., & Ambatipudi, K. (2019). Protein glycosylation: Sweet or bitter for bacterial pathogens? *Critical Reviews in Microbiology*, 45, 82–102.

Boerjan, W., Ralph, J., & Baucher, M. (2003). Lignin biosynthesis. *Annual Review of Plant Biology*, 54, 519–546.

Bogs, C., Battilani, P., & Geisen, R. (2006). Development of a molecular detection and differentiation system for ochratoxin A producing Penicillium species and its application to analyse the occurrence of Penicillium nordicum in cured meats. *International Journal of Food Microbiology*, 107, 39–47.

Bolton, M. D., Thomma, B. P. H. J., & Nelson, B. D. (2006). Sclerotinia sclerotiorum (Lib.) de Bary: Biology and molecular traits of a cosmopolitan pathogen. *Molecular Plant Pathology*, 7(1), 1–16.

Bolwell, G. P., & Wojtaszek, P. (1997). Mechanisms for the generation of reactive oxygen species in plant defence–A broad perspective. *Physiological and Molecular Plant Pathology*, 51(6), 347–366.

Bouizgarne, B., El-Maarouf-Bouteau, H., Frankart, C., Reboutier, D., Madiona, K., & Pennarun, A. M. (2006). Early physiological responses of Arabidopsis thaliana cells to fusaric acid: Toxic and signalling effects. *New Phytology*, 169, 209–218.

Bragulat, M. R., Martinez, E., Castella, G., & Cabanes, F. J. (2008). Ochratoxin A and citrinin producing species of the genus Penicillium from feedstuffs. *International Journal of Food Microbiology*, 126, 43–48.

Brown, D. W., Butchko, R. A. E., Busman, M., & Proctor, R. H. (2012). Identification of gene clusters associated with fusaric acid, fusarin, and perithecial pigment production in Fusarium verticillioides. *Fungal Genetics Biology*, 49, 521–532.

Caglayan, M. O., Şahin, S., & Üstündağ, Z. (2020). Detection strategies of Zearalenone for food safety: A review. *Critical Reviews in Analytical Chemistry*.

Campo Sánchez, S., Gilbert, K. B., & Carrington, J. C. (2016). Small RNA-based antiviral defense in the phytopathogenic fungus *Colletotrichum higginsianum*. *PLoS Pathogens*, 12, e1005640.

Caño-Delgado, A., Penfield, S., Smith, C., Catley, M., & Bevan, M. (2003). Reduced cellulose synthesis invokes lignification and defense responses in Arabidopsis thaliana. *The Plant Journal*, 34(3), 351–362.

CAST (Council for Agriculture Science and Technology). (2003). *Mycotoxins Risks in Plant, Animal, and Human Systems*, CAST: Ames, IA, USA.

Chain, E., Florey, H. W., & Jennings, M. A. (1942). An antibacterial substance produced by Penicillium claviforme. *British Journal of Experimental Pathology*, 23, 202–205.

Chanclud, E., & Morel, J.-B. (2016). Plant hormones: A fungal point of view. *Molecular Plant Pathology*, 17, 1289–1297.

Chang, J. H., Urbach, J. M., Law, T. F., Arnold, L. W., Hu, A., Gombar, S., et al. (2005). A high-throughput, near-saturating screen for type III effector genes from Pseudomonas syringae. *Proceedings of the National Academy of Sciences of the United States of America*, 102, 2549–2554.

Cheeke, P. R. (1998). *Natural Toxicant in Feeds, Forages, and Poisonous Plants*, Interstate Publishers Inc.: Danville, IL, USA.

Cheeke, P. R., & Shull, L. R. (1985). Mycotoxins. In *Natural Toxicants in Feeds and Poisonous Plants*, AVI Publishing Company Inc.: Westport, CT, USA, 393–477.

Chen, C. M., Gritzali, M., & Stafford, D. W. (1987). Nucleotide sequence and deduced primary structure of the cellobiohydrolase II from Trichoderma reesei. *Bio/Technology*, 5, 274–278.

Chen, J. S., Lin, B. R., Huang, Q. L., Hu, L. L., Zhuo, K., & Liao, J. L. (2017). A novel Meloidogyne graminicola effector, MgGPP, is secreted into host cells and undergoes glycosylation in concert with proteolysis to suppress plant defenses and promote parasitism. *PLoS Pathogens*, 13, e1006301.

Chen, X. L., Shi, T., Yang, J., Shi, W., Gao, X. S., Chen, D., Xu, X. W., Xu, J. R., Talbot, N. J., & Peng, Y. L. (2014). N-Glycosylation of effector proteins by an alpha-1,3-mannosyltransferase is required for the rice blast fungusto evade host innate immunity. *Plant Cell*, 26, 1360–1376.

Ciegler, A., Vesonder, R. F., & Jackson, L. K. (1977). Production and biological activity of patulin and citrinin from Penicillium expansum. *Applied and Environmental Microbiology*, 33, 1004–1006.

Ciuffetti, L. M., Manning, V. A., Pandelova, I., Betts, M. F., & Martinez, J. P. (2010). Host-selective Toxins, Ptr ToxA and Ptr ToxB, as necrotrophic effectors in the pyrenophora tritici-repentis—Wheat interaction. *New Phytology*, 187, 911–919.

Codex Committee on Food Additives and Contaminants (CCFAC). (2000). *Posting Date. Joint FAO/WHO Expert Committee on Food Additives: Position Paper on Zearalenone*, Publication CCFAC 00/19, Codex Alimentarius Commission: Rome, Italy.

Cohen, R., Suzuki, M. R., & Hammel, K. E. (2005). Processive endoglucanase active in crystalline cellulose hydrolysis by the brown rot basidiomycete Gloeophyllum trabeum. *Applied and Environmental Microbiology*, 71(5), 2412–2417.

Cole, R. J., Kirksey, J. W., Cutler, H. G., Doubnik, B. L., & Peckham, J. C. (1973). Toxin from Fusarium moniliforme: Effects on plants and animals. *Science*, 179, 1324–1326.

Cotty, P. J. (2007). Influence of field application of an atoxigenic strain of Aspergillus flavus on the populations of A. flavus infecting cotton bolls and on the aflatoxincontent of cottonseed. *Phytopathology*, 84, 1270.

Cundliffe, E., Cannon, M., & Davies, J. (1974). Mechanism of inhibition of eukaryotic protein synthesis by trichothecene fungal toxins. *Proceedings of the National Academy of Sciences of the United States of America*, 71, 30–34.

Cundliffe, E., & Davies, J. E. (1977). Inhibition of initiation, elongation, and termination of eukaryotic protein synthesis by trichothecene fungal toxins. *Antimicrobial Agents and Chemotherapy*, 11, 491–499.

Cutler, H. G. (1988). Trichothecenes and their role in the expression of plant disease. In *Biotechnology for Crop Protection*, Hedin, P. A., Menn, J. J., & Hollingworth, R. M. (eds.), Maple Press: York, PA, 50–72.

D'Alton, A., & Etherton, B. (1984). Effects of fusaric acid on tomato root hair membrane potentials and ATP levels. *Plant Physiology*, 74, 39–42.

Damodaran, C., Kathirvel-Pandian, S., & Seeni, S. (1975). Citrinin, a phytotoxin? *Cellular and Molecular Life Sciences*, 31, 1415–1417.

Danielsen, S., & Jensen, D. E. (1998). Relationships between seed germination, fumonisin content, and Fusarium verticillioides infection in selected maize samples from different regions of Costa Rica. *Plant Pathology*, 47, 609–614.

Datta, R., Kelkar, A., Baraniya, D., Molaei, A., Moulick, A., Meena, R. S., & Formanek, P. (2017). Enzymatic degradation of lignin in soil: A review. *Sustainability*, 9(7), 1163. doi: 10.3390/su9071163.

de Vries, R. P., Nadal, M., van den Brink, J., Vivas-Duarte, D. A., & Stålbrand, H. (2011). Fungal degradation of plant oligo—and polysaccharides. In *Carbohydrate Modifying Enzymes and Microorganisms*, Grunwald, P. (ed.), Pan Stanford Publishing Pte. Ltd.: Singapore.

De Vries, R. P., & Visser, J. (2001). Aspergillus enzymes involved in degradation of plant cell wall polysaccharides. *Microbiology and Molecular Biology Reviews*, 65, 497–522.

De Wit, P. J. G. M., Mehrabi, R., Van den Burg, H. A., & Stergiopoulos, I. (2009). Fungal effector proteins: Past, present and future. *Molecular Plant Pathology*, 10, 735–747.

Desjardins, A. E. (2006). *Fusarium Mycotoxins Chemistry, Genetics and Biology*, The American Phytopathological Society: St. Paul, Minnesota, USA.

Desjardins, A. E., & Hohn, T. M. (1997). Mycotoxins in plant pathogenesis. *Molecular Plant-Microbe Interactions*, 10, 147–152.

Doehlemann, G., & Hemetsberger, C. (2013). Apoplastic immunity and its suppression by filamentous plant pathogens. *New Phytologist*, 198(4), 1001–1016.

Doehlemann, G., Van Der Linde, K., Aßmann, D., Schwammbach, D., Hof, A., Mohanty, A., et al. (2009). Pep1, a secreted effector protein of Ustilago maydis, is required for successful invasion of plant cells. *PLoS Pathogens*, 5(2), e1000290.

Doehlert, D. C., Knutson, C. A., & Vesonder, R. E. (1994). Phytotoxic effects of fumonisin B1 on maize seedling growth. *Mycopathologia*, 127, 117–121.

Domijan, A. M., & Fumonisin, B. (2012). A neurotoxic mycotoxin. *Archives of Industrial Hygiene and Toxicology*, 63, 531–544.

Duplessis, S., Cuomo, C. A., Lin, Y.-C., Aerts, A., Tisserant, E., Veneault-Fourrey, C., et al. (2011). Obligate biotrophy features unraveled by the genomic analysis of rust fungi. *Proceedings of the National Academy of Sciences of the United States of America*, 108, 9166–9171.

Dutton, M. V., Evans, C. S., Atkey, P. T., & Wood, D. A. (1993). Oxalate production by Basidiomycetes, including the white-rot species Coriolus versicolor and Phanerochaete chrysosporium. *Applied Microbiology and Biotechnology*, 39(1), 5–10.

Eamens, A., Wang, M. B., Smith, N. A., & Waterhouse, P. M. (2008). RNA silencing in plants: Yesterday, today, and tomorrow. *Plant Physiology*, 147(2), 456–468.

El Gueddari, N., Rauchhaus, U., Moerschbacher, B., & Deising, H. (2002). Developmentally regulated conversion of surface-exposed chitin to chitosan in cell walls of plant pathogenic fungi. *New Phytologist*, 156, 103–112.

Ellendorff, U., Fradin, E. F., De Jonge, R., & Thomma, B. P. (2009). RNA silencing is required for Arabidopsis defence against Verticillium wilt disease. *Journal of Experimental Botany*, 60(2), 591–602.

Ellis, J. G., Rafiqi, M., Gan, P., Chakrabarti, A., & Dodds, P. N. (2009). Recent progress in discovery and functional analysis of effector proteins of fungal and oomycete plant pathogens. *Current Opinion in Plant Biology*, 12, 399–405.

Ellis, J. R., & McCalla, T. M. (1973). Effects of patulin and method of application on growth stages of wheat. *Applied Microbiology*, 25, 562–566.

Enoki, A., Itakura, S., & Tanaka, H. (1997). The involvement of extracelluar substances for reducing molecular oxygen to hydroxyl radical and ferric iron to ferrous iron in wood degradation by wood decay fungi. *Journal of Biotechnology*, 53(2–3), 265–272.

Fernandez-Fueyo, E., Ruiz-Dueñas, F. J., Ferreira, P., Floudas, D., Hibbett, D. S., Canessa, P., et al. (2012). Comparative genomics of Ceriporiopsis subvermispora and Phanerochaete chrysosporium provide insight into selective ligninolysis. *Proceedings of the National Academy of Sciences of the United States of America*, 109, 5458–5463.

Flor, H. H. (1942). Inheritance of pathogenicity in Melampsora lini. *Phytopathology*, 32, 653–669.

Foreman, P. K., Brown, D., Dankmeyer, L., Dean, R., Diener, S., Dunn-Coleman, N. S., Goedegebuur, F., Houfek, T. D., England, G. J., Kelley, A. S., Meerman, H. J., Mitchell, T., Mitchinson, C., Olivares, H. A., Teunissen, P. J., Yao, J., & Ward, M. (2003). Transcriptional regulation of biomass-degrading enzymes in the filamentous fungus Trichoderma reesei. *Journal of Biological Chemistry*, 278, 31988–31997.

Foroud, N. A., & Eudes, F. (2009). Trichothecenes in cereal grains. *International Journal of Molecular Sciences*, 10, 147–173. doi: 10.3390/ijms10010147.

Freeman, G. G., & Morrison, R. I. (1948). Trichothecin: An antifungal metabolic product of Trichothecium roseum Link. *Nature*, 162, 30.

Frisvad, J. C., & Filtenborg, O. (1983). Classification of terverticillate penicillia based on profiles of mycotoxins and other secondary metabolites. *Applied and Environmental Microbiology*, 46, 1301–1310.

Frisvad, J. C., Hubka, V., Ezekiel, C. N., Hong, S.-B., Nováková, A., Chen, A. J., Arzanlou, M., Larsen, T. O., Sklenář, F., Mahakarnchanakul. W., Samson, R. A., Houbraken, J. (2019). Taxonomy of *Aspergillus* section *Flavi* and their production of aflatoxins, ochratoxins and other mycotoxins. *Studies in Mycology,* 93, 1–63. doi: 10.1016/j.simyco.2018.06.001.

Frisvad, J. C., & Samson, R. A. (2004). Polyphasic taxonomy of Penicillium subgenus Penicillium. A guide to identification of food and air-borne terverticillate Penicillia and their mycotoxins. *Studies in Mycology*, 49, 1–174.

Frisvad, J. C., & Thrane, U. (1996). Mycotoxin production by food borne fungi. In *Introduction to Food-Borne Fungi*, Samson, R. A., Hoekstra, E. S., Frisvad, J. C., & Filtenborg, O. (eds.), Centraalbureau Voor Schimmelcultures: Baarn, The Netherlands, 251–260.

Fujikawa, T., Kuga, Y., Yano, S., Yoshimi, A., Tachiki, T., Abe, K., & Nishimura, M. (2009). Dynamics of cell wall components of Magnaporthe grisea during infectious structure development. *Molecular Microbiology*, 73, 553–570.

Fujikawa, T., Sakaguchi, A., Nishizawa, Y., Kouzai, Y., Minami, E., Yano, S., Koga, H., Meshi, T., & Nishimura, M. (2012). Surface alpha-1, 3-glucan facilitates fungal stealth infection by interfering with innate immunity in plants. *PLoS Pathogens*, 8, e1002882.

Galhano, R., Illana, A., Ryder, L. S., Rodriguez-Romero, J., Demuez, M., Badaruddin, M., et al. (2017). Tpc1 is an important Zn (II) 2Cys6 transcriptional regulator required for polarized growth and virulence in the rice blast fungus. *PLoS Pathogens*, 13(7), e1006516.

Gan, P., Ikeda, K., Irieda, H., Narusaka, M., O'Connell, R. J., Narusaka, Y., Takano, Y., Kubo, Y., & Shirasu, K. (2013). Comparative genomic and transcriptomic analyses reveal the hemibiotrophic stage shift of colletotrichum fungi. *New Phytology*, 197, 1236–1249.

Gathercole, P. S., Thiel, P. G., & Hofmeyr, J. H. (1986). Inhibition of pyruvate dehydrogenase complex by moniliformin. *Biochemistry Journal*, 233, 719–723.

Gedek, P., Bauer, J., & Schreiber, H. (1981). Mycotoxin production of silage deteriorating molds. *Wien Tierarztl Monatsschr*, 68, 299–301.

Gelderblom, W. C. A., Jaskiewicz, K., Marasas. W. F., Thiel, P. G., Horak, R. M., Vleggaar, R., & Kreik, N. P. J. (1988). Fumonisins-novel mycotoxins with cancer-promoting activity produced by Fusarium moniliforme. *Applied and Environmental Microbiology*, 54, 1806–1811.

Ghildiyal, M., & Zamore, P. D. (2009). Small silencing RNAs: An expanding universe. *Nature Reviews Genetics*, 10(2), 94–108.

Giraldo, M. C., Dagdas, Y. F., Gupta, Y. K., Mentlak, T. A., Yi, M., Martinez-Rocha, A. L., Saitoh, H., Terauchi, R., Talbot, N. J., & Valent, B. (2013). Two distinct secretion systems facilitate tissue invasion by the rice blast fungus magnaporthe oryzae. *Nature Communications*, 4, 1996.

Goodell, B., Jellison, J., Liu, J., Daniel, G., Paszczynski, A., Fekete, F., et al. (1997). Low molecular weight chelators and phenolic compounds isolated from wood decay fungi and their role in the fungal biodegradation of wood. *Journal of Biotechnology*, 53(2–3), 133–162.

Guo, M., Chen, Y., Du, Y., Dong, Y., Guo, W., Zhai, S., et al. (2011). The bZIP transcription factor MoAP1 mediates the oxidative stress response and is critical for pathogenicity of the rice blast fungus Magnaporthe oryzae. *PLoS Pathogens*, 7(2), e1001302.

Gupta, S., Krasnoff, S. B., Underwood, N. L., Renwick, J. A. A., & Roberts, D. W. (1991). Isolation of beauvericin as an insect toxin from Fusarium semitectum and Fusarium moniliforme var. subglutinans. *Mycopathologia*, 115, 185–189. doi: 10.1007/BF00462223.

Hamill, R. L., Higgens, C. E., Boaz, H. E., & Gorman, M. (1969). Structure of beauvericin, a new depsipeptide antibiotic toxic to artemia salina. *Tetrahedron Letters*, 49, 4255–4258.

Hammel, K. E., Kapich, A. N., Jensen, Jr., K. A., & Ryan, Z. C. (2002). Reactive oxygen species as agents of wood decay by fungi. *Enzyme and Microbial Technology*, 30(4), 445–453.

Hanumegowda, U. M., Dhulipala, V. C., & Reddy, C. S. (2002). Mechanism of secalonic acid D-induced inhibition of transcription factor binding to cyclic AMP response element in the developing murine palate. *Toxicological Sciences*, 70, 55–62. doi: 10.1093/toxsci/70.1.55.

Harris, L. J., Desjardins, A. E., Plattner, R. D., Nicholson, P., Butler, G., Young, J. C., Weston, G., Proctor, R. H., & Holn, T. M. (1999). Possible role of trichothecenes mycotoxins in virulence of Fusarium graminearum on maize. *Plant Disease*, 83, 954–960.

Haschek, W. M., Gumprecht, L. A., Smith, G., Tumbleson, M. E., & Constable, P. D. (2001). Fumonisin toxicosis in swine: An overview of porcine pulmonary edema and current perspectives. *Environmental Health Perspective*, 109(Suppl. S2), 251–257.

Heredia, A. (2003). Biophysical and biochemical characteristics of cutin, a plant barrier biopolymer. *Biochimica et Biophysica Acta (BBA)-General Subjects*, 1620(1–3), 1–7.

Hetherington, A. C., & Raistrick, H. (1931). Studies in the Biochemistry of Micro-organisms. Part XIV—On the production and chemical constitution of a new yellow colouring mater, citrinin, produced from glucose by Penicillium. *Philosophical Transactions of the Royal Society of London. Series B, Containing Papers of a Biological Character*, 220(468–473), 269–295.

Hooper, L. V., & Gordon, J. I. (2001). Glycans as legislators of host-microbial interactions: Spanning the spectrum fromsymbiosis to pathogenicity. *Glycobiology*, 11, 1R–10R.

Horbach, R., Navarro-Quesada, A. R., Knogge, W., & Deising, H. B. (2011). When and how to kill a plant cell: Infection strategies of plant pathogenic fungi. *Journal of Plant Physiology*, 168, 51– 62.

Huang, K., Czymmek, K. J., Caplan, J. L., Sweigard, J. A., & Donofrio, N. M. (2011). Suppression of plant-generated reactive oxygen species is required for successful infection by the rice blast fungus. *Virulence*, 2(6), 559–562.

Hutvagner, G., & Simard, M. J. (2008). Argonaute proteins: Key players in RNA silencing. *Nature Reviews Molecular Cell Biology*, 9(1), 22–32.

Ismaiel, A. A., & Papenbrock, J. (2014). The effects of patulin from Penicilllium vulpinum on seedling growth, root tip ultrastructure and glutathione content of maize. *European Journal of Plant Pathology*, 139, 497–509.

Ismaiel, A. A., & Tharwat, N. A. (2014). Antifungal activity of silver ion on ultrastructure and production aflatoxin B1 and patulin by two mycotoxigenic strains, Aspergillus flavus OC1 and Penicillium vulpinum CM1. *Journal of Medical Mycology*, 24, 193–204.

Jackowiak, H., Packa, D., Wiwart M., Perkowski, J., Busko, M., & Borusiewicz, A. (2002). Scanning electron microscopy of mature wheat kernels infected with Fusarium culmorum. *Journal of Applied Genetics*, 43, 167–176.

Jenkins, G. R., Tolleson, W. H., Newkirk, D. K., Roberts, D. W., Rowland, K. L., Saheki, T., Kobayashi, K., Howard, P. C., & Melchior, W. B., Jr. (2000). Identification of fumonisin B1 as an inhibitor of argininosuccinate synthetase using fumonisin affinity chromatography and in vitro kinetic studies. *Journal of Biochemical and Molecular Toxicology*, 14, 320–328.

Jennings, D. B., Ehrenshaft, M., Pharr, D. M., & Williamson, J. D. (1998). Roles for mannitol and mannitol dehydrogenase in active oxygen-mediated plant defense. *Proceedings of the National Academy of Sciences*, 95(25), 15129–15133.

Jensen, Jr., K. A., Ryan, Z. C., Vanden Wymelenberg, A., Cullen, D., & Hammel, K. E. (2002). An NADH: Quinone oxidoreductase active during biodegradation by the brown-rot basidiomycete Gloeophyllum trabeum. *Applied and Environmental Microbiology*, 68(6), 2699–2703.

Jones, J. D. G., & Dangl, J. L. (2006). The plant immune system. *Nature*, 444, 323–329.

Juan, C., Manyes, L., Font, G., & Juan-García, A. (2014). Evaluation of immunologic effect of Enniatin A and quantitative determination in feces, urine and serum on treated Wistar rats. *Toxicon*, 87, 45–53. doi: 10.1016/j.toxicon.2014.05.005.

Kahmann, R., & Basse, C. (2001). Fungal gene expression during pathogenesis-related development and host plant colonization. *Current Opinion in Microbiology*, 4(4), 374–380.

Kale, S. D., Gu, B., Capelluto, D. G., Dou, D., Feldman, E., Rumore, A., et al. (2010). External lipid PI3P mediates entry of eukaryotic pathogen effectors into plant and animal host cells. *Cell*, 142(2), 284–295.

Kameshwar, A. K. S., & Qin, W. (2016a). Lignin degrading fungal enzymes. In *Production of Biofuels and Chemicals from Lignin*, Springer, 81–130.

Kameshwar, A. K. S., & Qin, W. (2017). Gene expression metadata analysis reveals molecular mechanisms employed by Phanerochaete chrysosporium during lignin degradation and detoxification of plant extractives. *Current Genetics*, 1–18.

Kamoun, S. (2009). The secretome of plant-associated fungi and oomycetes. In *The Mycota*, H. B. Deising (ed.), Springer: Berlin, Heidelberg, 173–180.

Kawakubo, M., Ito, Y., Okimura, Y., Kobayashi, M., Sakura, K., Kasama, S., Fukuda, M. N., Fukuda, M., & Keen, N. T. (1990). Gene-for-gene complementarity in plant-pathogen interactions. *Annual Review of Genetics*, 24, 447–463.

Kerem, Z., Jensen, K. A., & Hammel, K. E. (1999). Biodegradative mechanism of the brown rot basidiomycete Gloeophyllum trabeum: Evidence for an extracellular hydroquinone-driven fenton reaction. *FEBS Letters*, 446(1), 49–54.

Keromnes, J., & Thouvenot, D. (1985). Role of penicillic acid in the phytotoxicity of Penicillium cyclopium and Penicillium canescens to the germination of corn seeds. *Applied and Environmental Microbiology*, 49, 660–663.

Kersten, P., & Cullen, D. (2007). Extracellular oxidative systems of the lignin-degrading Basidiomycete Phanerochaete chrysosporium. *Fungal Genetics and Biology*, 44(2), 77–87.

Kim, J.-E., Son, H., & Lee, Y.-W. (2018). Biosynthetic mechanism and regulation of zearalenone in Fusarium graminearum. *JSM Mycotoxins*, 68, 1–6.

Kirk, K., & Cullen, D. (1998). Enzymology and molecular genetics of wood degradation by white rot fungi. In *Environmental Friendly Technologies for Pulp and Paper Industry*, Young, R. A., & Akhtar, M. (eds.), Wiley: New York, 273–307.

Kluczek-Turpeinen, B., Tuomela, M., Hatakka, A., & Hofrichter, M. (2003). Lignin degradation in a compost environment by the deuteromycete Paecilomyces inflatus. *Applied Microbiology and Biotechnology*, 61(4), 374–379.

Kolattukudy, P. E. (2001). Polyesters in higher plants. *Biopolyesters*, 1–49.

Kolattukudy, P. T. (1981). Structure, biosynthesis, and biodegradation of cutin and suberin. *Annual Review of Plant Physiology*, 32(1), 539–567.

Korripally, P., Timokhin, V. I., Houtman, C. J., Mozuch, M. D., & Hammel, K. E. (2013). Evidence from Serpula lacrymans that 2, 5-dimethoxyhydroquinone is a lignocellulolytic agent of divergent brown rot basidiomycetes. *Applied and Environmental Microbiology*, 79(7), 2377–2383.

Koteswara Rao, V., Girisham, S., & Reddy, S. M. (2014). Influence of different species of Penicillium and their culture filtrates on seed germination and seedling growth of sorghum. *Journal of Biochemical Technology*, 5, 832–837.

Krasnoff, S. B., Sommers, C. H., Moon, Y.-S., Donzelli, B. G. G., Vandenberg, J. D., & Churchill, A. C. L. (2006). Production of mutagenic metabolites by Metarhizium anisopliae. *Journal of Agricultural and Food Chemistry*, 54, 7083–7088.

Kuhad, R. C., Singh, A., & Eriksson, K.-E. L. (1997). Microorganisms and enzymes involved in the degradation of plant fiber cell walls. *Advances in Biochemical Engineering/Biotechnology*, 57, 45–125.

Lamprecht, S. C., Marasas, W. E. O., Alberts, J. F., Cawood, M. E., Gelderblom, W. C. A., & Shephard, G. S. (1994). Phytotoxicity of fumonisins and TA-toxin to corn and tomato. *Phytopathology*, 84, 383–391.

Latge, J. P. (2007). The cell wall: A carbohydrate armour for the fungal cell. *Molecular Microbiology*, 66, 279–290.

Levasseur, A., Piumi, F., Coutinho, P. M., Rancurel, C., Asther, M., Delattre, M., Henrissat, B., Pontarotti, P., Asther, M., & Record, E. (2008). FOLy: An integrated database for the classification and functional annotation of fungal oxidoreductases potentially involved in the degradation of lignin and related aromatic compounds. *Fungal Genetics and Biology*, 45(5), 638–645.

Levine, A., Tenhaken, R., Dixon, R., & Lamb, C. (1994). H2O2 from the oxidative burst orchestrates the plant hypersensitive disease resistance response. *Cell*, 79(4), 583–593.

Li, S., Zhang, L., Yao, Q., Li, L., Dong, N., Rong, J., et al. (2013). Pathogen blocks host death receptor signalling by arginine GlcNAcylation of death domains. *Nature*, 501, 242–246.

Liebminger, S., Eberl, A., Sousa, F., Heumann, S., Fischer-Colbrie, G., Cavaco-Paulo, A., & Guebitz, G. M. (2007). Hydrolysis of PET and bis-(benzoyloxyethyl) terephthalate with a new polyesterase from Penicillium citrinum. *Biocatalysis and Biotransformation*, 25(2–4), 171–177.

Lin, B., Qing, X., Liao, J., & Zhuo, K. (2020). Role of protein glycosylation in Host-Pathogen interaction. *Cells*, 9(4), 1022.

Liu, B.-H., Wu, T.-S., Yu, F.-Y., & Wang, C.-H. (2005). Mycotoxin patulin activates the p38 kinase and JNK signaling pathways in human embryonic kidney cells. *Toxicology Science*, 89, 423–430. doi: 10.1093/toxsci/kfj049.

Lombard, V., Golaconda Ramulu, H., Drula, E., Coutinho, P. M., & Henrissat, B. (2014). The carbohydrate-active enzymes database (CAZy) in 2013. *Nucleic Acids Research*, 42(D1), D490–D495.

Lund, F., & Frisvad, J. C. (1994). Chemotaxonomy of Penicillium aurantiogriseum and related species. *Mycological Research*, 98, 481–492.

Luque, P., Bruque, S., & Heredia, A. (1995). Water permeability of isolated cuticular membranes: A structural analysis. *Archives of Biochemistry and Biophysics*, 317(2), 417–422.

Mach, R. L., Seiboth, B., Myasnikov, A., González, R., Strauss, J., Harkki, A. M., & Kubicek, C. P. (1995). The bgl1 gene of Trichoderma reesei QM 9414 encodes an extracellular, cellulose-inducible β-glucosidase involved in cellulase induction by sophorose. *Molecular Microbiology*, 16, 687–697.

Maćias, M., Ulloa, M., Gamboa, A., & Mata, R. (2000). Phytotoxic compounds from the new coprophilous fungus Guanomyces polythrix. *Journal of Natural Products*, 63, 757–761.
Majumdar, R., Rajasekaran, K., & Cary, J. W. (2017). RNA interference (RNAi) as a potential tool for control of mycotoxin contamination in crop plants: Concepts and considerations. *Frontiers in Plant Science*, 8, 200.
Mallebrera, B., Prosperini, A., Font, G., & Ruiz, M. J. (2018). In vitro mechanisms of Beauvericin toxicity: A review. *Food and Chemical Toxicology*, 111, 537–545.
Margis, R., Fusaro, A. F., Smith, N. A., Curtin, S. J., Watson, J. M., Finnegan, E. J., & Waterhouse, P. M. (2006). The evolution and diversification of Dicers in plants. *FEBS Letters*, 580(10), 2442–2450.
Marré, M. T., Vergani, P., & Albergoni, F. G. (1993). Relationship between fusaric acid uptake and its binding to cell structures by leaves of Egeria densa and its toxic effects on membrane permeability and respiration. *Physiological and Molecular Plant Pathology*, 42, 141–157.
Martín, I., Gálvez, L., Guasch, L., & Palmero, D. (2022). Fungal pathogens and seed storage in the dry state. *Plants (Basel)*, 11(22), 3167.
Martinez, D., Challacombe, J., Morgenstern, I., Hibbett, D., Schmoll, M., Kubicek, C. P., et al. (2009). Genome, transcriptome, and secretome analysis of wood decay fungus Postia placenta supports unique mechanisms of lignocellulose conversion. *Proceedings of the National Academy of Sciences*, 106(6), 1954–1959.
Martinez, D., Larrondo, L. F., Putnam, N., Gelpke, M. D. S., Huang, K., Chapman, J., et al. (2004). Genome sequence of the lignocellulose degrading fungus Phanerochaete chrysosporium strain RP78. *Nature Biotechnology*, 22(6), 695–700.
Martins, I., Garcia, H., Varela, A., Núñez, O., Planchon, S., Galceran, M. T., et al. (2014). Investigating Aspergillus nidulans secretome during colonisation of cork cell walls. *Journal of Proteomics*, 98, 175–188.
Masuda, D., Ishida, M., Yamaguchi, K., Yamaguchi, I., Kimura, M., & Nishiuchi, T. (2007). Phytotoxic effects of trichothecenes on the growth and morphology of Arabidopsis thaliana. *Journal of Experimental Botany*, 58, 1617–1626.
Matas, A., & Heredia, A. (1999). Molecular dynamics modellization and simulation of water diffusion through plant cutin. *Zeitschrift für Naturforschung C*, 54(11), 896–902.
McCormick, S. P., Stanley, A. M., Stover, N. A., & Alexander, N. J. (2011). Trichothecenes: From simple to complex mycotoxins. *Toxins*, 3, 802–814. doi: 10.3390/toxins3070802.
Mcgowan, J., & Fitzpatrick, D. A. (2017). Genomic, network, and phylogenetic analysis of the oomycete effector arsenal. *mSphere*, 2, e00408–e00417. doi: 10.1128/mSphere.00408–17.
McLaughlin, C. S., Vaughan, M. H., Campbell, I. M., Wei, C. M., Stafford, M. E., & Hansen, B. S. (1977). Inhibition of protein synthesis by trichothecenes. Mycotoxins in human and animal health. In *Mycotoxins in Human and Health*, Rodericks, J. V., Hesseltine, C. W., & Mehlman, M. A. (eds.), Pathotox: Park Forest, IL, USA, 263–275.
McLean, M. (1996a). The phytotoxicity of Fusarium metabolites: An update since 1989. *Mycopathologia*, 133, 163–179.
McLean, M. (1996b). The phytotoxicity of selected mycotoxins on mature, germinating Zea mays embryos. *Mycopathologia*, 132, 173–183.
Meena, M., Prasad, V., Zehra, A., Gupta, V. K., & Upadhyay, R. S. (2015). Mannitol metabolism during pathogenic fungal–host interactions under stressed conditions. *Frontiers in Microbiology*, 6, 1019.
Mendgen, K., & Hahn, M. (2002). Plant infection and the establishment of fungal biotrophy. *Trends Plant Science*, 7, 352–356.
Mentlak, T. A., Kombrink, A., Shinya, T., Ryder, L. S., Otomo, I., Saitoh, H., Terauchi, R., Nishizawa, Y., Merrill, A. H., Jr., Sullards, M. C., Wang, E., Voss, K. A., & Riley, R. T. (2001). Sphingolipid metabolism: Roles in signal transduction and disruption by fumonisins. *Environmental Health Perspectives*, 109, 283–289.

Merrill, A. H., Jr., Sullards, M. C., Wang, E., Voss, K. A., Riley, R. T. (2001). Sphingolipid metabolism: Roles in signal transduction and disruption by fumonisins. *Environ. Health Perspect.* 2001, 109(Suppl. S2), 283–289.

Mitchell, N. J., Bowers, E., Hurburgh, C., & Wu, F. (2016). Potential economic losses to the US corn industry from aflatoxin contamination. *Food Additives & Contaminants: Part A*, 33(3), 540–550.

Molina, L., & Kahmann, R. (2007). An Ustilago maydis gene involved in H2O2 detoxification is required for virulence. *The Plant Cell*, 19(7), 2293–2309.

Molnar, A., Melnyk, C., & Baulcombe, D. C. (2011). Silencing signals in plants: A long journey for small RNAs. *Genome Biology*, 12(1), 1–8.

Morgan, W., & Kamoun, S. (2007). RXLR effectors of plant pathogenic oomycetes. *Current Opinion in Microbiology*, 10(4), 332–338.

Mostrom, M. S., & Raisbeck, M. F. (2012). Trichothecenes. *Veterinary Toxicology*, 1239–1265, doi: 10.1016/b978-0-12-385926-6.00108-3.

Müller, H. M., & Amend, R. (1997). Formation and disappearance of mycophenolic acid, patulin, penicillic acid and PR toxin in maize silage inoculated with Penicillium roqueforti. *Archiv für Tierernährung*, 50, 213–225.

Munari. (2019). Enzymatic hydrolysis of poly (1,4-butylene 2,5-thiophenedicarboxylate) (PBTF) and poly (1,4-butylene 2,5-furandicarboxylate) (PBF) films: A comparison of mechanisms. *Environment International*, 130, 104852.

Murphy, C. A., Cameron, J. A., Huang, S. J., & Vinopal, R. T. (1996). Fusarium polycaprolactone depolymerase is cutinase. *Applied and Environmental Microbiology*, 62(2), 456–460.

Mutter, M. (1997). *New rhamnogalacturonan degrading enzymes from Aspergillus aculeatus*, Wageningen University and Research.

Mutter, M., Beldman, G., Pitson, S. M., Schols, H. A., & Voragen, A. G. (1998). Rhamnogalacturonan α-D-galactopyranosyluronohydrolase: An enzyme that specifically removes the terminal nonreducing galacturonosyl residue in rhamnogalacturonan regions of pectin. *Plant Physiology*, 117(1), 153–163.

Naveed, Z. A., Wei, X., Chen, J., Mubeen, H., & Ali, G. S. (2020). The PTI to ETI continuum in phytophthora-plant interactions. *Frontiers in Plant Science*, 11, 593905. doi: 10.3389/fpls.2020.593905.

Nelson, E. P., Dignani, M. C., & Anaissie, E. J. (1994). Taxonomy, biology, and clinical aspects of Fusarium species. *Clinical Microbiology Reviews*, 7, 479–504.

Nelson, T. S., Kirby, L. K., Beasley, J. N., Johnson, Z. B., & Ciegler, A. (1985). The effect of drying method and storage time on citrinin activity in corn. *Poultry Science*, 64, 464–468.

Nganje, W. E., Bangsund, D. A., Leistritz, F. L., Wilson, W. W., & Tiapo, N. M. (2004). Regional economic impacts of Fusarium head blight in wheat and barley. *Review of Agricultural Economics*, 26, 332–347.

Ngou, B. P. M., Ding, P., & Jones, J. D. G. (2022). Thirty years of resistance: Zig-zag through the plant immune system. *Plant Cell*, 34, 1447–1478.

Nutt, A., Sild, V., Pettersson, G., & Johansson, G. (1998). Progress curves—A mean for functional classification of cellulases. *European Journal of Biochemistry*, 258(1), 200–206.

Ochiai, A., Itoh, T., Kawamata, A., Hashimoto, W., & Murata, K. (2007). Plant cell wall degradation by saprophytic Bacillus subtilis strains: gene clusters responsible for rhamnogalacturonan depolymerization. *Applied and environmental microbiology*, 73(12), 3803–3813. Chicago.

O'Neill, A., Araújo, R., Casal, M., Guebitz, G., & Cavaco-Paulo, A. (2007). Effect of the agitation on the adsorption and hydrolytic efficiency of cutinases on polyethylene terephthalate fibres. *Enzyme and Microbial Technology*, 40(7), 1801–1805.

Ofong, A. U., & Pearce, R. B. (1994). Suberin degradation by Rosellinia desmazieresii. *European Journal of Forest Pathology*, 24(6–7), 316–322.

Oliveira, C. A. F., Ivanova, L., Solhaug, A., & Fæste, C. K. (2020). Enniatin B1-induced lysosomal membrane permeabilization in mouse embryonic fibroblasts. *Mycotoxin Research*, 36, 23–30. doi: 10.1007/s12550-019-00366-8.

Osman, S. F., Gerard, H. C., Fett, W. F., Moreau, R. A., & Dudley, R. L. (1995). Method for the production and characterization of tomato cutin oligomers. *Journal of Agricultural and Food Chemistry*, 43, 2134–2137.

Paciolla, C., Dipierro, N., Mule, G., Logrieco, A., & Dipierro, S. (2004). The mycotoxins beauvericin and T-2 induce cell death and alteration to the ascorbate metabolism in tomato protoplasts. *Physiological and Molecular Plant Pathology*, 65, 49–56.

Panstruga, R., & Dodds, P. N. (2009). Terrific protein traffic: The mystery of effector protein delivery by filamentous plant pathogens. *Science*, 324, 748–750.

Paszczynski, A., Crawford, R., Funk, D., & Goodell, B. (1999). De novo synthesis of 4, 5-dimethoxycatechol and 2, 5-dimethoxyhydroquinone by the brown rot fungus Gloeophyllum trabeum. *Applied and Environmental Microbiology*, 65(2), 674–679.

Peng, X. L., Xu, W. T., Wang, Y., Huang, K. L., Liang, Z. H., Zhao, W. W., & Luo, Y. B. (2010). Mycotoxin ochratoxin A-induced cell death and changes in oxidative metabolism of Arabidopsis thaliana. *Plant Cell Report*, 29, 153–161. doi: 10.1007/s00299-009-0808-x.

Penttilä, M., Lehtovaara, P., Nevalainen, H., Bhikhabhai, R., & Knowles, J. (1986). Homology between cellulase genes of Trichoderma reesei: Complete nucleotide sequence of the endoglucanase I gene. *Gene*, 45, 253–263.

Pérez, J., Munoz-Dorado, J., De la Rubia, T. D. L. R., & Martinez, J. (2002). Biodegradation and biological treatments of cellulose, hemicellulose and lignin: An overview. *International Microbiology*, 5, 53–63.

Pérez, S., Mazeau, K., & du Penhoat, C. H. (2000). The three-dimensional structures of the pectic polysaccharides. *Plant Physiology and Biochemistry*, 38(1–2), 37–55.

Perincherry, L., Lalak-Kańczugowska, J., & Stępień, Ł. (2019). Fusarium-produced mycotoxins in plant-pathogen interactions. *Toxins (Basel)*, 11(11), 664.

Pestka, J. (2008). Mechanisms of deoxynivalenol-induced gene expression and apoptosis. *Food Additives & Contaminants: Part A*, 25, 1128–1140. doi: 10.1080/02652030802056626.

Petre, B., Joly, D. L., & Duplessis, S. (2014). Effector proteins of rust fungi. *Frontiers in Plant Science*, 5, 416.

Pitson, S. M., Voragen, A. G., & Beldman, G. (1996). Stereochemical course of hydrolysis catalyzed by arabinofuranosyl hydrolases. *FEBS Letters*, 398(1), 7–11.

Plattner, P. A., & Nager, U. (1947). Über die Chemie des Enniatins. *Experientia*, 3, 325–326. doi: 10.1007/BF02164246.

Proctor, R. H., McCormick, S. P., Alexander, N. J., & Desjardins, A. E. (2009). Evidence that a secondary metabolic biosynthetic gene cluster has grown by gene relocation during evolution of the filamentous fungus Fusarium. *Molecular Microbiology*, 74, 1128–1142.

Qi, F. F., Zhang, W. X., Zhang, F. J., Chen, G. J., & Liu, W. F. (2014). Deciphering the effect of the different n-glycosylationsites on the secretion, activity, and stability of cellobiohydrolase I from Trichoderma reesei. *Applied and Environmental Microbiology*, 80, 3962–3971.

Qi, T., Guo, J., Liu, P., He, F., Wan, C., Islam, M. A., et al. (2019). Stripe rust effector PstGSRE1 disrupts nuclear localization of ROS-promoting transcription factor TaLOL2 to defeat ROS-induced defense in wheat. *Molecular Plant*, 12(12), 1624–1638.

Ralph, J., Lundquist, K., Brunow, G., Lu, F., Kim, H., Schatz, P. F., et al. (2004). Lignins: Natural polymers from oxidative coupling of 4-hydroxyphenyl-propanoids. *Phytochemistry Reviews*, 3(1), 29–60.

Ramachandran, S. R., Yin, C., Kud, J., Tanaka, K., Mahoney, A. K., Xiao, F., & Hulbert, S. H. (2017). Effectors from wheat rust fungi suppress multiple plant defense responses. *Phytopathology*, 107(1), 75–83.

Reese, E. T., & Levinson, H. S. (1952). A comparative study of the breakdown of cellulose by microorganisms. *Physiologia Plantarum*, 5(3), 345–366.

Reiss, J. (1978). Effects of mycotoxins on higher plants, algae, fungi and bacteria. In *Mycotoxic Fungi, Mycotoxins, Mycotoxicoses: An Encyclopaedic Handbook, Volume 3: Mycotoxicoses of Man and Plants: Mycotoxin Control and Regulatory Practices*, Wyllie, T. D., Morehouse, L. G. (eds.), Marcel Dekker Inc.: New York, NY, USA, 119–143.

Rheeder, J. P., Marasas, W. E. O., & Vismer, H. E. (2002). Production of fumonisin analogs by Fusarium species. *Applied and Environmental Microbiology*, 68, 2101–2105.

Richard, J. L., Payne, G. A., Desjardin, A. E., Maragos, C., Norred, W. P., & Pestka, J. J. (2003). Mycotoxins, risks in plant, animal and human systems. In *CAST Task Force Report 139*, Council for Agricultural Science and Technology: Ames, IA, USA, 101–103.

Ridley, B. L., O'Neill, M. A., & Mohnen, D. (2001). Pectins: Structure, biosynthesis, and oligogalacturonide-related signaling. *Phytochemistry*, 57(6), 929–967.

Rodriguez-Moreno, L., Ebert, M. K., Bolton, M. D., & Thomma, B. P. H. J. (2018). Tools of the crook-infection strategies of fungal plant pathogens. *The Plant Journal*, 93(4), 664–674.

Romero-Contreras, Y. J., Ramírez-Valdespino, C. A., Guzmán-Guzmán, P., Macías-Segoviano, J. I., Villagómez-Castro, J. C., & Olmedo-Monfil, V. (2019). Tal6 from Trichoderma atroviride is a LysM effector involved in mycoparasitism and plant association. *Frontiers in Microbiology*, 10, 2231.

Ryder, L. S., & Talbot, N. J. (2015). Regulation of appressorium development in pathogenic fungi. *Current Opinion in Plant Biology*, 26, 8–13.

Saitoh, H., Fujisawa, S., Mitsuoka, C., Ito, A., Hirabuchi, A., Ikeda, K., et al. (2012). Large-scale gene disruption in magnaporthe oryzae identifies MC69, a secreted protein required for infection by monocot and dicot fungal pathogens. *PLoS Pathoges*, 8, e1002711.

Saloheimo, A., Henrissat, B., Hoffrén, A. M., Teleman, O., & Penttilä, M. (1994). A novel, small endoglucanase gene, egl5, from Trichoderma reesei isolated by expression in yeast. *Molecular Microbiology*, 13, 219–228.

Saloheimo, M., Lehtovaara, P., Penttilä, M., Teeri, T. T., Ståhlberg, J., Johansson, G., Pettersson, G., Claeyssens, M., Tomme, P., & Knowles, J. K. (1988). EGIII, a new endoglucanase from Trichoderma reesei: The characterization of both gene and enzyme. *Gene*, 63, 11–22.

Samuels, L., Kunst, L., & Jetter, R. (2008). Sealing plant surfaces: Cuticular wax formation by epidermal cells. *Annual Review of Plant Biology*, 59, 683–707.

Sanderson, P. G., & Spotts, R. A. (1995). Postharvest decay of winter pear and apple fruit caused by species of Penicillium. *Phytopathology*, 85, 103–110.

Sato, S., Liu, F., Koc, H., & Tien, M. (2007). Expression analysis of extracellular proteins from Phanerochaete chrysosporium grown on different liquid and solid substrates. *Microbiology*, 153(Pt 9), 3023.

Schechter, L. M., Vencato, M., Jordan, K. L., Schneider, S. E., Schneider, D. J., & Collmer, A. (2006). Multiple approaches to a complete inventory of Pseudomonas syringae pv. tomato DC3000 type III secretion system effector proteins. *Molecular Plant-Microbe Interactions*, 19, 1180–1192.

Scholl, P. F., & Groopman, J. D. (2008). Long-term stability of human aflatoxin B1 albumin adducts assessed by isotope dilution mass spectrometry and high-performance liquid chromatography-fluorescence. *Cancer Epidemiology, Biomarkers & Prevention*, 17, 1436–1439. doi: 10.1158/1055–9965.epi-07–2926.

Scott, P. M., van Walbeek, W., Kennedy, B., & Anyeti, B. (1972). Mycotoxins (ochratoxin A, citrinin and sterigmatocystin) and toxigenic fungi in grains and other agricultural products. *Journal of Agricultural and Food Chemistry*, 20, 1103–1109.

Segal, L. M., & Wilson, R. A. (2018). Reactive oxygen species metabolism and plant-fungal interactions. *Fungal Genetics and Biology*, 110, 1–9.

Shah, P., Gutierrez-Sanchez, G., Orlando, R., & Bergmann, C. (2009). A proteomic study of pectin-degrading enzymes secreted by Botrytis cinerea grown in liquid culture. *Proteomics*, 9(11), 3126–3135.

Sharma, P., Jha, A. B., Dubey, R. S., & Pessarakli, M. (2012). Reactive oxygen species, oxidative damage, and antioxidative defense mechanism in plants under stressful conditions. *Journal of Botony*, 217037.

Shilina, S. G., Bekker, Z. E., & Goshaev, M. G. (1973). Isolation and comparative characterization of vivotoxin from wilt-infected cotton plants and of fusaric acid. *Ekologo-Fiziologlcheskle Metody Bor'be Fuzarioznym Viltom Khlop*, 2, 219–230.

Shoemaker, S., Schweickart, V., Ladner, M., Gelfand, D., Kwok, S., Myambo, K., & Innis, M. (1983). Molecular cloning of exocellobiohydrolase I derived from Trichoderma reesei strain L27. *Bio/Technology*, 1, 691–696.

Sieber, C. M. K., von Bargen, K. W., Studt, L., Niehaus, E.-M., & Espino, J. J. (2013). Deciphering the cryptic genome: Genome-wide analyses of the rice pathogen Fusarium fujikuroi reveal complex regulation of secondary metabolism and novel metabolites. *PLoS Pathology*, 9, e1003475.

Silva, I. R., Jers, C., Meyer, A. S., & Mikkelsen, J. D. (2016). Rhamnogalacturonan I modifying enzymes: An update. *New Biotechnology*, 33(1), 41–54.

Singh, V. K., Singh, H. B., & Upadhyay, R. S. (2017). Role of fusaric acid in the development of 'Fusarium wilt' symptoms in tomato: Physiological, biochemical and proteomic perspectives. *Plant Physiology and Biochemistry*, 118, 320–332.

Steinbuchel, A. (2001). *Biopolymers*, Vol. 4, Wiley-VCH, 636.

Stergiopoulos, I., & de Wit, P. J. G. M. (2009). Fungal effector proteins. *Annual Review of Phytopathology*, 47, 233–263.

Sun, W. X., Jia, Y. J., Feng, B. Z., O'Neill, N. R., Zhu, X. P., Xie, B. Y., & Zhang, X. G. (2009). Functional analysis of Pcipg2From the straminopilous plant pathogen Phytophthora capsici. *Genesis*, 47, 535–544.

Suzuki, M. R., Hunt, C. G., Houtman, C. J., Dalebroux, Z. D., & Hammel, K. E. (2006). Fungal hydroquinones contribute to brown rot of wood. *Environmental Microbiology*, 8(12), 2214–2223.

Sy-Cordero, A. A., Pearce, C. J., & Oberlies, N. H. (2012). Revisiting the enniatins: A review of their isolation, biosynthesis, structure determination and biological activities. *The Journal of Antibiotics*, 65, 541–549. doi: 10.1038/ja.2012.71.

Takao, S. (1965). Organic acid production by Basidiomycetes: I. screening of acid-producing strains. *Applied Microbiology*, 13(5), 732–737.

Takashima, S., Nakamura, A., Hidaka, M., Masaki, H., & Uozumi, T. (1999). Molecular cloning and expression of the novel fungal beta-glucosidase genes from Humicola grisea and Trichoderma reesei. *Journal of Biochemistry*, 125, 728–736.

Tanaka, K., Sago, Y., Zheng, Y., Nakagawa, H., & Kushiro, M. (2007). Mycotoxins in rice. *International Journal of Food Microbiology*, 119, 59–66.

Tashiro, F., Hiral, K., & Ueno, Y. (1979). Inhibitory effects of carcinogenic mycotoxins on deoxyribonucleic acid-dependent ribonucleic acid polymerase and ribonuclease H. *Applied and Environmental Microbiology*, 38, 191–196. doi: 10.1128/aem.38.2.191–196.1979.

Teeri, T. (1997). Crystalline cellulose degradation: New insight into the function of cellobiohydrolases. *Trends Biotechnology*, 15, 160–167.

Teeri, T., Salovuori, I., & Knowles, J. (1983). The molecular cloning of the major cellulase gene from Trichoderma reesei. *Bio/Technology*, 1, 696–699.

Teeri, T. T., Lehtovaara, P., Kauppinen, S., Salovuori, I., & Knowles, J. (1987). Homologous domains in Trichoderma reesei cellulolytic enzymes: Gene sequence and expression of cellobiohydrolase II. *Gene*, 51, 43–52.

Telles-Pupulin, A. R., Diniz, S. P. S. S., Bracht, A., & Ishii-Iwamoto, E. L. (1996). Effects of fusaric acid on respiration in maize root mitochondria. *Biologia Plantarum*, 38, 421–429.

Temme, N., & Tudzynski, P. (2009). Does Botrytis cinerea ignore H2O2-induced oxidative stress during infection? Characterization of Botrytis activator protein 1. *Molecular Plant-Microbe Interactions*, 22(8), 987–998.

Thiel, P. G. (1978). A molecular mechanism for the toxic action of moniliformin, a mycotoxin produced by Fusarium moniliforme. *Biochemical Pharmacology*, 27(4), 483–486.

Thiel, P. G., Marasas, W. E. O., Sydenham, E. W., Shephard, G. S., Gelderblom, W. C. A., & Nieuwenhuis, J. J. (1991). Survey of fumonisin production by Fusarium species. *Applied and Environmental Microbiology*, 57, 1089–1093.

Tripathi, R. K., & Misra, R. S. (1981). Effect of aflatoxin B1 on chromatin-bound ribonucleic acid polymerase and nucleic acid and protein synthesis in germinating maize seeds. *Applied and Environmental Microbiology*, 42, 389–393. doi: 10.1128/aem.42.3.389–393.1981.

Tronchet, M., Balague, C., Kroj, T., Jouanin, L., & Roby, D. (2010). Cinnamyl alcohol dehydrogenases-C and D, key enzymes in lignin biosynthesis, play an essential role in disease resistance in Arabidopsis. *Molecular Plant Pathology*, 11(1), 83–92.

Ueno, Y. (1983). *Trichothecenes: Chemical, Biological, and Toxicological Aspects*, Elsevier: Amsterdam, The Nerthlands.

Urbaniak, M., Stepien, Ł., & Uhlig, S. (2019). Evidence for naturally produced beauvericins containing N-methyl-tyrosine in hypocreales fungi. *Toxins*, 11, 182.

Van Asch, M. A. J., Rijkenberg, F. H. F., & Coutinho, T. A. (1992). Phytotoxicity of fumonisin B1, moniliformin and T-2 toxin in corn callus cultures. *Phytopathology*, 82, 1330–1332.

Van Der Merwe, K. J., Steyne, P. S., Fourie, L., Scott, D. B., & Theron, J. J. (1965). Ochratoxin A, a toxic metabolite produced by Aspergillus ochraceus. *Nature*, 205, 1112–1113.

Van der Vlugt-Bergmans, C. J. B., Meeuwsen, P. J. A., Voragen, A. G. J., & van Ooyen, A. J. J. (2000). Endo-xylogalacturonan hydrolase, a novel pectinolytic enzyme. *Applied and Environmental Microbiology*, 66, 36–41.

Vesonder, R. F., Labeda, D. P., & Peterson, R. E. (1992). Phytotoxic activity of selected water-soluble metabolites of Fusarium against Lemna minor L. (duckweed). *Mycopathologia*, 118, 185–189.

Vianello, A., & Macri, F. (1978). Inhibition of plant cell membrane transport phenomena induced by zearalenone (F-2). *Planta*, 143, 51–57.

Wakulinski, W. (1989). Phytotoxicity of the secondary metabolites of fungi causing wheat head fusariosis (head blight). *Acta Physiologiae Plantarum*, 11, 301–306.

Walton, T. J. (1990). Waxes, cutin and suberin. *Methods in Plant Biochemistry*, 4(11), 5–158.

Wan, Q., Wu, G., He, Q., Tang, H., & Wang, Y. (2015). The toxicity of acute exposure to T-2 toxin evaluated by the metabonomics technique. *Molecular BioSystems*, 11, 882–891. doi: 10.1039/c4mb00622d.

Wang, H., Hwang, S. F., Eudes, F., Chang, K. F., Howard, R. J., & Turnbull, G. D. (2006). Trichothecenes and aggressiveness of Fusarium graminaearum causing seeding blight and root rot in cereals. *Plant Pathology*, 55, 224–230.

Wang, W.-H., Lin, C.-Y., Chang, M. R., Urbina, A. N., Assavalapsakul, W., Thitithanyanont, A., Chen, Y.-H., Wang, Y., Peng, X., Xu, W., Luo, Y., & Zhao, W. (2011a). Transcript and protein profiling analysis of OTA-induced cell death reveals the regulation of the toxicity response process in Arabidopsis thaliana. *Journal of Experimental Botany*, 29, 1–17.

Wang, Y., Peng, X., Xu, W., Luo, Y., & Zhao, W. (2011b). Transcript and protein profiling analysis of OTA-induced cell death reveals the regulation of the toxicity response process in Arabidopsis thaliana. *Journal of Experimental Botany*, 29, 1–17.

Wang, Y., Peng, X., Xu, W., Luo, Y., Zhao, W., Hao, J., Liang, Z., Zhang, Y., & Huang, K. (2012). Transcript and protein profiling analysis of OTA-induced cell death reveals the regulation of the toxicity response process in Arabidopsis thaliana. *J Exp Bot.*, 63(5), 2171–2187.

Ward, M., Wu, S., Dauberman, J., Weiss, G., Larenas, E., Bower, B., Rey, M., Clarkson, K., & Bott, R. (1993). Cloning, sequnece and preliminary structural analysis of a small, high pI endoglucanase (EGIII) form Trichoderma reesei. In *Trichoderma reesei Cellulases and Other Hydrolases (TRICEL 93)*, Vol. 8, Suominen, P., & Reinikainen, T. (eds.), Foundation for Biotechnical and Industrial Fermentation Research, Kirkkonummi, Finland. pp. 153–158.

Weiberg, A., Wang, M., Lin, F. M., Zhao, H., Zhang, Z., Kaloshian, I., et al. (2013). Fungal small RNAs suppress plant immunity by hijacking host RNA interference pathways. *Science*, 342(6154), 118–123.

Werner, S., Sugui, J. A., Steinberg, G., & Deising, H. B. (2007). A chitin synthase with a myosin-like motor domain is essential for hyphal growth, appressorium differentiation, and pathogenicity of the maize anthracnose fungus Colletotrichum graminicola. *Molecular Plant–Microbe Interactions*, 20, 1555–1567.

Whisson, S. C., Boevink, P. C., Moleleki, L., Avrova, A. O., Morales, J. G., Gilroy, E. M., et al. (2007). A translocation signal for delivery of oomycete effector proteins into host plant cells. *Nature*, 450(7166), 115–118.

White, A. G., & Truelove, B. (1972). The effects of aflatoxin B1, citrinin, and ochratoxin A on amino acid uptake and incorporation by cucumber. *Canadian Journal of Botany*, 50, 2659–2664.

Wiebe, L. A., Bjeldanes, L. F., & Fusarin C. (1981). A mutagen from Fusarium moniliforme grown on corn. *Journal of Food Science*, 46, 1424–1426.

Williamson, G., Kroon, P. A., & Faulds, C. B. (1998). *Microbiology*, 144, 2011–2023.

Wilson, D. M. (1976). Patulin and penicillic acid. *Advances in Chemistry*, 149, 90–109.

Wingard, S. A. (1928). Hosts and symptoms of ring spot, a virus disease of plants. *Journal of Agriculture Research*, 37, 127–153.

Worrall, J. J., Anagnost, S. E., & Wang, C. J. K. (1991). Conditions for soft rot of wood. *Canadian Journal of Microbiology*, 37, 869–874.

Wortman, J. R., Fedorova, N., Crabtree, J., Joardar, V., Maiti, R., Haas, B. J., Amedeo, P., Lee, E., Angiuoli, S. V., Jiang, B., Anderson, M. J., Denning, D. W., White, O. R., & Nierman, W. C. (2006). Whole genome comparison of the *A. fumigatus* family. *Medical Mycology*, 44(Suppl 1), S3–S7. https://doi.org/10.1080/13693780600835799.

Wu, F. (2006). Mycotoxin reduction in Bt corn: Potential economic, health, and regulatory impacts. *Transgenic Research*, 15, 277–289.

Wu, G., Shortt, B. J., Lawrence, E. B., Leon, J., Fitzsimmons, K. C., Levine, E. B., et al. (1997). Activation of host defense mechanisms by elevated production of H2O2 in transgenic plants. *Plant Physiology*, 115(2), 427–435.

Wu, H. S., Yin, X. M., Liu, D. Y., Ling, N., Bao, W., Ying, R. R., et al. (2008). Effect of fungal fusaric acid on the root and leaf physiology of watermelon (Citrullus lanatus) seedlings. *Plant Soil*, 308, 255–266.

Wu, T.-S., Yu, F.-Y., Su, C.-C., Kan, J.-C., Chung, C.-P., & Liu, B.-H. (2005). Activation of ERK mitogen-activated protein kinase in human cells by the mycotoxin patulin. *Toxicology and Applied Pharmacology*, 207, 103–111. doi: 10.1016/j.taap.2004.12.006.

Wubben, J. P., ten Have, A., van Kan, J. A., & Visser, J. (2000). Regulation of endopolygalacturonase gene expression in Botrytis cinerea by galacturonic acid, ambient pH and carbon catabolite repression. *Current Genetics*, 37(2), 152–157.

Wymelenberg, A. V., Minges, P., Sabat, G., Martinez, D., Aerts, A., Salamov, A., et al. (2006). Computational analysis of the Phanerochaete chrysosporium v2. 0 genome database and mass spectrometry identification of peptides in ligninolytic cultures reveal complex mixtures of secreted proteins. *Fungal Genetics and Biology*, 43(5), 343–356.

Wymelenberg, A. V., Sabat, G., Martinez, D., Rajangam, A. S., Teeri, T. T., Gaskell, J., et al. (2005). The Phanerochaete chrysosporium secretome: Database predictions and initial mass spectrometry peptide identifications in cellulose-grown medium. *Journal of Biotechnology*, 118(1), 17–34.

Yamaji, K., Fukushi, Y., Hashidoko, Y., & Tahara, S. (2005). Penicillium frequentans from Picea glehnii seedling roots as a possible biological control agent against damping-off. *Ecological Research*, 20, 103–107.

Yamaji, K., Fukushi, Y., Hashidoko, Y., Yoshida, T., & Tahara, S. (1999). Characterization of antifungal metabolites produced by Penicillium species isolated from seeds of Picea glehnii. *Journal of Chemical Ecology*, 25, 1643–1654.

Yin, C., Ramachandran, S. R., Zhai, Y., Bu, C., Pappu, H. R., & Hulbert, S. H. (2019). A novel fungal effector from Puccinia graminis suppressing RNA silencing and plant defense responses. *New Phytologist*, 222(3), 1561–1572.

Yoon, J. J., Cha, C. J., Kim, Y. S., Son, D. W., & Kim, Y. K. (2007). The brown-rot basidiomycete Fomitopsis palustris has the endo-glucanases capable of degrading microcrystalline cellulose. *Journal of Microbiology and Biotechnology*, 17(5), 800–805.

Zahid, M. S., Nawaz, M. A., Tanveer, M. H., Shahzadi, S., Fayyaz, M., Shafique, I., & Abbas, A. (2023). Expression analysis and characterization of resistance (R) genes in contrasting Fusarium wilt resistant chickpea genotypes. *Integrative Plant Biotechnology*, 1(1), 9–16. https://doi.org/10.55627/pbiotech.001.01.0666.

Zandleven, J., Beldman, G., Bosveld, M., Benen, J., & Voragen, A. (2005). Mode of action of xylogalacturonan hydrolase towards xylogalacturonan and xylogalacturonan oligosaccharides. *Biochemical Journal*, 387, 719–725.

4 The Cytogenomics and Cytomics of Fungi

Muhammad Waris, Zobia Jabeen, Syed Abdul Sadiq, Ayesha Lutf, and Zakia Habib

4.1 INTRODUCTION TO CYTOGENOMICS

In ecosystems, fungi are a varied group of eukaryotic microorganisms that perform key roles in nutrient cycling, decomposition, and symbiotic interactions with plants. Unveiling fungi's biology, evolution, and possible uses in numerous industries requires a thorough understanding of their genetic composition and chromosomal architecture. An effective way to dive into the complexity of fungal genomes at the chromosomal level is to use the developing science of cytogenomics, which combines cytology with genomics (Hochstenbach et al., 2019). Revelations into fungal biology, notably genome architecture, genetic variability, and adaption tactics, are provided by this multidisciplinary approach. Understanding the intricate life cycles, reproductive techniques, and adaption processes in fungi has been made possible because of cytogenomics (Taga et al., 2015; Haq & Ijaz, 2022).

Cytogenetics is often defined as the study of chromosomes under a microscope. In a larger sense, however, it is a division of genetics dealing with the biological components involved in heredity, such as the chromosome, DNA, gene, episome, etc. Since high-throughput technology has recently made strides, cytogeneticists can now analyze chromosomal abnormalities without being constrained by them (Haq et al., 2022). Arrays provide a more accurate way to detect copy number variations and imbalanced chromosomal rearrangements than karyotyping. Array-based research is frequently utilized as a first line of testing to uncover tiny and major modifications since it provides larger genome coverage with higher resolution. The neologism "cytogenomics" was created solely to represent technological developments in the industry, giving it a tenuous justification for displacing the preeminent word "cytogenetics" (Liehr, 2021b). It is recommended that the term "cytogenomics" be used to refer to a broad area of study in genomics and genetic diagnostics. In this sense, "cytogenomics" is considered identical to "chromosomics," a word Professor Uwe Claussen coined in 2005 (Claussen, 2005; Liehr, 2019). "Cytogenomics" or "chromosomics" refers to all chromosome-related research focusing on the three-dimensional morphology of chromosomes and their significance for gene regulation. This strategy aims to integrate and unify research in biology and medicine, in contrast to the majority of "omics" topics, which divide disciplines (Claussen, 2005). It is crucial to keep in mind that every single "cytogenomic approach" accessible offers distinct and complementary ways to extract data from a genome; this information can be

DOI: 10.1201/9781003452362-4

recovered at the level of a single cell or millions of cells and various resolutions (Hochstenbach et al., 2021).

4.2 FUNGAL GENOME ARCHITECTURE

The fungal kingdom is thought to include a great diversity of species within the eukaryotic domain, with estimates ranging from 2.2 to 3.8 million distinct fungal species. As a result, the kingdom of fungi is the most varied in the eukaryotic realm (Möller & Stukenbrock, 2017). Presently, fungi are classified into eight phyla that are each taxonomically different. The phyla Ascomycota and Basidiomycota, which together comprise the subkingdom Dikarya, are home to most of the most researched and thoroughly characterized fungi. Comparatively less study has been done on the six oldest phyla (Blastocladiomycota, Chytridiomycota, Mucoromycota, Zoopagomycota, Cryptomycota, and Microsporidia), often known as early diverging fungi, than on the Dikarya fungi (Naranjo-Ortiz & Gabaldón, 2019). From genetic, morphological, and ecological standpoints, the variety within the fungal kingdom is quite evident. Fungal morphologies range from macroscopic, multicellular filamentous fungi to ubiquitous single-celled yeasts found throughout the environment and obligatory intracellular diseases linked to their hosts (Habib, 2022; Ijaz et al., 2022). The range of ecological functions that fungi perform, their significance as model organisms in scientific study, and the multitude of uses for fungi in industrial contexts are other examples of this diversity. The variety found within fungal genomes serves as a partial reflection of the phenotypic as well as ecological diversity present throughout the fungal kingdom (Mohanta & Bae, 2015).

The generation of both permanent and transient genetic and phenotypic variety, which in turn aids in speciation, depends on variation in fungal genomes, whether on a small or large scale (Haq & Ijaz, 2020a). Genomic microevolution, or variations in allelic frequencies across species, is frequently influenced by small-scale genomic alterations, which include modifications at the nucleotide and gene levels. The evolution of fungal genomes is also influenced by high mutation rates, genomic positions, and environments (Stajich, 2017). Other means for small-scale fungal genome evolution include horizontal gene transfer across different fungal species, cross-kingdom genetic transmission from prokaryotes, meiotic drivers, and introgression. The process of genomic macroevolution, which refers to evolutionary changes happening at and above the species level, can be facilitated by a combination of variables linked to small-scale genomic mutations (Haq et al., 2023). Fungal genomes' macroevolutionary routes have also been significantly and extensively altered by genomic alterations. These alterations include karyotype variances as well as structural alterations and reorganizations within certain chromosomes. These changes can be attributed to events like aneuploidy, the amplification of chromosomes, and polyploidy (Priest et al., 2020).

4.3 CYTOGENOMICS OF FUNGI

Fundamentally, cytogenomics is the junction of genomics, which is the study of all genes and their roles, and cytology, which is the study of cell structure and function.

Cytogenomics in the context of fungal research refers to the thorough chromosomal study of fungal genomes (Haq & Ijaz, 2020b). This method includes research on chromosomal organization, variations, and functional consequences in fungal species (Pires et al., 2016). Furthermore, it offers insightful information on chromosomal abnormalities, gene localization, and genomic structure. Some of the main methods in cytogenomics are discussed in the following sections.

4.4 FLUORESCENCE IN SITU HYBRIDIZATION (FISH)

Fluorescence in situ hybridization (FISH) is a method used in traditional cytogenetic research to examine chromosomes (Liehr, 2019). Major chromosomal abnormalities such monosomies, trisomies, chromosomal rearrangements, and significant deletions or duplications have all been successfully identified thanks to this method. With the use of two fundamental components (a DNA probe and a target sequence) and the banding method of FISH, particular DNA sequences on chromosomes may be physically mapped. The presence or lack of certain DNA sequences on genomes can be found and localized using FISH. It makes use of fluorescent or hapten-labeled probes that only attach to chromosomal regions with which they have a striking degree of similarity in sequence (Liehr et al., 2017). To see one or more specific areas of the genome, DNA probes are labeled with various colored fluorescent tags. Confocal microscopy or fluorescence microscopy may be used to determine where the fluorescent probe is linked to the chromosome. FISH can be carried out either directly on mitotic or meiotic chromosomes and interphase nuclei or indirectly by using comparative genomic hybridization analysis to check the complete genome for chromosomal rearrangements. FISH is a method that is frequently used in labs for cytogenetic as well as other biological investigations (Liehr, 2021a).

FISH is often performed on a microscope slide that has a chromosomal or nuclear spread. Chromosomes or nuclei produced with this technique have a defined shape and are well separated. However, using cytological methods to investigate mitotic chromosomes in fungi has proven challenging. With many fungi, blocking mitosis during metaphase, when chromosomes are compact and simple to distinguish based on their physical characteristics, is unsuccessful due to the coenocytic architecture of hyphae for Zygomycota and Glomeromycota. Furthermore, in contrast to plants and animals, fungi offer very few observations of the mitotic process since their chromosomal sizes are substantially smaller than those of meiotic bivalents, and metaphase chromosomes are seldom observed during somatic divisions.

Expertise in molecular biology, cytogenetics, immunocytochemistry, microscopy, and cellular imaging analysis is necessary. Meanwhile, these techniques' limitations include their low resolution and restricted target coverage. FISH is a potent tool for eukaryotic chromosome and genome research using a combined cytogenetics and molecular biology perspective, although filamentous fungi have not seen much of its application. Previously, only a few glomalean species of mycorrhizae have been subjected to DNA FISH, including *Coprinus cinereus*, *Botrytis cinerea*, *Alternaria alternata*, *Nectria haematococca*, *Cochliobolus heterostrophus*, and *Ustilago maydis*. Among them, *C. cinereus* was analyzed using meiotic pachytene chromosomes as a target, whereas other species were assessed using interphase nuclei, mitotic chromosomes, and DNA fibers from somatic cells. Given that fungi have significantly lower specimen sizes than higher plants and animals, FISH is undoubtedly more challenging to carry out on them. The insufficient recognition of FISH's viability in filamentous fungus by researchers, however, may also be a contributing factor to the limited use of FISH in filamentous fungi (Tsuchiya & Taga, 2010).

Moreover, a number of examples from ascomycetes, including *Alternaria alternata* (Hatta et al., 2002), *Fusarium* (Ma et al., 2010), *Leptosphaeria maculans* (Balesdent et al., 2013), and *Mycosphaerella graminicola* (Stukenbrock et al., 2010), have shown links between pathogenicity traits and genome structure or chromosomal composition. Supernumerary chromosomes, which are normally smaller in size and are commonly found in fungi, are extremely important in the context of plant disease. The evolution of fungal genomes, the dynamics of fungal populations, and the development of particular pathogenic traits in fungi are all influenced by mechanisms that vary the overall size of the fungal genome, particularly those involving polyploidy, aneuploidy, and the activity of mobile genetic elements. There are currently no known relationships between the pathogenicity of *Colletotrichum* fungus and its cytogenetic or cytogenomic characteristics. Despite differences in the genome size and the presence of extra chromosomes in *C. lindemuthianum* and *C. gloeosporioides*, respectively, no correlations between these genetic traits and the pathogenicity of these fungi have been demonstrated.

Molecular cytogenetics methods are essential for deciphering the chromosomal intricacies of fungi and providing information about their genomic architecture. Molecular cytogenetics, particularly fluorescence in situ hybridization (FISH), permits the analysis of several sample types, including tissue sections and interphase nuclei, in contrast to traditional banding cytogenetics, which requires chromosomal preparations. A target DNA (either metaphases, interphases, or DNA fibers for molecular combing) and a probe DNA, the latter of which is labeled with a detectable hapten for visualization under a fluorescence microscope, are essential components of this technique (Liehr, 2021a). The approach modified for fungal cytogenomics is outlined in the following steps:

Denaturation of targeted DNA and probe: First, the target (from fungal metaphases, interphase nuclei, or DNA fibers) and the probe (from other sources) undergo denaturation in order to prepare the DNA strands for later hybridization.

Hybridization: The targeted DNA and probe can be joined more easily by allowing them to stay in a hybridization solution at 37°C. Stable DNA hybrids are formed in this regulated environment, ensnaring certain genomic sequences of interest.

Repetitive DNA sequence blockage: It is possible to reduce background signals caused by repeating DNA sequences using a blocking step. This safety measure guarantees specificity in the next detection, improving the accuracy of the results.

Washing off non-specific DNA probe: To remove any unnecessary or non-specifically bound probe DNA, use the appropriate buffers. Molecular cytogenetic analysis's specificity is improved by this rigorous washing procedure, which also raises the signal-to-noise ratio.

Hapten detection: Moreover, introduce a fluorophore-labeled antibody if the hapten labeling necessitates further detection. This stage improves chromosomal localization accuracy by helping to see the precise genomic areas bound by the probe DNA. An antifade solution and a coverslip can also be added if the probe DNA has already been fluorescently labeled. The fluorescence signals are maintained for further analysis in this stage.

Fluorescence microscopy: Attain high-resolution pictures to examine the geographic distribution of the labeled genomic areas, which will provide important information about the fungal chromosomal architecture.

It is crucial for FISH research that samples contain target DNA with a high molecular weight that has not been degraded. FISH methods have a wide range of applications, including analyzing bacterial and fungal samples. Thus, the entire target DNA can be obtained from a range of multicellular organism materials, including native cells, separated nuclei, tissue sections, metaphase chromosomes, and purified DNA. Following are the probes that are applied in FISH techniques:

- Locus-specific probes
- Repetitive probes
- Partial chromosome painting probes
- Whole-chromosome paints
- Whole-genome probes

Locus-specific probes: Most locus-specific probes (LSPs) originate from molecular cloning. Genetic vectors, which include plasmids, artificial chromosomes from bacteria and yeast, and other vector types, can serve as hosts if they contain the necessary insert of species-specific DNA that is meant to be targeted with FISH. The inserts should be at least 12 kb in size for best results.

Repetitive probes: In FISH investigations, repetitive DNA may be easily visualized. Therefore, strong and easily evaluable signals are produced by repeating probes that target centromeres, telomeres, or other repetitive regions, such as interspersed satellite DNAs. Remarkably, at least one repetitive DNA (D4Z4) located in 4q24 can be tracked using molecular computing and has some significance for human genetic diagnosis.

Partial chromosomal paints: Partial chromosomal paints (PCPs) are developed using glass needle-based chromosomal microdissection (midi). These PCPs usually do not enlarge to the size of a chromosome arm and can stain at least one or two euchromatic chromosomal subbands concurrently.

Whole-chromosome paint: There are two basic methods for producing a whole-chromosome paint (WCP) that dyes a complete chromosome. It was suggested that sequencing by chromosomal flow sorting allows a WCP to be established for a particular chromosome. Furthermore, interspecies hybrids such as fungal hybrids that include genetic material from many fungal species have been investigated as potential sources for producing WCP probes that are particular to individual species. These approaches provide useful resources for thorough chromosomal analysis in the study of fungi (Liehr et al., 2017; Liehr, 2021a).

4.5 KARYOTYPING

To thoroughly investigate the diversity of fungi, there is still considerable work to be done in fungal research. As significant resources in cutting-edge biotechnology, fungi have enormous promise. A potential option for detecting fungus is molecular-based methods. The genome size of filamentous fungi is tiny, yet their gene count is impressively constant (Mehrabi et al., 2017). Their genomes are between 30 and 100 Mbp in size and are anticipated to include between 10,000 and 13,000 genes (Ellwood et al., 2010). Fungi are potential candidates for determining the suitability of cutting-edge sequencing methods because of their simplicity and modest size compared to many other eukaryotes. The nuclei of fungi are too tiny, and there have been very few successful attempts to see fungus chromosomes. In most cases, there are no more than 10 chromosomes per nucleus as determined by conventional light imaging and stained with industry-standard dyes like Giemsa or aceto-orcein. Furthermore, condensed chromosomes may be released from the hyphal cell and distributed across the surface of a plate using a technique known as "germ tube burst," which was invented in the late 1980s. This technique often results in chromosomal resolution. It is used in fluorescence microscopy, in situ hybridization, and traditional light microscopy dyes (Wieloch, 2006).

4.5.1 Electrophoretic Karyotyping

According to electrophoretic karyotyping of many fungi, chromosomal length polymorphisms (CLPs) are a frequent and noticeable characteristic of these species. These polymorphisms have been linked to two different mechanisms: mitotic and meiotic recombination processes (Mehrabi et al., 2017), as well as raising the copy numbers of certain sequences such as ribosomal DNA (rDNA) along with subtelomeric repeats (Mehrabi et al., 2012). Since numerous distinct karyotypes may be found in a given species, these CLPs frequently have very minimal genetic effects. Chromosome recombination has been found to cause CLPs within a specific strain to finally vanish in various organisms. Chromosomal separation on gel permits the detection of supernumerary chromosomes, which are often too tiny to be seen in nuclei, as well as the assessment of genome size, chromosomal length polymorphism, etc. (Halder et al., 2004).

4.5.2 Pulse-Field Gel Electrophoresis

For the separation and study of big DNA molecules, such as chromosomes or other genomic fragments, PFGE is a specialized kind of gel electrophoresis. It is a crucial technique in cytogenomics since it is frequently used to research the structural organization of genomes, including measuring chromosomal sizes, detecting DNA fragmentation, and identifying genetic variants (Lopez-Canovas et al., 2019). Chromosome-sized DNA fragments are separated using pulsed-field gel electrophoresis (PFGE), a technique also known as electrophoretic karyotyping, on agarose gel where the electric field's orientation is regularly changed. This method has shown that many fungi exhibit chromosomal length polymorphisms. Since a single clone frequently gives rise to sole isolates with identical karyotypes, PFGE gives these isolates a "genetic fingerprint." Since the size range and chromosome count within isolates of a given species are often stable, PFGE can discriminate between fungi with similar morphologies. This method could track the inheritance of chromosomal length polymorphisms and demonstrates that meiosis produces novel-sized chromosomes in some fungal species (Gil-Lamaignere et al., 2003).

PFGE provides benefits in investigating karyotype polymorphisms, identifying tiny chromosomes, and measuring precise chromosomal size. To separate fungal chromosomes up to several megabases, PFGE is a useful method for karyotyping different types of fungi. This method is based on a direct current electric field, the direction or strength of which is regularly changed (Soll, 2000). A 120-degree reorientation field is produced by the PFGE device's hexagonal electrode array, or CHEF (contour-clamped homogenous electric field). The electric field's intensity and direction change regularly due to this field. Large DNA molecules can migrate in a zig-zag pattern across the agarose matrix thanks to these dynamic circumstances. Short pulse periods matched with the electric field direction can efficiently move small DNA molecules inside the gel matrix, but longer pulse times help bigger DNA molecules migrate. Nuclear chromosomal and chromosome length variations can be assessed very well using PFGE. PFGE is an accessible technique that may be used to examine somatic chromosomes in interphase nuclei, making a variety of fungal species appropriate for it. This ground-breaking technology makes it feasible to evaluate the karyotypes of imperfect fungi, which were previously difficult to research using conventional karyotypic analysis methods. Regarding chromosomal size estimation and the identification of small-sized chromosomes, PFGE offers a comparatively high level of accuracy. Our knowledge of structural genomics in fungal biology has greatly benefited from the karyotyping of various fungal species. The detection of cases of horizontal chromosomal transfer across closely related fungus species has also been made easier by this research (Mohamed Mahmoud Ahmed, 2012).

4.6 APPLICATIONS OF PFGE

There is a wide range of applications for PFGE, including karyotyping, strain identification of closely related species, characterization of transformed strains, creation of linkage maps, and DNA preparation for genomic research.

4.6.1 Genetic Mapping

Genetic mapping has been one of PFGE's most significant possibilities. Traditional blotting and hybridization procedures may be modified to assign cloned genes and other probes to linkage groups using Southern blot or fluorescence in situ hybridization. Linkage studies in fungi without sexual cycles benefit from this method since it is significantly quicker than parasexual analysis. DNA must be broken by acid depurination or ultraviolet (UV) irradiation for effective Southern blot transfer of DNA bigger than 2 Mb from conventional agarose gels onto membranes. The UV technique is more trustworthy for nicking DNA. However, it should be emphasized that DNA will not be nicked or transferred unless it is stained with ethidium bromide.

Rare-cutting endonucleases may help localize genes on chromosomal fragments when numerous chromosomes are co-migrating or when big chromosomes are not effectively resolved. In a previous research study, *Aspergillus flavus*, *Aspergillus parasiticus*, and *Aspergillus nidulans* were used to examine the locations of conserved genes (argB, benA, trpC, niaD, and pyrG). The placement of these aflatoxigenic species in a distinct taxonomic section from *A. nidulans* is supported by the similarity in the map locations of gene probes in *A. flavus* and *A. parasiticus* (Chiou et al., 2002). Another use of the hybridization method was establishing the disomy of RFLP loci in *Septoria tritici* (Beadle et al., 2003).

4.6.2 Strain Identification

Chromosome length polymorphisms can be used as a supplement for the taxonomic categorization of certain species or strains within species. For instance, Vaughan-Martini et al. (1993) employed CHEF analysis to look at the type strains of 10 species of the genus *Saccharomyces* (Vaughan-Martini et al., 1993). Six species had consistent karyotypes, but *S. dairensis*, *S. exiguus*, *S. sarvazzii*, and *S. unisporus* had heterogeneous karyotypes. Further, even though the individual strains from the various races were sexually compatible, separate karyotypes were discovered for each of the 14 races investigated in *Ustilago hordei* (McCluskey et al., 1994).

4.6.3 Transformed Strain Analysis

Recombinant DNA may be found directly on individual chromosomes using PFGE. For instance, Southern hybridization in *Sordaria macrospora* revealed that all vector molecules had been incorporated into a single chromosome at a single locus; that is, vector copies were incorporated at a single locus in each transformant. The structural alterations of the altered chromosomes were linked to the integration of the changing DNA. *Aspergillus niger*, *Ophiostoma ulmi*, and *Acremonium chrysogenum*'s transformation of chromosomes reportedly increased in size. Inserting a plasmid containing a hygromycin resistance gene led to transformants with modified karyotypes in the echinocandin B-producing strain of *A. nidulans*. Several chromosomal modifications were seen, such as apparent chromosome loss, size changes, and the emergence of major chromosomes. These karyotype alterations were stabilized during mitosis (Pöggeler et al., 2000).

4.7 COMPARATIVE GENOMIC HYBRIDIZATION

The development of comparative genomic hybridization (CGH) was a major breakthrough, which provided the first efficient way to systematically scan entire genomes for differences in DNA copy number (Pinkel & Albertson, 2005). Genomic DNA from the test and reference cell populations is used in a typical CGH analysis. This DNA is then hybridized with a genome representation after being differentially labeled, allowing sequences at different genomic locations to be distinguished. Comparative genomic hybridization, or CGH, makes it possible to compare more than two fungal genomes simultaneously. This is accomplished by assigning unique labels in the genomic DNA of various species or strains of fungi. Unlabeled Cot-1 DNA is frequently added to reactions to stop highly repetitive sequences from hybridizing. To enable a more precise and focused comparison of the genomic variations among fungi, Cot-1 DNA suppresses the hybridization of these repetitive elements.

The genome was first represented using metaphase chromosomes, and copy number differences between test and reference genomic DNA were mapped to the actual locations on the chromosomes. However, DNA microarrays, which are made up of elements directly mapped to the genome sequence, have largely replaced the use of chromosomes (Pinkel et al., 1998; Pollack et al., 1999). The relative copy number of those sequences in the test and reference genomes is then ideally proportional to the relative hybridization intensity of the test and reference signals at that particular location. Variations in the reference genome's copy number within the test cells' genome are directly indicated by rises and falls in signal intensity ratios. Typically, data are normalized by setting the genome's modal ratio to a standard value, which is usually 0.0 on a logarithmic scale or 1.0 on a linear scale. Additional methods, such as flow cytometry or fluorescence in situ hybridization (FISH), can be utilized to determine the true copy number associated with a particular ratio level (Vági et al., 2014).

Several approaches have been employed with array CGH, including large-insert genomic clones, like BACs (bacterial artificial chromosomes). Since obtaining a significant amount of BAC DNA with the necessary purity for the construction of arrays is difficult, different amplification techniques are adopted for limited starting material. These techniques include rolling circle amplification, degenerate primer PCR with one or more primer sets, and ligation-mediated polymerase chain reaction (PCR). Currently, there are BAC arrays that provide complete genome tiling pathways. Moreover, arrays of less intricate nucleic acids such as cDNAs, selected PCR products, and oligonucleotides are employed. Although most CGH techniques use hybridization with whole genomic DNA, some methods use simplified genome representations produced by PCR. Computational analysis of the genome sequence is utilized to create an array of elements that complement the sequences in the representation. Currently, several single nucleotide polymorphism (SNP) genotyping systems are being evaluated for their ability to determine DNA copy number and allelic content throughout the genome; some of these systems use simplified genomic representations (Pinkel & Albertson, 2005).

Butler et al. (2009) compared the genomes of seven species of *Candida* with those of other free-living and parasitic yeasts, with the exception of *C. dubliniensis*. This thorough investigation covered the diversity of *Candida* species worldwide and identified genetic alterations linked to the development of parasitism. Expansions of cell wall and secreted proteins were among these alterations; these were most pronounced in the families of agglutinin-like sequence (ALS) and secreted aspartyl proteinase (SAP) (Butler et al., 2009). The study could not identify the particular characteristics of *Candida albicans* that explain its unique virulence and medical significance by considering historical events connected to the origins of *Candida* and candidosis. Comparative genome hybridization (CDH) was used by Moran et al. (2004) to examine the genomes of *C. dubliniensis* and *C. albicans*. A comparison study may shed light on the genetic components influencing the reduced virulence of *C. dubliniensis*, as the researchers found a multitude of genes unique to that species (Moran et al., 2004).

4.8 INTRODUCTION TO CYTOMICS

Proteomics, metabolomics, and genomics have made great strides in biology. However, their effectiveness is limited when examining single or uncommon cells or monitoring fast cellular processes over extended periods of time. This also emphasizes how important it is to consider epigenetic modifications and the cellular environment's altering effects. To get around these restrictions and obtain a more thorough understanding, functional analyses should be added to "omic" studies. Through the use of an integrated approach, scientists can determine the molecular components and clarify their functional roles, as well as how the cellular context and epigenetic factors affect them (Bernas et al., 2006). As a result, cell-based assays are becoming more and more in demand for drug screening and fundamental research. These assays tackle complex cell-to-cell and cell-to-environment interactions, providing results that are rich in information and are both useful and technically accessible.

Cytomics emerges as an innovative analytical approach in the context of the intersection of systems biology and omics. Its goal is to detect various biochemical characteristics or, more specifically, molecular phenotypes, that are present in individual cells. A thorough grasp of single-cell molecular interactions and cellular dynamics is possible through cytometry of complex cellular systems, which is known as cytomics. By delving into the minute details of cellular behavior, this method enables researchers to gain a more comprehensive understanding of biological systems (Herrera et al., 2007). Like other omics fields such as proteomics and genomics, cytogenomics aims to clarify the details of diverse cellular systems known as cytomes (Gomase & Tagore, 2008). The various cellular systems and functional elements present in multicellular organisms are represented by these cytomes. The interaction between the extracellular environment and the genome gives rise to the functional diversity of cytomes. Consequently, cytomics acts as a link between the functional features of cells and tissues that are impacted by outside influences and genomics and proteomics (Bernas et al., 2006). Cytomics analysis is distinguished by its cell-by-cell foundation, which allows for the analysis of heterogeneous systems without the information loss associated with bulk technologies. In contrast to methods that yield mean values from large sample sizes or tissue homogenates, cytomics provides a comprehensive understanding of cellular diversity and function down to the single-cell level (Valet et al., 2004).

To understand the complexities of tissues and organisms, cytogenomics fundamentally depends on using extremely sensitive and minimally invasive fluorescence-based techniques in conjunction with the general idea of examining individual cell behavior. Multiple fluorescence detectors can be integrated into specialized instruments, allowing cytomic assays to be multiparametric, polychromatic, and multiplexed. This is made possible by the abundance of fluorescent markers available. Fluorescence-based measurements may originate from kinetic, sequential, or single end-point measurements, which can be quantitative and qualitative (Eils & Athale, 2003). These features are common to all cytomic technologies. Still, there are important differences depending on whether quantitative cell fluorescence data are acquired through fluorescence-pulse analysis in flow-based cytomics or in conjunction with cell morphology in image-based cytomics (O'Connor et al., 2001).

Currently available cytomic technologies that are significant include flow cytometry (FCM), confocal laser scanning microscopy (CLSM), laser scanning cytometry (LSC), and spinning-disk confocal microscopy (SDCM) (Herrera et al., 2007).

4.9 FLOW CYTOMETRY

An automated, high-throughput technique called flow cytometry enables extremely quick investigations of several chemical and physical properties of individual cells within a cell population (McKinnon, 2018). A stream of individual cells is moved ahead of one or more light sources (usually lasers), and various detectors are used to detect various optical signals, notably fluorescence and scattered light. To determine a cell's internal complexity or granularity, visible light dispersion is measured in two distinct directions: forward (forward scatter/FSC), which can reveal the cell's relative size, and at 90° (side scatter/SSC), which reveals the cell's internal complexity or granularity (Ibrahim & van den Engh, 2007). Fluorescence and light scattering are two different characteristics of flow cytometry. Sample preparation techniques like transfection to express fluorescent proteins (like green fluorescent protein), staining with fluorescent dyes like propidium iodide (PI) for DNA, or labeling with fluorescently conjugated antibodies like CD3 FITC make it easier to measure fluorescence. These methods make it possible to precisely identify fluorescence in samples (McKinnon, 2018).

Genome size: The first FACS devices, which allowed for the sorting of individual cells based on their fluorescence properties, were launched in the 1970s (Fulwyler, 1974) after the first fluorescence-based flow cytometry device was created and commercialized in the 1960s. When yeast cell growth was initially studied regarding its relationship to DNA content, protein synthesis, and cell viability, flow cytometry was not employed by microbiologists for around 10 years after its invention (Hutter & Eipel, 1978). Numerous research goals are met by the broad use of flow cytometry to analyze nuclei and evaluate their DNA content. It is a useful tool for figuring the genome size, among other things. For instance, preserved spores can be immediately labeled with PI (Anderson et al., 2010) for examination, whereas separated nuclei can be stained with dyes like PI or Hoechst (Sędzielewska et al., 2011; Talhinhas et al., 2017). This research uses strains with known genome sizes as reference points. PI fluorescence peaks may be used as histogram markers to determine the genome size for each sample. These outcomes are frequently analogous to those seen in genome sequencing initiatives. It was discovered that there was genome size variation in *Cronartium quercuum* and *Colletotrichum lindemuthianum* among various isolates (Veselská et al., 2014).

Ploidy: The use of flow cytometry to stain spores or nuclei with propidium iodide (PI) allows for the quick and easy observation of ploidy alterations in a wide variety of fungi (Hara et al., 2012; Runa et al., 2015). De Lucas et al. (1998) showed that using flow cytometry, it is simple to distinguish between haploid and diploid conidia of *A. fumigatus* and *A. nidulans*. There were distinct single (haploid) and double (diploid) peaks in the fluorescence of the conidia. The level of PI fluorescence in a small (10%) subset of haploid conidia was doubled, which may indicate that conidia were halted after their S-phase (De Lucas et al., 1998).

DNA content: Furthermore, using the fluorescent DNA dye acriflavine hydrochloride, Bridge et al. (1987) assessed the conidial DNA content of several *Penicillium* species and discovered variability both within and across isolates (Bridge et al., 1987). Conidial size and DNA content were shown to be correlated. PI staining was used to evaluate DNA degradation in *A. nidulans* (Katayama et al., 2012). It was reported that the 1C peak was shown to diminish by around half during a five hour period while a mutant underwent germination. The number of cells exhibiting decreased levels of red fluorescence increased along with this decrease in DNA content throughout time. The cell cycle was studied using flow cytometry, making detecting DNA content possible (Lobo et al., 2007; Heimel et al., 2010).

Nuclei: Dormant conidia can differ in their number of nuclei, with the precise count variable depending on the fungus species. The number of nuclei may be determined by flow cytometry, either by labeling the DNA (Hara et al., 2002) or by expressing histone–fluorescent protein fusions (Maruyama et al., 2001; Runa et al., 2015). In their research, Maruyama et al. (2001) showed that a single conidium in *A. oryzae* can have up to seven nuclei. According to unique peaks in the intensity of the H2B-EGFP fluorescence, Maruyama et al. (2001) showed that they could separate conidia with one or two nuclei using FACS. After allowing the uni-nucleated conidia to go through sporulation, the spores that emerged once more showed a variety of nuclear numbers in their progeny. A thorough examination into nuclear heterogeneity was done by Runa et al. (2015). The protoplasts of these strains were fused together using complementary auxotrophic mutants of *A. flavus* that expressed histone fusions with YFP and CFP (fluorescent proteins). They discovered that fused cells mostly had either YFP or CFP fluorescence in their nuclei by using both FACS and microscopy. The co-localized fluorescence signals were only present in fewer than 10% of the nuclei, which is interesting. It was demonstrated that the latter merged cells were either diploid or heterokaryotic (Runa et al., 2015).

A recent speciation event occurred in the developing fungal pathogen *Colletotrichum kahawae*. Due to its specific propensity to infect green coffee berries, it has separated from the larger "*C. gloeosporioides* species complex" and given rise to coffee berry disease. If the infection were to spread to Asia or the Americas, Arabica coffee growing would be immediately at risk, as it is now a serious and widespread problem in Africa. Although there is little genetic variability within *C. kahawae*, there are notable differences in the pathogen's aggressiveness. In the preceding study, a variety of *C. kahawae* isolates were evaluated for two cytogenomic traits, including genome size and minichromosome profiles. This collection includes closely related species for comparison and spans a broad spectrum of genetic variation and various levels of aggression. These results represented the first time flow cytometry was used to calculate genome sizes within the *Colletotrichum* genus. The results showed that the genome size of *C. kahawae* was around eight megabases (Mb) larger on average than that of its nearest relatives, which averaged 71.3 Mb. These findings suggest that *C. kahawae*, which contains the viruses that cause coffee berry disease, should be recognized as a different species. Additionally, it has been discovered that *C. kahawae* has two to five minichromosomes. This finding suggests a possible relationship between the number of minichromosomes and the degree of aggression among the isolates under investigation. However, aggressiveness and genome size were not shown to be significantly correlated (Pires et al., 2016).

Flow cytometry, a technique frequently used to gauge how sensitive eukaryotic cells or bacteria are to therapeutic treatments, is the tool of choice for evaluating the effects of antifungal medications and therapies on fungi and their spores in fungal cytomics research. For the purpose of quickly identifying cellular features based on the intensity of fluorescence released upon tagging, fluorescence-based excitation and detection technologies are frequently integrated with flow cytometry in fungal research. For instance, researchers can employ fluorescent tags attached to antibodies that target specific components of fungal cells, including mitochondria. This makes it possible to identify cells with different mitochondrial contents. This process produces fluorescent signals that provide data on cellular responses, medication efficaciousness, and possible mechanisms of action against fungal infections and spores. These signals are highly important in understanding the impacts of antifungal therapies. This advanced method in fungal cytomics makes it easier to comprehend in great depth the molecular and cellular interactions that medicinal drugs have with fungal cells.

Flow cytometry provides a higher degree of quantitative information in fungal cytomics by simultaneously labeling fungal cells with a mixture of fluorophores. This combination of fluorophores offers a detailed comprehension of the relative amounts of every component that is being targeted. The obtained data offer clues on the processes behind the activities of antifungal treatments and enables the evaluation of their efficacy. Furthermore, flow cytometry's adaptability includes the capacity to separate particular fungal cell types in preparation for a more thorough investigation. This sorting feature makes it easier to thoroughly examine the side effects that antifungal medications cause. As a result, researchers studying fungal cytomics may explore the complex cellular reactions and changes brought about by antifungal therapy, improving our understanding of the therapeutic

4.10 CONFOCAL FLUORESCENCE MICROSCOPY

Confocal fluorescence microscopy reduces the field depth, removes out-of-focus blur, and makes it possible to create successive optical stacks from thick specimens, which improves z-axis resolution over conventional fluorescence microscopy. Confocality is a useful tool for imaging single cells labeled with appropriate fluorescent probes in both fixed and living preparations (Rubart, 2004). This technique allows for time- and space-resolved fluorescence measurements, which provides detailed information about cells and is increasingly used in scientific and technological applications (Ivorra et al., 2006). Although CLSM usually offers better image quality, it has a slow imaging frame rate. In contrast, SDCM can achieve video rate imaging, which is important for effective dynamic observations. On the other hand, the newest confocal systems are hybrids, allowing for superior resolution and fast speed.

4.11 LASER SCANNING CYTOMETRY

LSC is a scanning cytofluorimeter that combines the advantages of image and flow cytometry, all under the microscope. Measuring the fluorescence of individual cells in solid-state preparations, such as monolayers, smears, imprints, cytospins, or tissue sections on different supports like slides, dishes, and multiwell plates, allows for multiparametric analysis. LSC provides improved specificity and sensitivity over conventional microscopic techniques. Despite operating at a slower data acquisition speed than flow cytometry, it has a similar structure for data analysis. Furthermore, proceeding with the coordinates after slide restaining enables the repositioning of examined cells of interest. Finally, unique single cells identified by their fluorescence measurements can be displayed as single-cell images or assembled into image galleries for further examination (Cyclin, 2001).

4.12 HIGH-CONTENT SCREENING BIOIMAGING

Typically, a high-content screening bioimaging (HCS-B) system comprises a motorized fluorescence microscope, a CCD camera for capturing images, and a digitizing system for storing images in a high-capacity computer. Software intended for image analysis and acquisition oversees all these processes. This innovative technical idea enables large-scale cell research employing multiwell plates and bulk analysis of millions of cells per experiment when used in high-throughput screening (HTS) cell-based assays (Abraham et al., 2004). High temporal and high spatial resolution may be combined in these systems, which can analyze a single well every second. For complicated studies including combinations of various cell types and agonist and antagonist concentrations, as well as diverse exposure durations and the kinds and numbers of parameters examined for each condition, HCS-B is highly effective. Furthermore, functional assays using live cells are made possible by the ability to maintain constant CO_2 levels and temperature. Additional insights into the behaviors of diverse subpopulations, which may differ in developmental stage, cell cycle phase, transfection status, or other parameters, can be gained by identifying responses in individual cells (Herrera et al., 2007). Bioimaging of various fungal species, such as *Candida albicans*, is potentially involved in biomedical research (Evans & Matsudaira, 2006).

4.13 LASER CAPTURE MICRODISSECTION

Utilizing homogenous cell populations for performing significant proteomic and genomic investigations is imperative. Working with solid tissue samples presents a challenge for flow cytometry cell sorting, even if it is useful for purifying particular cell types in suspension. Many techniques for microdissection have been developed as a result of improving cell isolation. However, they are often inaccurate and time-consuming. A method for automating and standardizing the microdissection procedure has emerged: laser capture microdissection (LCM). This has improved precision and repeatability when picking out certain cells from complicated tissues for further genetic research (Curran et al., 2000). Typically, LCM systems have an inverted microscope with a low-power near-infrared laser attached to it. A transparent 100 μm thermoplastic film is placed over tissue slices on regular glass slides. The laser's focused energy application melts the film in particular spots, which bonds it to the intended cells. This makes it possible to choose specific cells or clusters with accuracy. The film holding the selected cells is removed once the desired cells have been selected, leaving the non-selected tissue in contact with the glass slide. After being separated by LCM, the cells can be collected and subjected to several molecular analysis techniques. Therefore, the comparison of loss of heterozygosity, mutation detection, and gene expression patterns across various cell types within a tissue is made possible by mRNA measurements and cDNA microarrays of cells purified by LCM from micro-dissected tissues (Kunz & Chan, 2004). Proteins of interest have been identified using mass spectrometric sequencing, peptide mass fingerprinting, zymography, and Western blot methods. These techniques are particularly useful for differentiating between expression levels in developing, sick, and normal tissues (Zieziulewicz et al., 2003).

4.14 FLUORESCENT DYES IN FUNGAL CYTOLOGY

In fungal cytology, fluorescent dyes are essential because they allow researchers to precisely visualize and analyze various cellular structures and activities. These dyes glow when exposed to particular light wavelengths, making it possible to precisely identify and observe different fungal components (Hickey et al., 2004). Fluorescent dyes that are often employed in fungal cytology include the following.

***Plasma membrane-selective dyes*:** The plasma membrane and other organelle membranes within fungal cells may be stained by FM dyes, which are specialized probes with a predilection for membrane structures (Fischer-Parton et al., 2000). These dyes, including FM4-64 and FM1-43, belong to the class of amphiphilic styryl dyes, which was pioneered by Betz and colleagues (Betz et al., 1996; Bolte et al., 2004). In hydrophobic scenarios like membranes, their fluorescence is very noticeable. One important benefit is that these colors do not need to be removed from the growing medium the fungus is grown in. The FM dyes are mostly internalized by fungal cells via endocytosis, which makes them useful markers for this biological process. After integration, these dyes diffuse across distinct organelle membranes and stain different organelles in a time-dependent mode (Atkinson et al., 2002). The dye's intracellular migration is primarily linked to the endomembrane system's vesicle trafficking network and the physical continuity of several membranes (Hickey, 2001).

FM4-64 produces unique organelle staining patterns that differ from those produced by FM1-43. Both dyes work well as general membrane-specific stains, especially when low-magnification cell imaging is involved. On the other hand, FM4-64 has shown to be better at staining vesicles and vacuolar membranes within the Spitzenkörper at higher magnification applications, whereas FM1-43 is better at staining mitochondria. While neither dye stains the nuclear envelope, both show substantial plasma membrane staining, making them useful for visualizing septum formation. Though numerous organelles are labeled with FM4-64 and FM1-43, only a small number are recognized (Hickey et al., 2004). Moreover, cell type and growth circumstances influence when to stain organelles. Broad, swiftly expanding hyphae, like the vegetative hyphae toward the edge of the colony, often stain more quickly than short, slowly expanding hyphae like germ tubes. Filamentous fungi use the cationic cyanine dye DIOC6 (3), which accumulates in the nuclear envelope, mitochondria, and endoplasmic reticulum (ER). In plants, the ER and nuclear envelope can be effectively stained by greater dye concentrations, whereas lower dye concentrations preferentially stain mitochondria. The nuclear envelope, ER, and mitochondria in yeast cells and filamentous fungi have all been successfully stained with this dye (Hickey, 2001). DiOC6(3) (3,3′-Dihexyloxacarbocyanine Iodide) dye has been employed to visualize fungal cell structure, including the mitochondria, vacuoles, and Spitzenkörper of *Phanerochaete velutina* hyphal tips (Zhuang et al., 2009).

Nuclear dyes: In fungal cytology, nuclear dyes are vital for visualizing and examining the nucleus, an important organelle containing genetic information. The fluorescent dye DAPI (4′,6-diamidino-2-phenylindole), which binds selectively to DNA sections rich in adenine and thymine (AT), is one nuclear dye frequently utilized within fixed fungal cells. Nevertheless, DAPI can be employed as an essential probe if the cell permits dye penetration. DAPI absorption by vegetative hyphae is occasionally restricted while germinating spores are considered more responsive. DAPI is normally activated at UV wavelengths, which might not be the best consideration for imaging living cells. However, it can also be stimulated at 405 nm, a wavelength presently reached with a cost-effective blue diode laser for confocal microscopy (Haugland, 2002). This dye has been used to visualize organelles and the genome sizes in *Cyanidioschyzon merolae* and *Chlamydomonas reinhardtii* (Kuroiwa et al., 2012).

Moreover, the staining of other cellular components, including mitochondrial DNA and polyphosphates, has made it difficult to achieve effective live-cell nuclear imaging with DAPI utilizing CLSM (confocal laser scanning microscopy). Molecular Probes Inc. offers a variety of cyanine SYTO dyes that can stain DNA and RNA. Upon analyzing many green fluorescing SYTO dyes, it was observed that they readily penetrate cells and exhibit varying degrees of nuclei staining in vegetative hyphae (Hickey, 2001). Unfortunately, these dyes have not proven appropriate for live-cell imaging studies because of their high phototoxicity, which causes organelle destruction when subjected to a laser beam in CLSM scans. DIOC6 (3) can stain the nuclear envelope, mitochondria, and ER. In cells stained with FM4-64 for an extended period, nuclei can also be seen as negatively stained structures. However, using a nuclear-targeted fluorescent protein like GFP is the most efficient way to visualize nuclei in fungal species (Hickey et al., 2004).

Mitochondrial dyes: Mitochondrial dyes are luminous substances that are particularly synthesized to highlight and stain the mitochondria in cells. These dyes are vital resources for investigating the shape, dynamics, and function of mitochondria in cell biology and fluorescence microscopy. The mitochondria, essential cellular organelles for energy production and other cellular functions, can be labeled and studied using dyes specific to mitochondria. In fungal cytomics, the shape of mitochondria in apical hyphal compartments is typically tubular and elongated, while species-specific differences may exist in the size of individual mitochondria. Mitochondria are found numerously in cells and closely spaced in actively developing vegetative hyphae. Shorter and more widely distributed mitochondria can be seen in sub-apical areas. Furthermore, the mitochondria in slowly developing germ tubes are shorter in length and have fewer numbers. Rhodamine 123, DASPMI, FM1-43, and MitoTracker green are commonly used dyes for efficiently labeling the mitochondria in filamentous fungus cells (Fischer-Parton et al., 2000).

A fluorescent dye, Rhodamine 123, has cationic characteristics and can diffuse into cells easily and be quickly absorbed by active mitochondria in hyphae. It is a potentiometric dye, the fluorescence of which is directly proportional to the electrochemical gradient across the mitochondrial membrane; the greater the membrane potential the stronger the fluorescence and vice versa. The fluorescence of mitochondria at growing tips is often brighter than that of mitochondria, which suggests that the tip areas are more active. This increased activity might be explained by the fact that more ATP is required, which is necessary to maintain tip development (Swift et al., 2001). It was identified that using the fungicide azoxystrobin to collapse the electrochemical gradient of mitochondrial membranes might reduce Rhodamine 123's fluorescence in the case of *Botrytis cinerea* (Haq et al., 2021). This fungicide blocks the cytochrome bc1 complex, the site of mitochondrial membrane transport. It is important to remember that Rhodamine 123 occasionally causes background staining of the surrounding media and the hyphal plasma membrane. Compared to Rhodamine 123, DASPMI takes a longer staining time (about 20 minutes) to mark mitochondria. Although it has the same potentiometric properties as Rhodamine 123, its effects are much less noticeable. DASPMI prevents background staining of the media and does not stain the plasma membrane. One benefit of DASPMI is that it has a large Stokes shift, meaning that it emits red light when it is excited with blue light. Due to this characteristic, it is a good choice for dual labeling experiments, such as GFP, which fluoresces green when stimulated by blue light (Freitag et al., 2004).

The FM dyes derived from DASPMI, specifically FM1-43, stain *Neurospora crassa* mitochondria more quickly (about 20 minutes) than FM4-64 (about 45 minutes). FM1-43 is the favored mitochondrial stain of the two dyes because it is noticeably more selective for mitochondria and produces stronger staining than FM4-64 (Bolte et al., 2004). FM1-43 also stains the plasma membrane, but not as strongly as Rhodamine 123 does. MitoTracker probes are cell permeable and specifically target mitochondria; thereafter, they diffuse across the plasma membrane and accumulate within functioning mitochondria. DIOC6 (3) at low doses (usually 0.5 mM) can stain mitochondria in hyphae and spores (Fischer-Parton et al., 2000).

Vacuolar dyes: Filamentous fungi have a complex vacuolar system that includes several tiny spherical organelles and a dynamic network of tubular components in juvenile hyphal compartments and enormous spherical organelles in older compartments. Three dyes, DFFDA, MDY-64, and FM4-64, are mostly used for staining the vacuolar system (Hickey et al., 2004).

FM4-64, a vital fluorescent dye to identify vacuolar membranes, indicates the endocytic process's completion in fungal cells. When the dye is applied to *Neurospora* vegetative hyphae, it becomes visible in the tubular elements around 20 minutes after the dye is first applied and in the large spherical vacuoles about 10 minutes afterward. The vacuolar compartments of *Magnaporthe* germ tubes stain later, after around 45 minutes. This staining pattern's temporal correlation makes a more sophisticated knowledge of fungal cytomics possible, which offers insightful information on the dynamics of endocytosis in various fungal structures. Moreover, another vacuolar dye, MDY-64, can stain the membranes within the spores and hyphae and the membranes inside the vacuolar membranes. Despite its efficiency in membrane staining, it is still unclear exactly what mechanism drives the staining process (Ashford et al., 2001).

A derivative of 6-carboxyfluorescein diacetate (CFDA), Oregon Green 488 carboxylic acid (carboxy-DFFDA), has better photostability than its equivalent. The ability of both of these lipophilic probes to pass through the plasma membrane of *Pisolithus tinctorius* hyphae was confirmed by Cole et al. (1998). After entering the cytoplasm, they are hydrolyzed to produce 6-carboxyfluorescein and Oregon Green 488 carboxylic acid, respectively. Subsequently, with the help of an organic anion transporter, these derivatives show quick passage across the vacuolar membrane and into vacuoles. Notably, adding the anion transport inhibitor probenecid may obstruct this transport phase. One particularly efficient dye for coloring the lumens of spherical and tubular vacuoles inside hyphae is carboxy-DFFDA (Roca et al., 2003).

***Cell wall dyes*:** Calcofluor White M2R potentially serves as a staining dye for cell walls via its ability to bind to chitin, intercalate into developing chitin chains, and impede the formation of chitin microfibrils in filamentous fungi (Gonçalves et al., 2006; Siqueira et al., 2013). It is often utilized at doses of around 25 mM, which can suppress cell growth and increase chitin production in both yeast and hyphal cells, whether fixed or alive. But Calcofluor can be useful as a cell stain at extremely low concentrations, below 1.5 mM. This stain plays an important role in visualizing cell structures in fungal cytology (Hickey et al., 2004).

***Dyes for cell viability*:** It is frequently crucial to evaluate the vitality of fungal cells within a population. Combining propidium iodide with fluorescein diacetate (FDA) stains is a very successful method. At 488 nm, both stains can be stimulated. FDA is easily absorbed by living cells and hydrolyzes to produce fluorescein, which releases green light. In contrast, only cells with a damaged or weakened plasma membrane can absorb propidium iodide. Propidium iodide is thus only incorporated by dead or injured cells, which is why red fluorescence is produced. Living cells (green) and dead cells (red) can be distinguished by simultaneous imaging at the proper emission wavelengths. The live/dead cell test should be carried out when the dyes are added since cells do not maintain fluorescein. BCECF (2′-7′-bis (carboxyethyl)-5(6)-carboxyfluorescein) can be utilized as a substitute for FDA since it shows improved cell retention. This dual-staining method may assess fungal cell populations' vitality (Haugland, 2002). The vitality of fungi may be assessed using this dual-staining method.

REFERENCES

Abraham, V.C., D.L. Taylor and J.R. Haskins. 2004. High content screening applied to large-scale cell biology. *Trends Biotechnol.* 22:15–22.

Anderson, C.L., T.L. Kubisiak, C.D. Nelson, J.A. Smith and J.M. Davis. 2010. Genome size variation in the pine fusiform rust pathogen Cronartium quercuum f. sp. fusiforme as determined by flow cytometry. *Mycologia* 102:1295–1302.

Ashford, A.E., L. Cole and G.J. Hyde. 2001. Motile tubular vacuole systems. In *Biology of the Fungal Cell.* Springer. pp. 243–265.

Atkinson, H.A., A. Daniels and N.D. Read. 2002. Live-cell imaging of endocytosis during conidial germination in the rice blast fungus, Magnaporthe grisea. *Fungal Genet. Biol.* 37:233–244.

Balesdent, M.H., I. Fudal, B. Ollivier, P. Bally, J. Grandaubert, F. Eber, . . . & T. Rouxel. 2013. The dispensable chromosome of L eptosphaeria maculans shelters an effector gene conferring avirulence towards B rassica rapa. *New Phytologist*, 198(3):887–898.

Beadle, J., M. Wright, L. McNeely and J.W. Bennett. 2003. Electrophoretic karyotype analysis in fungi. *Adv. Appl. Microbiol.* 53:243–270.

Bernas, T., G. Gregori, E.K. Asem and J.P. Robinson. 2006. Integrating cytomics and proteomics. *Mol. Cell. Proteomics* 5:2–13.

Betz, W.J., F. Mao and C.B. Smith. 1996. Imaging exocytosis and endocytosis. *Curr. Opin. Neurobiol.* 6:365–371.

Bolte, S., C. Talbot, Y. Boutte, O. Catrice, N.D. Read and B. Satiat-Jeunemaitre. 2004. FM-dyes as experimental probes for dissecting vesicle trafficking in living plant cells. *J. Microsc.* 214:159–173.

Bridge, P.D., L. Hudson, Z. Kozakiewicz, A.H.S. Onions and R.R.M. Paterson. 1987. Investigation of variation in phenotype and DNA content between single-conidium isolates of single Penicillium strains. *Microbiology* 133:995–1004.

Butler, G., M.D. Rasmussen, M.F. Lin, M.A.S. Santos, S. Sakthikumar, C.A. Munro, E. Rheinbay, M. Grabherr, A. Forche and J.L. Reedy. 2009. Evolution of pathogenicity and sexual reproduction in eight Candida genomes. *Nature* 459:657–662.

Chiou, C.-H., M. Miller, D.L. Wilson, F. Trail and J.E. Linz. 2002. Chromosomal location plays a role in regulation of aflatoxin gene expression in Aspergillus parasiticus. *Appl. Environ. Microbiol.* 68:306–315.

Claussen, U. 2005. Chromosomics. *Cytogenet. Genome Res.* 111:101–106.

Cole, L., D. A. Orlovich, and A. E. Ashford 1998. Structure, function, and motility of vacuoles in filamentous fungi. *Fungal Genet. Biol.* 24:86–100.

Curran, S., J.A. McKay, H.L. McLeod and G.I. Murray. 2000. Laser capture microscopy. *Mol. Pathol.* 53:64.

Cyclin, E. 2001. Intranuclear Compartmentalization of Cyclin E during the cell. *Cancer Res.* 61:1220–1226.

De Lucas, J.R., A.I. Dominguez, A. Mendoza and F. Laborda. 1998. Use of flow-cytometry to distinguish between haploid and diploid strains of Aspergillus fumigatus. *Fungal Genet. Rep.* 45:7–9.

Eils, R. and C. Athale. 2003. Computational imaging in cell biology. *J. Cell Biol.* 161:477–481.

Ellwood, S.R., Z. Liu, R.A. Syme, Z. Lai, J.K. Hane, F. Keiper, C.S. Moffat, R.P. Oliver and T.L. Friesen. 2010. A first genome assembly of the barley fungal pathogen Pyrenophora teres f. teres. *Genome Biol.* 11:1–14.

Evans, J.G. and P. Matsudaira. 2006. Linking microscopy and high content screening in large-scale biomedical research. High content screen. A powerful approach to syst. *Cell Biol. Drug Discov.* 33–38.

Fischer-Parton, S., R.M. Parton, P.C. Hickey, J. Dijksterhuis, H.A. Atkinson and N.D. Read. 2000. Confocal microscopy of FM4-64 as a tool for analysing endocytosis and vesicle trafficking in living fungal hyphae. *J. Microsc.* 198:246–259.

Freitag, M., P.C. Hickey, N.B. Raju, E.U. Selker and N.D. Read. 2004. GFP as a tool to analyze the organization, dynamics and function of nuclei and microtubules in Neurospora crassa. *Fungal Genet. Biol.* 41:897–910.

Fulwyler, M.J. 1974. Status quo in flow-through cytometry. *J. Histochem. Cytochem.* 22:605–606.

Gil-Lamaignere, C., E. Roilides, J. Hacker and F.-M. Müller. 2003. Molecular typing for fungi—a critical review of the possibilities and limitations of currently and future methods. *Clin. Microbiol. Infect.* 9:172–185.

Gomase, V.S. and S. Tagore. 2008. Cytomics. *Curr. Drug Metab.* 9:263–266.

Gonçalves, A.B., I.M. Santos, R.R.M. Paterson and N. Lima. 2006. FISH and Calcofluor staining techniques to detect in situ filamentous fungal biofilms in water. *Rev. Iberoam. Micol.* 23:194–198.

Habib, Z. 2022. DNA-binding One Zinc Finger (DOF) transcription factors in Vigna species: A review. *Phtopathogenom. Dis. Control* 1:37–54.
Halder, A., S. Halder, A. Fauzdar and A. Kumar. 2004. Molecular approaches of chromosome analysis: An overview. *Proc. Indian Nat. Sci. Acad.* 153–221.
Hatta, R., K. Ito, Y. Hosaki, T. Tanaka, A. Tanaka, M. Yamamoto, K. Akimitsu, and T. Tsuge. 2002. The conditionally dispensable chromosome controls host-specific pathogenicity in the fungal plant pathogen Alternaria alternata. *Genetics*, 161, 59–70.
Haq, I.U. and S. Ijaz. 2020a. *Plant Disease Management Strategies for Sustainable Agriculture Through Traditional and Modern Approaches*. Springer Nature.
Haq, I.U. and S. Ijaz. 2020b. *Etiology and Integrated Management of Economically Important Fungal Diseases of Ornamental Palms*. Springer.
Haq, I.U. and S. Ijaz. 2022. *Trends in Plant Disease Assessment*. Springer Nature.
Haq, I.U., S. Ijaz, A. Faraz and N.A. Khan. 2021. Characterization of Curvularia buchloes causing leaf spots on Medicago sativa L. (alfalfa) and its management through fungicides. *J. Plant Dis. Prot.* 128:493–500.
Haq, I.U., S. Ijaz and I.A. Khan. 2022. *Phytomycology and Molecular Biology of Plant Pathogen Interactions*. CRC Press.
Haq, I.U., S. Ijaz, M.Z. Latif, I.A. Khan, H.M. Ali and S. Kaur. 2023. Phylogenomics resolves the etiology of dieback disease and deciphers Ceratocystis dalbergicans sp. nov., causal agent of Dalbergia sissoo decline. *Front. Genet.* 14:1136688.
Hara, S., F.J. Jin, T. Takahashi and Y. Koyama. 2012. A further study on chromosome minimization by protoplast fusion in Aspergillus oryzae. *Mol. Genet. Genom.* 287:177–187.
Hara, S., R.F. Tsuji, O. Hatamoto and T. Masuda. 2002. A simple method for enrichment of uninucleate conidia of Aspergillus oryzae. *Biosci. Biotechnol. Biochem.* 66:693–696.
Haugland, R.P. 2002. *Handbook of Fluorescent Probes and Research Products*. Molecular Probes.
Heimel, K., M. Scherer, M. Vranes, R. Wahl, C. Pothiratana, D. Schuler, V. Vincon, F. Finkernagel, I. Flor-Parra and J. Kämper. 2010. The transcription factor Rbf1 is the master regulator for b-mating type controlled pathogenic development in Ustilago maydis. *PLoS Pathog*. 6:e1001035.
Herrera, G., L. Diaz, A. Martinez-Romero, A. Gomes, E. Villamón, R.C. Callaghan and J.-E. O'Connor. 2007. Cytomics: A multiparametric, dynamic approach to cell research. *Toxicol. Vitr*. 21:176–182.
Hickey, P.C. 2001. Imaging vesicle trafficking and organelle dynamics in living fungal hyphae. KB thesis scanning project 2015.
Hickey, P.C., S.R. Swift, M.G. Roca and N.D. Read. 2004. Live-cell imaging of filamentous fungi using vital fluorescent dyes and confocal microscopy. *Methods Microbiol.* 34:63–87.
Hochstenbach, R., T. Liehr and R.J. Hastings. 2021. Chromosomes in the genomic age. Preserving cytogenomic competence of diagnostic genome laboratories. *Eur. J. Hum. Genet*. 29:541–552.
Hochstenbach, R., E. van Binsbergen, H. Schuring-Blom, A. Buijs and H.K.P. van Amstel. 2019. A survey of undetected, clinically relevant chromosome abnormalities when replacing postnatal karyotyping by whole genome sequencing. *Eur. J. Med. Genet.* 62:103543.
Hutter, K.-J. and H.E. Eipel. 1978. Flow cytometric determinations of cellular substances in algae, bacteria, moulds and yeasts. *Antonie Van Leeuwenhoek* 44:269–282.
Ibrahim, S.F. and G. van den Engh. 2007. Flow cytometry and cell sorting. *Cell Sep. Fundam. Anal. Prep. Methods* 19–39.
Ijaz, S., I.U. Haq, S. Mukhtar and Z. Habib. 2022. Molecular Phytopathometry. *Trends Plant Dis. Assess.* 167–201.
Ivorra, C., M. Kubicek, J.M. González, S.M. Sanz-González, A. Álvarez-Barrientos, J.-E. O'Connor, B. Burke and V. Andrés. 2006. A mechanism of AP-1 suppression through interaction of c-Fos with lamin A/C. *Genes Dev.* 20:307–320.

Katayama, T., H. Uchida, A. Ohta and H. Horiuchi. 2012. Involvement of protein kinase C in the suppression of apoptosis and in polarity establishment in Aspergillus nidulans under conditions of heat stress. *PLoS One* 7:e50503.

Kunz, G.M. and D.W. Chan. 2004. The use of laser capture microscopy in proteomics research–a review. *Dis. Markers* 20:155–160.

Kuroiwa, T., M. Ohnuma, Y. Imoto, O. Misumi, T. Fujiwara, S. Miyagishima, N. Sumiya and H. Kuroiwa. 2012. Lipid droplets of bacteria, algae and fungi and a relationship between their contents and genome sizes as revealed by BODIPY and DAPI staining. *Cytologia* 77:289–299.

Liehr, T. 2019. From human cytogenetics to human chromosomics. *Int. J. Mol. Sci.* 20:826.

Liehr, T. 2021a. Molecular cytogenetics in the era of chromosomics and cytogenomic approaches. *Front. Genet.* 12:720507.

Liehr, T. 2021b. International system for human cytogenetic or cytogenomic nomenclature (ISCN): Some thoughts. *Cytogenet. Genome Res.* 161(5):223–224.

Liehr, T., K. Kreskowski, M. Ziegler, K. Piaszinski and K. Rittscher. 2017. The standard FISH procedure. *Fluoresc. Situ Hybrid. Appl. Guid.* 109–118.

Lobo, D.S., I.B. Pereira, L. Fragel-Madeira, L.N. Medeiros, L.M. Cabral, J. Faria, M. Bellio, R.C. Campos, R. Linden and E. Kurtenbach. 2007. Antifungal Pisum sativum defensin 1 interacts with Neurospora crassa cyclin F related to the cell cycle. *Biochemistry* 46:987–996.

Lopez-Canovas, L., M.B.M. Benitez, J.A.H. Isidron and E.F. Soto. 2019. Pulsed field gel electrophoresis: Past, present, and future. *Anal. Biochem.* 573:17–29.

Ma, L.J., H.C. Van Der Does, K.A. Borkovich, J.J. Coleman, M.J. Daboussi, A. Di Pietro, M. Dufresne, M. Freitag, M. Grabherr, B. Henrissat and P.M. Houterman. 2010. Comparative genomics reveals mobile pathogenicity chromosomes in *Fusarium*. *Nature* 464:367–373. https://doi.org/10.1038/nature08850

Maruyama, J., H. Nakajima and K. Kitamoto. 2001. Visualization of nuclei in Aspergillus oryzae with EGFP and analysis of the number of nuclei in each conidium by FACS. *Biosci. Biotechnol. Biochem.* 65:1504–1510.

McCluskey, K., J. Agnan and D. Mills. 1994. Characterization of genome plasticity in Ustilago hordei. *Curr. Genet.* 26:486–493.

McKinnon, K.M. 2018. Flow cytometry: An overview. *Curr. Protoc. Immunol.* 120:1–5.

Mehrabi, R., A. Mirzadi Gohari and G.H.J. Kema. 2017. Karyotype variability in plant-pathogenic fungi. *Annu. Rev. Phytopathol.* 55:483–503.

Mehrabi, R., M. Taga, M. Aghaee, P.J.G.M. De Wit and G.H.J. Kema. 2012. Karyotyping methods for fungi. *Plant Fungal Pathog. Methods Protoc.* 591–602.

Mohamed Mahmoud Ahmed, A. 2012. Molecular and cytological karyotype of Nectria haematococca mating population I and inheritance mode of a 410-kb supernumerary chromosome in the genome. PhD thesis.

Mohanta, T.K. and H. Bae. 2015. The diversity of fungal genome. *Biol. Proced. Online* 17:1–9.

Möller, M. and E.H. Stukenbrock. 2017. Evolution and genome architecture in fungal plant pathogens. *Nat. Rev. Microbiol.* 15:756–771.

Moran, G., C. Stokes, S. Thewes, B. Hube, D.C. Coleman and D. Sullivan. 2004. Comparative genomics using Candida albicans DNA microarrays reveals absence and divergence of virulence-associated genes in Candida dubliniensis. *Microbiology* 150:3363–3382.

Naranjo-Ortiz, M.A. and T. Gabaldón. 2019. Fungal evolution: Diversity, taxonomy and phylogeny of the Fungi. *Biol. Rev.* 94:2101–2137.

O'Connor, J., R.C. Callaghan, M. Escudero, G. Herrera, A. Martínez, M. Monteiro and H. Montolíu. 2001. The relevance of flow cytometry for biochemical analysis. *IUBMB Life* 51:231–239.

Pinkel, D. and D.G. Albertson. 2005. Comparative genomic hybridization. *Annu. Rev. Genom. Hum. Genet.* 6:331–354.

Pinkel, D., R. Segraves, D. Sudar, S. Clark, I. Poole, D. Kowbel, C. Collins, W.-L. Kuo, C. Chen and Y. Zhai. 1998. High resolution analysis of DNA copy number variation using comparative genomic hybridization to microarrays. *Nat. Genet.* 20:207–211.

Pires, A.S., H.G. Azinheira, A. Cabral, S. Tavares, D. Tavares, M. Castro, V. Várzea, M.C. Silva, R. Abranches and J. Loureiro. 2016. Cytogenomic characterization of Colletotrichum kahawae, the causal agent of coffee berry disease, reveals diversity in minichromosome profiles and genome size expansion. *Plant Pathol.* 65:968–977.

Pöggeler, Masloff, Jacobsen and Kück. 2000. Karyotype polymorphism correlates with intraspecific infertility in the homothallic ascomycete Sordaria macrospora. *J. Evol. Biol.* 13:281–289.

Pollack, J.R., C.M. Perou, A.A. Alizadeh, M.B. Eisen, A. Pergamenschikov, C.F. Williams, S.S. Jeffrey, D. Botstein and P.O. Brown. 1999. Genome-wide analysis of DNA copy-number changes using cDNA microarrays. *Nat. Genet.* 23:41–46.

Priest, S.J., V. Yadav and J. Heitman. 2020. Advances in understanding the evolution of fungal genome architecture. *F1000Research* 9.

Roca, M.G., L.C. Davide, M.C. Mendes-Costa and A. Wheals. 2003. Conidial anastomosis tubes in Colletotrichum. *Fungal Genet. Biol.* 40:138–145.

Rubart, M. 2004. Two-photon microscopy of cells and tissue. *Circ. Res.* 95:1154–1166.

Runa, F., I. Carbone, D. Bhatnagar and G.A. Payne. 2015. Nuclear heterogeneity in conidial populations of Aspergillus flavus. *Fungal Genet. Biol.* 84:62–72.

Sędzielewska, K.A., J. Fuchs, E.M. Temsch, K. Baronian, R. Watzke and G. Kunze. 2011. Estimation of the Glomus intraradices nuclear DNA content. *New Phytol.* 192:794–797.

Siqueira, V.M., H.M.B. Oliveira, C. Santos, R.R.M. Paterson, N.B. Gusmão and N. Lima. 2013. Biofilms from a Brazilian water distribution system include filamentous fungi. *Can. J. Microbiol.* 59:183–188.

Soll, D.R. 2000. The ins and outs of DNA fingerprinting the infectious fungi. *Clin. Microbiol. Rev.* 13:332–370.

Stajich, J.E. 2017. Fungal genomes and insights into the evolution of the kingdom. *Microbiol. Spectr.* 5:10–1128.

Stukenbrock, E.H., F.G. Jorgensen, M. Zala, T.T. Hansen, B.A. McDonald, and M.H. Schierup 2010. Whole-genome and chromosome evolution associated with host adaptation and speciation of the wheat pathogen. *Mycosphaerella graminicola. PLoS Genetics* 6:e1001189.

Swift, S.R., C.A. Hart, D.W. Bartlett and N.D. Read. 2001. Interactions between azoxystrobin and Puccinia recondita, Erysiphe graminis, and Botrytis cinerea on the microscale. *Scanning* 23:153–154.

Taga, M., K. Tanaka, S. Kato and Y. Kubo. 2015. Cytological analyses of the karyotypes and chromosomes of three Colletotrichum species, C. orbiculare, C. graminicola and C. higginsianum. *Fungal Genet. Biol.* 82:238–250.

Talhinhas, P., D. Tavares, A.P. Ramos, S. Gonçalves and J. Loureiro. 2017. Validation of standards suitable for genome size estimation of fungi. *J. Microbiol. Methods* 142:76–78.

Tsuchiya, D. and M. Taga. 2010. Fluorescence in situ hybridization for molecular cytogenetic analysis in filamentous fungi. *Mol. Cell Biol. Methods Fungi* 235–257.

Vági, P., D.G. Knapp, A. Kósa, D. Seress, Á.N. Horváth and G.M. Kovács. 2014. Simultaneous specific in planta visualization of root-colonizing fungi using fluorescence in situ hybridization (FISH). *Mycorrhiza* 24:259–266.

Valet, G., J.F. Leary and A. Tarnok. 2004. Cytomics—New technologies: Towards a human cytome project. *J. Int. Soc. Anal. Cytol.* 59(2):167–171.

Vaughan-Martini, A., A. Martini and G. Cardinali. 1993. Electrophoretic karyotyping as a taxonomic tool in the genus Saccharomyces. *Antonie Van Leeuwenhoek* 63:145–156.

Veselská, T., J. Svoboda, Ž. Růžičková and M. Kolařík. 2014. Application of flow cytometry for genome size determination in Geosmithia fungi: A comparison of methods. *Cytom. Part A* 85:854–861.

Wieloch, W. 2006. Chromosome visualisation in filamentous fungi. *J. Microbiol. Methods* 67:1–8.

Zhuang, X., M. Tlalka, D.S. Davies, W.G. Allaway, S.C. Watkinson and A.E. Ashford. 2009. Spitzenkörper, vacuoles, ring-like structures, and mitochondria of Phanerochaete velutina hyphal tips visualized with carboxy-DFFDA, CMAC and DiOC6 (3). *Mycol. Res.* 113:417–431.

Zieziulewicz, T.J., D.W. Unfricht, N. Hadjout, M.A. Lynes and D.A. Lawrence. 2003. Shrinking the biologic world—nanobiotechnologies for toxicology. *Toxicol. Sci.* 74:235–244.

5 Fungal Transcriptomics

Ahmad Nisar, Waqar Ahmad,
Nabeeha Aslam Khan, Nasir Ahmad Khan,
Amer Habib, Anjum Faraz, and
Hayssam Mohamad Ali

5.1 INTRODUCTION

5.1.1 Molecular Plant Pathology

Molecular plant pathology is a comprehensive field that explores the complex molecular interactions between pathogens that infect plants and plants. This field integrates molecular biology, genetics, biochemistry, and bioinformatics elements to comprehend and treat plant diseases. To quickly and effectively identify infections, molecular diagnostics methods such as next-generation sequencing (NGS), qPCR (quantitative PCR), and PCR (polymerase chain reaction) are employed (Sloots et al., 2015). The identification of particular genetic markers that allow the differentiation of various disease strains or species (Goswami et al., 2022) is crucial for developing targeted disease management strategies. Additionally, examining pathogen-produced proteins that influence host cellular functions to suppress plant defenses and promote infection, known as effector proteins, provides insights into pathogen virulence mechanisms (Tomczynska et al., 2020). Furthermore, understanding how immune responses are triggered in plants by pathogen-associated molecular patterns (PAMPs) is essential for enhancing plant resistance through breeding or biotechnological approaches (Rathore and Ghosh, 2018). Resistance genes (R-genes) are analyzed to determine and describe the plant genes that provide resistance to specific pathogens (Hammond-Kosack and Kanyuka, 2007). Whole-genome sequencing involves identifying genes important in pathogenesis and resistance by sequencing the genomes of pathogens and host plants (He et al., 2021). Transcriptomics involves understanding the dynamics of host–pathogen interactions by analyzing gene expression patterns during infection (Sackton, 2019). Proteomics and metabolomics involve investigating every protein and metabolite implicated in the pathogenesis of the disease (Chen and Gerszten, 2020). Defense signaling involves investigating how plants detect pathogen invasion and transduce signals to trigger defense mechanisms (Wu et al., 2014). This research explores the functions of plant hormones, such as ethylene, salicylic acid, and jasmonic acid, in mediating defense responses (hormonal cross-talk) (Yang et al., 2019). Biotechnology is described as developing crops that have been genetically modified to resist disease effectively (Kumar et al., 2020). Silencing critical genes in pathogens or pests using RNA interference (RNAi) technology has emerged as a promising strategy for pest and disease control by disrupting essential

DOI: 10.1201/9781003452362-5

biological processes in target organisms (Zotti and Smagghe, 2015). Similarly, using genome-editing techniques such as CRISPR-Cas to modify plant genomes enables the introduction of resistant traits or the elimination of susceptibility genes, thereby enhancing crop resilience against various pathogens (Talakayala et al., 2022).

Incorporating molecular methods into conventional breeding programs will accelerate the emergence of resistant varieties (Ahmar et al., 2020). CRISPR-Cas9 is used for precise genome editing to study gene function and develop disease-resistant plants (Erdoğan et al., 2023). Agrobacterium-mediated transformation involves introducing foreign genes into plants (Hwang et al., 2017). High-throughput sequencing enables comprehensive analysis of genomes (genomics), allowing researchers to identify genetic variations, study evolutionary relationships, and understand the genetic basis of traits, including disease resistance (De Wit et al., 2012). Transcriptomics involves RNA sequencing for studying gene expression profiles (Dong and Chen, 2013). Proteomics involves mass spectrometry and is used to analyze the protein composition of samples (Mann et al., 2001). Metabolomics uses techniques like gas chromatography-mass spectrometry (GC-MS) and liquid chromatography-mass spectrometry (LC-MS) for profiling metabolites (Zeki et al., 2020). Sequence analysis uses tools for aligning and comparing genetic sequences (Tippmann, 2004). Integrative approaches to model and simulate biological networks and pathways provide valuable insights into complex biological interactions, enabling researchers to predict system behaviors, identify key regulatory components, and developing targeted strategies for disease control and crop improvement (Machado et al., 2011). Databases like GenBank, TAIR, and others are utilized for genetic information (Lamesch et al., 2010). Molecular diagnostic kits are developed for the early detection of plant pathogens in the field (Donoso and Valenzuela, 2018). Surveillance systems are used for implementing molecular tools for monitoring and managing disease outbreaks (Sabat et al., 2013). Marker-assisted selection (MAS) involves using molecular markers to select disease-resistant traits in breeding programs (Devi et al., 2017). Transgenic plant breeding involves creating plants with introduced genes that confer resistance to multiple pathogens (Low et al., 2018). Integrated pest management (IPM) involves combining molecular insights with traditional pest control methods to achieve sustainable agriculture (Schmidt et al., 2021). Developing disease-resistant crops is a sustainable approach to minimizing reliance on chemical pesticides, reducing environmental impact, and enhancing agricultural productivity (Gossen and McDonald, 2020).

5.1.2 Omics in Plant Pathology

5.1.2.1 Genomics

In recent years, developments in molecular techniques have opened doors to understanding how biocontrol agents prevent plant pathogens (Massart and Jijakli, 2007). Next-generation sequencing (NGS) technology has sped up the sequencing of microbial genomes (Knief, 2014). Advancements in high-throughput sequencing have reduced the expense of NGS analysis for biocontrol agents. Eight BCAs have a genome database available out of the numerous published genomes of BCAs. As with draft genomes, certain genes might be skipped, and contigs or scaffolds' relative positions could be incorrectly placed or misinterpreted. However, by characterizing genes essential for biocontrol, the genome's preliminary availability enables improved high-throughput analysis. Kloet et al. (2011) have proposed a new intersecting field termed "eco genomics" explaining how to distinguish this growing subject from existing methodologies like the more comprehensive metagenomics. Our knowledge of the intricate interactions between hosts, pathogens, and environments has improved because of a concentrated application of ecological genomics (eco genomics) approaches (Martin, 2013). This has improved our understanding of how pathogens are controlled through biological means. Metagenomics and transcriptomics may provide insight into a location's entire microbial community. Pathogen activity can be regulated by identifying these microbiomes and their subconstituents. Information can also aid in the development of new consortiums for biocontrol agents. The genomics "revolution," which has reduced sequencing costs and offered higher-throughput sequencing, has also aided in the genome sequencing of significant plant pathogens. The gene information content of these pathogens has enhanced our knowledge of the mechanisms involved in pathogenesis, potential imparted resistance, and the interactions between pathogenesis-reducing agents and helpful microbes. Genome editing and RNA interference (RNAi) are becoming important genetic tools for modifying the genomes of pathogens, host plants, and beneficial microbes (Klosterman et al., 2016; Collinge, 2018). Extracellular enzymes secreted by biocontrol agents that work through myco-parasitism break down the cell walls of fungal pathogens. The genes responsible for producing antagonistic metabolites can be identified through genome analysis of BCAs. These genes could be extremely different or the same in different BCA strains. Utilizing microbiological and biochemical methods to assess the features of MBCA and further understand the interpretation of novel pathways revealed by genome sequencing could be beneficial. The identification of genes and the processes involved in plant–BCA interactions has been made possible by genome sequencing. The Kyoto Encyclopedia of Genes and Genomes (KEGG) database describes host–pathogen interactions, and comparative genomics research can be used to identify these genes (Kanehisa et al., 2002). This type of genome sequencing makes it easier to identify genes, like chitinase genes, that exhibit functional similarities associated with biocontrol attributes (Mukherjee et al., 2013). Significant differences were found in the comparative genomics of *P. fuorescens* strains, indicating the existence of interspecific and inter-strain variants (Loper et al., 2012). Greater similarities, however, might suggest a common function and make mining such genes easier. Using a genome mining approach, novel cluster genes for orphan biosynthesis in biocontrol strains have been observed to identify comparable secondary metabolites (Eyiwumi et al., 2015). There have been reports of genomic variations between *Pseudomonas aeruginosa* M18 and other strains of the human pathogen (Wu et al., 2011). An outstanding overview of the functions of certain sequenced genomes of biocontrol agents has been given by (Srivastava et al., 2014). Since next-generation sequencing tools are now widely available, metagenomics and meta-transcriptomics research are now simpler, are being used by labs all over the world, and lead to quicker identification of microbiomes and their interactions with both the pathogen and the host (Massart et al., 2015a, 2015b).

Understanding the interactions between the resident and introduced microbiota will be made simpler by combining the fields of metabolomics and signalomics and other additional emerging fields that concentrate on omics techniques targeting signaling molecules (Mhlongo et al., 2018). Determining the ideal dosages and application timing for pathogen-controlling agents will help better understand microbial BCA and microbiota interactions. This all-encompassing approach will enhance the microbial suppression of plant pathogens through the alteration of indigenous microbiota (Mazzola and Freilich, 2017) and the utilization of helper strains to support MBCAs in their hostile activity and survival (Massart et al., 2015a). Additionally, various microbiomes with various microbial inhabitants and modes of action will help to provide a workable solution for the sustainable management of plant diseases (Gopal et al., 2013; Massart et al., 2015a; Syed et al., 2018).

5.1.2.2 Proteomics

In vitro studies of genes and proteins were the main focus of biological research in the early 1990s. However, as technology developed and evolved, the focus of the research switched to other molecular aspects, such as transcriptomics analysis, proteomics, metabolomics, and structural genomics. Integrative analysis and a multidisciplinary approach are essential for understanding the complexity of microorganisms. Proteomics and genomics are quickly becoming recognized as necessary in the post-genomic era. Research has demonstrated that the protein complement is dynamic and complex in response to environmental cues. At this point, the cell proteome is anticipated to address the most significant and pertinent queries about biocontrol agents and the metabolites they produce. The proteome is the whole set of proteins in a cell, tissue, organ, living organism, or environment at a given time. Research based on proteomics offers the chance to examine the function of individual proteins throughout an organism's life cycle. Understanding how different protein types interact to support the coordinated growth and development of every individual organism can be applied to other taxa. Despite their utility, proteomics techniques have several limitations and complications. Among the most crucial ones are the following: a top-down proteomics technique that might miss proteins of significance, a lack of deeper proteome coverage, the difficulty in comparing proteomics data of unsequenced orphan organisms, and effective protein quantification. Proteomics, when combined with other methods, can clarify details about the pathogenicity of plant infections, leading to the more efficient identification and management of plant diseases. There are two methods for quantifying proteins: a targeted method and a discovery-driven method. The discovery-driven approach focuses on investigating all proteins expressed during the biocontrol process and their underlying mechanisms of action.

However, a specific approach designed for biocontrol is used to assess the activity of the enzymes (proteases, chitinases, and cellulases) (Aegerter and Gordon, 2006). Gel-based techniques are used for the majority of proteome research related to biocontrol. These methods have improved protein analysis over the previous gel-based approaches (Otto et al., 2012). Tandem mass spectrometry (MS) and liquid chromatography are commonly used for BCA investigations. Proteomics of a BCA's biocontrol characteristics can provide insight into the strain's basic biological reactions by revealing its proteome fingerprint.

Proteins associated with biocontrol mechanisms that have biotechnological significance and are expressed differently can be discovered using a proteomics approach (Alfonso et al., 2016). To accommodate their saprotrophic phase, several fungi secret proteins. Fungal pathogens depend upon proteins for various functions, including adhesion to plant surfaces, host penetration, and other virulence-determining processes. A sophisticated workflow of proteomic approaches has been shown by González-Fernández et al. (2010) *to* clarify MS-based tools for examining plant pathogenic fungi and analyzing the events they trigger during the infection process. According to research by Kang et al. (2019), the proteome responds differently based on whether the host was colonized by a helpful or antagonistic pathogen alone or by both. However, when both pathogens colonized a host, there was a greater transcriptional and translational response than when each pathogen was inoculated alone, suggesting that the pathogen was more likely to trigger the response than the beneficial microorganism.

5.1.2.3 Transcriptomics in Plant Defense Mechanisms

The field of transcriptomics has made significant progress toward advancing our knowledge of fungal plant diseases. Examining gene expression during infection and expression profiling under various in vitro growth circumstances, such as starvation and other stresses, has improved our understanding of fungal phyto-pathogenicity. The advancement of omics technologies has significantly enhanced our comprehension of gene expression. The most effective method for characterizing potential transcripts involved in numerous biological processes is transcriptome analysis. ESTs, subtractive libraries, and DNA microarrays can be used to correlate the enormous transcriptome response to different variables with a biological activity (Herrera-Estrella, 2014). Several genes are translated at various stages throughout biological control and infection processes, displaying temporal and geographic variations and controls from environmental stimuli (Sharma et al., 2017). Researchers have proposed that a comprehensive integration of data from multiple distinct omics techniques is necessary to provide a complete picture of the events occurring during the infection process and the actions of BCAs. Trichoderma–plant pathogen interactions have been the subject of numerous transcriptome studies. *T. atroviride* IMI206040's transcriptome has been examined at various phases of its interactions with *R. solani*, *Phytophthora capsici*, and *Botrytis cinerea* (Reithner et al., 2011). Compared to the transcribed genes predicted by genomic data to be active for *T. atroviride*, only a small proportion (38.4%) of genes were active during the host plant's interaction with *R. solani* before contact (Reithner et al., 2011). Transcriptome-based research has been particularly helpful in serving as a reference for investigations into the benefits that *Trichoderma* strains impart on their hosts and the identification of key genes that induce mycoparasitism. As an option that solely includes active mRNA species, translatome-based studies are supported by the differential gene expression and artifact appearance during transcription (Meteignier et al., 2017). Sharma et al. (2017) propose that understanding the molecular mechanisms underlying BCAs can be achieved by utilizing proteome, transcriptome, and genome data. Lysee et al. (2017) have shown a similar series of events in a study of the interactions between the pathogen *Helminthosporium solani*, the biocontrol agent *Clonostachys rosea*, and the host *Solanum tuberosum* in three dimensions. When the pathogen was present, it was found that *C. rosea*'s gene expression was increased. Host genes that were down-regulated in reaction to the pathogen became upregulated and expressed when the biocontrol agent was present. The authors proposed that the reduction in illness incidence could have been attributed to a confluence of elements, such as mycoparasitism, competition, and the impact of the biocontrol agent. In a novel study that uses transcriptomics and genomes to comprehend mycoparasitism-based biocontrol strategies, Zhao et al. (2020) analyzed the transcriptome and genome of the plant pathogen *Sclerotinia sclerotium*'s mycoparasite *Coniothyrium minitans*. The information gathered demonstrated that *C. minitans* uses various protein classes to parasitize its host. It has been proposed that several *C. minitans* secondary metabolites have promise as biocontrol agents.

5.1.2.4 Transcriptomics in Fungus for Studying Virulence Factors and Fungal Adaptation

Transcriptomics quantifies the transcriptome, or all of the transcripts in a cell, and their abundance in a particular physiological state or developmental stage (Dong and Chen, 2013). It offers the ability to detect quantitative and qualitative variations in gene expression when contrasting various mRNA populations, and it has exerted a noteworthy influence on almost every aspect of biological science. A homologous chitinase (nagl) promoter was used to create a transgenic strain of *Aspergillus niger* that encodes the glucose oxidase gene (goxA). This strain showed enhanced capacity to suppress many fungal infections, including *Pythium ultimum*, *B. cinerea*, and *R. solani*. The rapid growth of the altered *A. niger* strain led to the lysis of *P. ultimum* and *R. solani*. Therefore, heterologous genes with pathogen-inducible promoters can augment the systemically induced resistance features of biocontrol agents. An altered genetic strain of *Erwinia ananas* (NR1) was created. Instead of producing lytic enzymes or antibiotics, the typical rice-indigenous bacterium colonizes rice leaves. Genes of strain B2 of *Serratia marcescens* were present in the genetically altered variant.

5.2 APPLICATIONS OF TRANSCRIPTOMICS PROFILING TECHNIQUES IN FUNGAL PLANT PATHOLOGY

5.2.1 Differential Display (DD) and cDNA-AFLP

With open-ended techniques like DD and cDNA-AFLP, the transcriptional response is profiled during fungus–plant interactions, pathogenesis, and the fungal colonization of plants. This section provides a brief overview of cDNA-AFLP (Bachem et al., 1996) and mRNA DD (DDRT-PCR) (Liang and Pardee, 1992). These techniques are amplified cDNA fragment electrophoretic fingerprinting-based gel-based transcriptome profiling tools. A PCR-based technique called DDRT-PCR allows for the side-by-side comparison of many RNA samples, helping to identify activated and suppressed genes.

There are two main processes in DDRT-PCR: Using a set of degenerate, anchored oligo (dT) primers, RNAs extracted from different cell types are first reverse transcribed (RT) to produce cDNA pools. Second, using an upstream arbitrary primer and the original attached dT primer, random partial sequences from the cDNA pools are amplified by PCR. The PCR reaction's annealing temperature should be low to get the most amplified mRNA species. During the amplification procedure, a35S-dATP is added to allow the short sequences to be seen on a sequencing gel. The same primer sets are used for DDRT-PCR in various cell groups. To get the most amplified mRNA species, the PCR annealing temperature should be low (Liang and Pardee, 1992). An advancement of DD techniques is cDNA-AFLP. Using random hexamers as primers, cDNA is synthesized from whole RNA or mRNA in this manner. After two restriction enzymes—typically a 4- and a 6-cutter—digest the resulting fragments, adapters are ligated to the ends of the fragments. The first amplification step amplifies only the fragments broken down by both restriction enzymes and having distinct adapters on both ends. Using primers on the adapters that include one or more additional nucleotides, selective PCR amplification is used in the subsequent amplification steps to fractionate the complicated initial mixture of cDNA into smaller subsets. The sensitivity of the analysis can be raised by the stringency of the PCR amplification (adding more nucleotides to the primers). Genes with low levels of expression can also be found in this manner. The amplified fragments are isolated on high-resolution gels with a bp range of approximately 100–400.

A reliable indicator of the relative variations in gene expression levels is the variation in visible band intensities. Finding the matching full-length cDNA is typically necessary to further characterize interesting transcripts. Transcriptomics' primary goal is to generate and compare snapshots of mRNA populations. To be useful for comparisons with new data sets, these transcription profiles need to include quantitative values that are digitally maintained and allow for normalization methods to account for changes in the efficacy of mRNA extraction, cDNA synthesis, and detection. Currently, cDNA-AFLP and DDRT-PCR do not meet these requirements (Stein and Liang, 2002). Venkatesh et al. (2005) developed a data processing method that creates quantitative transcriptome profiles from cDNA differential data.

Amplified cDNA fragments are separated into DNA sequences, and detector signals are transformed into virtual gel images that may be analyzed semi-automatically. The following steps comprise data processing: Densitometry is used to determine the band intensity; densitometric values are normalized, and identifying fragments originating from the same transcript involves matching cDNA bands in lanes corresponding to samples treated with the same primer combination. For each pair of corresponding bands, an intensity ratio is determined. Transcription patterns are demonstrated using intensity ratios (control versus treatment) for cDNA fragments based on primer combination and DNA mobility. Researchers were able to explain the procedure by looking at DDRT-PCR data on the effects of oilseed rape *Brassica napus* secondary metabolites on the transcriptome of the pathogenic fungus *Leptosphaeria maculans*.

It has been shown that cDNA-AFLP patterns are more reproducible than DDRT-PCR. CDNA-AFLP can produce a comprehensive picture of gene expression, although it requires many PCR cycles. Additionally, unlike SAGE and microarray data, individually collected data sets are not easily comparable. On the other hand, the quantitative examination of band intensities using cDNA-AFLP allows for determining reliable gene expression profiles.

5.2.1.1 Suppression Subtractive Hybridization (SSH)

SSH is a widely used method for separating closely linked DNA samples by dividing their DNA molecules. It is particularly effective for generating subtracted cDNA or genomic DNA libraries, integrating subtraction and normalization into a single step based on the suppression PCR effect. The normalization stage balances the target population's DNA fragment abundance, and the subtraction step eliminates sequences shared by the populations under comparison. According to Rebrikov et al. (2004), this makes it easier to investigate the subtracted library and greatly increases the possibility of finding low-abundance differentially expressed cDNA or genomic DNA fragments. The "tester" cDNA contains the differentially expressed cDNAs (targets), but the driving cDNAs either lack them entirely or have them at lower concentrations. First, a four-base cutting restriction enzyme is used to digest the double-stranded cDNAs of the driver and tester, producing blunt ends.

The two populations of testers, Tester I and Tester II, are created when the tester cDNA fragments are divided into two samples and ligated to two different adapters, 1 and 2. Since the adapters do not have phosphate groups at their ends, only the longer strand of each adapter can form a covalent bond with the 5′-ends of the cDNA. The SSH approach takes advantage of two hybridizations. At first, each tester sample is added with an excess of driver. After being heat-denatured, the samples are given time to anneal. The single-stranded cDNA tester fraction (A) is normalized. Normalization occurs due to the second-order kinetics of hybridization, which speeds up the re-annealing process that produces homo-hybrid cDNAs (B) for the more abundant molecules.

Furthermore, the single-stranded cDNAs in the tester fraction (A) are greatly enriched in cDNAs for differentially expressed genes as "common" non-target cDNAs form heterohybrids (C) with the driver. The two hybridization samples from the first hybridization are mixed in the second hybridization. Only the remaining normalized and subtracted single-stranded tester cDNAs are capable of re-association and the creation of new (B), (C), and (E) hybrids. This step further enriches fraction (E) for differentially expressed genes by adding a second portion of the denatured driver. An important characteristic sets apart the newly generated hybrids (E) from the hybrids (B) and (C) that were created during the first and second hybridizations. At the 5′-ends, hybrids (E) have distinct adapter sequences. One comes from sample 1, and the other from sample 2. The two sequences allow the preferred amplification of the subtracted and normalized fraction (E) using PCR and a set of primers, P1 and P2, corresponding to the outer part of adapters 1 and 2, respectively. To enable this kind of selective amplification, an extension reaction is performed before the PCR process fills in the sticky ends of the molecules for primer annealing. Only type (E) molecules are capable of causing exponential amplification in any PCR cycle. Long inverted repeats on the ends of type (B) molecules allow them to stabilize into "panhandle-like" structures following each denaturation–annealing PCR process. However, the resulting "panhandle-like" structure cannot be employed as a template for exponential PCR because intramolecular annealing of the longer adapter sequences is more stable and much preferred over intermolecular annealing of the much shorter PCR primers. This is the result of PCR suppression.

Additionally, only a linear rate of amplification is possible for type (C) molecules, while primer-binding sites are absent from type (A) and (D) molecules. Only type (E) molecules can be amplified exponentially by PCR because they have a distinct adapter sequence at the ends of their molecules (Diatchenko et al., 1996). Because several filamentous fungi enter plant tissues through this specialized cellular structure, understanding the pattern of genes expressed during appressorium development is essential for understanding the molecular mechanisms underlying pathogenesis, fungal colonization of plants, and plant–pathogen interactions in fungal plant pathology applications. A fungal suppression subtraction library containing many genes expressed by *Magnaporthe oryzae* throughout the appressorium's maturation has been developed (Lu et al., 2005). Of the 105 genes analyzed, 34 were expressed in conidia/mycelia and appressoria, whereas 71 were expressed exclusively in appressoria. SSH found genes specifically expressed in the appressorium, including GAS1 (MAS3), GAS3, and PTH11, all of which have been linked to the development or function of appressorium in the past.

Downy mildew is caused by *Peronospora parasitica* in *Brassica* species and *Arabidopsis thaliana*. Bittner-Eddy et al. (2003) employed SSH to identify *P. parasitica* genes implicated in pathogenicity to study the molecular mechanism of pathogenesis. One thousand three hundred forty-five clones were sequenced, corresponding to cDNA fragments from 618 Arabidopsis genes and 25 putative *P. parasitica* (At) genes (Ppat 1–25). Expression pattern analyses revealed that 15 Ppats were expressed exclusively in planta. Peptides encoded by 11 Ppats have homology (BLASTP values ole05) to proteins involved in phosphorylation, protein secretion, osmoregulation, amino acid metabolism, membrane or cell wall biosynthesis, and cation transport. None of the other 14 genes, represented as potentially unique oomycete genes, have homologs in the large *Phytophthora* species EST database. These findings highlight the usefulness of SSH in identifying new *P. parasitica* genes expressed in plants, which supports using other gene discovery techniques like EST sequencing. Xiong et al. (2001) used SSH screening to identify 22 rice genes that showed differential expression in response to blast infection from a cDNA library prepared by the blast fungus *Pyricularia grisea*. Nine shared high sequence similarity with previously reported genes associated with pathogenesis or stress-inducible pathways. Some of them were rice chitinase class III, glutathione S-transferase (GST), pathogenesis-related protein class I (PR-1), metallothionein-like protein, lipid transfer protein, rice probenazole-induced protein (PBZ1), and late-embryogenesis abundant (LEA)5. Four primary disadvantages of SSH exist: Transcript isolation and sequencing are time-consuming and laborious, possibly leading to multiple clones with identical sequences. Additionally, northern blot or real-time PCR must confirm all transcripts and unidirectional comparisons, meaning that differential expression in only one of the two populations can be assessed. For these reasons, it is inappropriate for studies comparing more than two cell populations.

5.2.1.2 Serial Analysis of Gene Expression (SAGE)

Thousands of transcripts can be analyzed simultaneously using SAGE, an efficient gene expression profiling method (Datson et al., 1999). The SAGE approach is based on the idea that the entire transcript may be identified with just a 10 bp messenger

RNA oligonucleotide sequence tag. The unique sequence cleaves the transcript with a specific restriction endonuclease (anchoring enzyme) at a predefined distance from the poly(A) tail. Identifying every projected transcript from fungal genomes using 410 distinct sequence combinations that can be produced from a 10 bp sequence tag is more than possible. In conclusion, a biotinylated oligo (dT) primer converts mRNA isolated from the targeted tissue or tissues into double-stranded cDNA. Next, an anchoring enzyme like NlaIII, which can identify the CTAG sequence and cut every 256 bp, restricts the cDNA. The 3′-most segment of the cleaved cDNA with a common NlaIII cohesive end at its 5′-terminus is then recovered by adhering to streptavidin-coated beads. Following the division of the reaction mixture into two parts, two distinct linkers are ligated to each component using the NlaIII cohesive termini. These linkers are designed to have a type IIS enzyme site recognized as the tagging enzyme and located near (or partially overlapped by) the 3′-NlaIII sequence (often FokI or BsmFI). Reaction mixtures recover released sections once the type IIS enzyme has broken them down. T4 DNA polymerase then blunt-ends the products' resultant staggered ends. After another mixing, two parts are ligated. Since the amino group prevents the linkers' 5′-ends from being ligated, only the mRNA-derived termini can be ligated in a tail-to-tail orientation. After PCR amplification, polyacrylamide gel electrophoresis (PAGE) separates the products after the anchoring enzyme NlaIII cleaves them. Ditag segments flanked by NlaIII cohesive termini on both ends are separated and ligated to produce concatemers. PAGE is used to recover highly concatenated products, which are then cloned into a plasmid vector for sequencing (Velculescu et al., 1995). SAGE has many drawbacks despite being a helpful method for transcriptomics. The main disadvantage of SAGE is that it needs a lot of input RNA—2.5–5 mg poly(A)+ RNA. Many technological advancements have been made to get around this restriction. The principal alterations are SAGE-Lite, MicroSAGE, and serial analysis of reduced extracts (SADE). Using 500–5000 times less RNA as the starting material (1–5 ng poly(A)+ RNA), MicroSAGE is a modified version of SAGE. From RNA isolation until tag release, every step is completed in a single tube reaction. The single-tube method is simpler to execute and also reduces material loss in subsequent phases (Datson et al., 1999). Ditags are amplified for 28 cycles using MicroSAGE's two-step ditag amplification method. After gel purification, the ditags are amplified again for a maximum of 18 cycles. The fact that MicroSAGE produces a lot of duplicate ditags is a drawback. Similar in design, SAGE-Lite is another approach that enables global transcription analysis from less than 100 ng of total RNA (Peters et al., 1999). SAGE adaptation for reduced extracts was also developed, allowing for a 1000-fold decrease in the initial material. We call this modification "serial analysis of downsized extracts" (Virlon et al., 1999). The size of the SAGE tag is another issue with traditional SAGE. Often, a 15 bp tag cannot definitively identify the origin gene. This tag size is not as problematic in organisms with fewer genes (like yeast), but in creatures with more complicated genomes, it is (Matsumura et al., 2005). Saha et al. (2002) substituted the Type IIS enzyme MmeI for the tagging enzyme BsmFI to isolate 20 bp tags, overcoming the challenge posed by conventional SAGE. We refer to this SAGE variation as LongSAGE. It is challenging to fill in the two-nucleotide recessed 5′-terminus that results from the digestion of a DNA fragment with MmeI. However, this reduces the

LongSAGE tag size to 18 bp because the SAGE approach necessitates blunting the 5′-termini of the tags. Just the annotation of the 15 bp tag acquired during the initial SAGE process is done using the information from the resulting 20 bp tag. Matsumura et al. (2003) developed a SuperSAGE method to get around the issues with SAGE and LongSAGE. This method uses EcoP15I, a type III restriction endonuclease tagging enzyme. Finally, 26 bp tags were successfully isolated from transcripts using EcoP15I. Filling in the two nucleotide recessions that EcoP15I digestion generates at the 3′-terminus is simple. Based on the 26 bp tags, SuperSAGE can provide accurate gene expression profiles in light of this. SuperSAGE is a powerful tool for transcriptomics analyses of host–pathogen interactions. The interaction transcriptome, referring to the simultaneously monitored quantitative gene expression of a host and one of its eukaryotic pathogens, can be interpreted using this form of SAGE (Matsumura et al., 2005). Matsumura et al. (2003) applied SuperSAGE to rice leaves infected with *M. oryzae*. The most actively transcribed *M. oryzae* gene in blast-infected rice leaves is the hydrophobin gene, which was discovered by simultaneously observing the gene expression profiles of the blast fungus and rice host using the fully sequenced genomes of both species.

5.3 APPLICATIONS IN FUNGAL PLANT PATHOLOGY

The SAGE technique is not as often applied in researching plant pathogenic fungi as it is in studying animals and plants. SAGE has examined a few phytopathogenic fungi, including *Ustilago maydis*, *Blumeria graminis*, and *M. oryzae*. Developing methods to control diseases mostly depends on understanding the molecular mechanisms underpinning host–pathogen interactions. Gene expression analysis has been widely used for this purpose. The most destructive rice disease in the world, rice blast, is caused by *M. oryzae*. With the availability of whole genome sequences for rice and the blast fungus, the rice–blast fungus interaction system tests the usefulness of genomic data for understanding host–pathogen interactions at a molecular level. This highly tractable fungal pathogen acts as a foundational model for research on plant pathology. Irie et al. (2003) performed SAGE using mRNA from conidia germinating with or without cAMP to determine the genes involved in appressorium formation and host invasion. cAMP-treated conidia yielded 5087 tags, including 2889 unique tags, whereas untreated conidia yielded 2342 unique tags out of 3938. In response to cAMP treatment, up- and down-regulated genes correlated with 57 and 53 unique tags, respectively. Several genes known to be implicated in pathogenicity, including MPG1, MAS1, and MAC1, were elevated after receiving cAMP treatment. This suggests that cAMP is necessary for fungal pathogenicity. When plants detect invasive pathogens, they spread their defenses along several fronts (Maleck and Dietrich, 1999). One of the most striking responses is the so-called hypersensitive response (HR), which entails apoptosis-like cell death around the foci of pathogen invasion. HR contributes to the containment of the germs and the prevention of disease dissemination. INF1, a well-known elicitor that causes unsuitable HR in hosts like *Nicotiana benthamiana*, is secreted by *Phytophthora infestans* (Kamoun et al., 1997, 1998). Matsumura et al. (2003) employed *N. benthamiana*, a “non-model” organism for which there is no DNA database, treated with an INF1 elicitor to examine gene expression changes before the so-called HR using the SuperSAGE approach. SuperSAGE facilitated the quick identification of genes that the elicitor up- or down-regulated. Remarkably, a large number of the down-regulated genes encoded photosynthesis-related proteins. Genes encoding photosystem II (PSII) proteins were markedly suppressed in *N. benthamiana* leaves treated with INF1. This, in turn, led to a buildup of reactive oxygen species (ROS) in the chloroplast, resulting in HR in leaf tissues. Following INF1 treatment, several photosynthetic genes were systematically down-regulated, which strongly implies that *N. benthamiana*'s shared transcriptional regulatory factors were responding to INF1. It is crucial to characterize mRNA levels worldwide in filamentous fungi to comprehend the mechanics behind the infection process. For this purpose, Thomas et al. (2002) analyzed approximately 60, 000 tags that have been isolated from the appressoria, germinated conidia, and ungerminated conidia of barley powdery mildew (*Bacillus graminis* f. sp. *hordei*). In total, 1274 (20%) out of the 1669 unique *B. graminis* ESTs were found to be SAGE tags. In all three stages, most tags showed high abundance. A putative metallothionein, a protein linked to oxidative and heavy metal stress, was the most abundant tag (Andrews, 2000).

Given that other genes with significant transcript levels are involved in protein synthesis, modification, and destruction, it is plausible that early germling development involves fast protein production and degradation (McCafferty and Talbot, 1998). In several fungi, the cyclic AMP (cAMP)/protein kinase A (PKA)-dependent pathway and the mitogen-activated protein kinase (MAPK) signaling cascade control fungal development and pathogenicity (D'Souza and Heitman, 2001; Lee et al., 2003) including *U. maydis* (Martinez-Espinoza et al., 2002), the causal agent of corn smut disease. Larraya et al. (2005) used the SAGE method to analyze the transcriptomes of two mutants of *U. maydis*: Dadr1 (which lacks the catalytic subunit of PKA) and Dubc1 (which lacks the regulatory subunit of PKA encoded by ubc1). These mutants showed changes in the transcript levels of genes encoding ribosomal proteins (RPs), genes controlled by b mating-type proteins, and genes involved in metabolic activities. Increased transcript levels for phosphate acquisition and storage genes were seen in the ubc1 mutant, suggesting a connection between cAMP and phosphate metabolism. Elevated phosphate levels stimulated the filamentous development of wild-type cells in response to lipids, consistent with PKA modulating *U. maydis* morphogenesis.

5.3.1 Microarray Techniques (cDNA Microarrays, DNA Chips)

DNA microarray technology could greatly advance our understanding of plant–microbe interaction systems. It has fundamentally altered our ability to monitor global variations in gene expression. Affymetrix arrays, oligonucleotide microarrays, and DNA chips are examples of the two primary types of microarrays used for expression profiling: those based on the in situ synthesis of oligonucleotide probes and those based on the deposition of preassembled DNA probes (cDNA microarrays).

An mRNA molecule's unique binding or hybridization to its original DNA coding sequence, which takes the form of a cDNA template spotted on an array, is the basis for how a cDNA microarray works. To put it briefly, before the spotting process, random cDNA libraries or PCR-amplified DNA fragments are spotted on a glass microscope slide coated with polylysine. Samples containing both healthy and fungus-infected leaves have their mRNAs isolated.

After that, messenger RNA is transformed into cDNA via reverse transcription. The first sample's DNA is now labeled with the green fluorescent dye Cyanine 3 (Cy3), and the second sample's DNA is labeled with the red fluorescent dye Cyanine 5 (Cy5). The probes, a matrix of spotted single-strand DNAs, are covered with the target, a combination of cDNAs tagged with Cy3 and Cy5. The array is then scanned using laser emission scanners, either confocal or non-confocal.

Programs like ArrayPlot are used to display variations in gene expression. Subsequently, the amount of each labeled target associated with each place on the array is quantified. The ratio of signal intensities between control and test cDNA targets indicates whether mRNA species or genes are expressed, stimulated, repressed, or unaffected. Following the acquisition of gene expression profiles, a hierarchical clustering procedure displays the outcomes as a matrix with a gene for every row and an experiment for every column.

In addition to cDNA clones, oligonucleotides with a length of approximately 20 nucleotides can also be employed as probes on an array (Affymetrix arrays or DNA chips). Since these microarrays are typically hybridized with a single sample, high-density oligonucleotide arrays offer immediate information about the expression levels of an mRNA. For each gene, an oligo-microarray contains two sets of probe pairs instead of one or more cDNA clones. Affymetrix arrays are utilized in gene expression profiling and gene polymorphism detection, including single nucleotide polymorphisms (SNP) and genomic mutations (Nakanishi et al., 2001). Clone collections frequently contain a high degree of redundancy; moreover, incomplete gene representation, incorrect clone annotation, and legal restrictions on the distribution and use of cDNA libraries are just a few of the fundamental limitations of cDNA microarrays (Brigand et al., 2006).

Additionally, the comparatively large size of the cDNA amplicons may indicate the presence of repetitive sequences or similarity to related genes, which could unpredictably impair the probes' specificity (Handley et al., 2004). An alternative approach to address these problems is the use of long (50–70 nucleotides) oligonucleotides for probe manufacturing (long oligonucleotide microarrays) (Kane et al., 2000; Hughes et al., 2001). With the filamentous fungus *Neurospora crassa*, Kasuga et al. (2005) used an extended oligomer microarray (using immobilized 70 mer oligonucleotide probes) to measure the relative expression of genes involved in the physiological and biochemical processes related to conidial germination. Microarray expression studies are producing an immense amount of functional genomics data, including gene expression, and these data may be vital in providing insights into the functions and interactions of specific genes inside and within metabolic pathways (Brown and Botstein, 1999; Lockhart and Winzeler, 2000; Young, 2000). A paper titled "Minimum Information about a Microarray Experiment (MIAME)" was published (Brazma et al., 2001). This article describes the minimal data needed to ensure that microarray data can be interpreted and that analysis results can be independently verified.

Our understanding of the pathogenesis, etiology, and interactions between fungi and plants could be accelerated by applying microarrays. Variations in a pathogen's gene expression could explain why closely related disease strains differ in their pathogenicity. Candidate virulence factors are genes that express themselves selectively during infection or in situations that resemble infection. At least several pathogens have established gene disruption methods, which can make it easier to analyze potential genes effectively. Afterward, virulence assays can verify a role in pathogenesis (Kazan et al., 2001). DNA array technology has become one of the most important techniques in functional genomics for analyzing the genome-wide patterns of gene expression in an organism since its inception in the mid-1990s. The first indication of the DNA array technique's applicability for monitoring the gene expression of plant–microbe interactions came from studies on Arabidopsis defense responses. The expression of genes in Arabidopsis plants exposed to ethylene, methyl jasmonate (MJ), salicylic acid (SA), or the incompatible fungal pathogen *Alternaria brassicicola* was examined (Schenk et al., 2000). Investigations of a few genes at a time have not previously revealed the existence of a significant network of regulatory linkages and coordinated signaling pathways, according to analysis of the expression data obtained. Utilizing microarray technology, researchers have examined global gene expression in over 20 species of filamentous fungus, including a wide range of scientific subjects. Research on the regulation of gene expression in key metabolic pathways has revealed notable differences between filamentous fungi and yeast concerning transcriptional control. Genes considered essential for host environmental adaptation have been associated with multiple stages of pathogenicity (Breakspear and Momany, 2007).

5.4 TRANSCRIPTOMICS IN THE BIOLOGICAL CONTROL OF PLANT PATHOGENS

5.4.1 Characterization of Biocontrol Agent Colonization

The study of how biocontrol agents, usually microorganisms like bacteria, fungi, or viruses, settle down in a particular setting—typically a host plant or soil—and start to have positive effects is known as colonization characterization. Because of their antagonistic characteristics against several phytopathogenic fungi, a filamentous fungi called *Trichoderma* is frequently employed in plant biocontrol. Additionally, certain *Trichoderma* species infiltrate plant roots, triggering defense mechanisms in the plant that protect against various diseases, such as bacteria, viruses, and fungi (Harman et al., 2004). Usually, root tissue penetration is restricted to the first or second cell layers and only the intercellular spaces (Yedidia et al., 1999). Of the hydrophobins from *Trichoderma* species that have been characterized, only a small number are class II hydrophobins. Two hydrophobins have been identified in *T. reesei*: The primary function of HFBI is cell wall binding and hyphal growth, whereas the primary function of HFBII is as an extracellular protein secreted into the medium and involved in sporulation (Askolin et al., 2005; Bailey et al., 2002). During sporulation, *T. harzianum* exhibits a significant expression of the hydrobhobin srh1 (Munoz et al., 1997). A protein called QID3 was identified in *T. harzianum* (Lora et al., 1995). This protein has two distinct domains that share structural similarities with the N-terminal half of plant bimodular proteins and the C-terminus of fungal hydrophobins. While conducting a differential mRNA display study on *Trichoderma asperellum* mycelia interacting with plant roots, it was found a clone encoding for a protein similar to hydrophobin.

5.4.2 Understanding the Mechanisms of Plant–Pathogen Interactions

A comprehensive understanding of plant–pathogen interactions has already been attained through a multitude of research from the perspectives of both the pathogen and the plant. These findings demonstrate that the specific communication that occurs between the invasive pathogen and the plant is what initiates the plant–pathogen interaction (Dodds and Rathjen, 2010; Lodha and Basak, 2012; Hammond-Kosack and Jones, 2000; Lucas, 1998). Various disease agents, including bacteria, fungi, nematodes, oomycetes, viruses, and insects, use plants as hosts (Schenk et al., 2000). Plants cannot mount effective anti-infectious defensive responses in compatible interactions, which prevents pathogens from finishing their life cycle (Lodha and Basak, 2012; Schenk et al., 2000). Plants initiate intricate defense mechanisms against pathogenic contacts to prevent pathogen growth during incompatible interactions (Lodha and Basak, 2012; Hammond-Kosack and Jones, 2000). Plants generally develop two types of defenses when they interact with pathogens: systemic defense responses and local defense responses (Lodha and Basak, 2012; Hammond-Kosack and Jones, 2000; Schenk et al., 2009). Recognition is the initial stage of a local response, and it takes place when a plant uses pathogen-associated molecular patterns (PAMPs) on its cell surface to identify pathogens or pathogen isolates (such as pattern recognition receptors) (Zipfel, 2008). Plant resistance proteins mediate this type of recognition, which triggers far more dependable, high-impact defense responses. These responses result in rapid ion fluxes across the plasma membrane, cell wall reinforcement, activation of MAP kinase, active production of reactive oxygen species (ROS), and ultimately, the hypersensitive response, or programmed cell death (PCD) (Zipfel, 2008; Hammond-Kosack and Jones, 2000; Schenk et al., 2009; Mur et al., 2008). A systemic or long-term response that reduces pathogen infection locally and generates signals that strengthen defense responses to infections in distal systemic organs is known as systemic acquired resistance (SAR). SAR is essential for both building disease resistance and healing from illness-related harm. Endogenous salicylic acid (SA) is necessary for SAR activation, which also increases ethylene, jasmonic acid (JA), nitric oxide (NO), and pathogenesis-related (PR) proteins. These actions then trigger several downstream processes. (Lodha and Basak, 2012; Andrew et al., 2002; Torres et al., 2006; Vallad and Goodman, 2004; Dong, 2001; Kreps et al., 2002). Numerous studies have been conducted on the signaling pathways set off by various stressors, such as pathogen invasion and herbivory/wounding (Pieterse and Van Loon, 2004). Plant responses to various stressors under similar circumstances must be investigated to decipher such complicated defense mechanisms.

5.4.2.1 Biocontrol Efficacy

Studying entomopathogenic fungi ultimately aims to advance the economical use of mycoinsecticides for the management of various insect pests, in addition to providing a deeper understanding of their distinct biology (Wang and Feng, 2014). Since fungal biocontrol agents kill slowly and cause problems with environmental stability, there are still obstacles to their widespread use (Ortiz-Urquiza et al., 2015; St. Leger and Wang, 2010). The main goals of genetic engineering efforts are to increase the virulence of a fungus against certain pests and improve the fungal environmental stabilities against stress factors like UV irradiation and high temperatures. These efforts have been made in addition to formulation techniques.

5.4.2.2 Genetic Modification of Fungal Virulence

The first experiment using a genetic improvement in fungal virulence was genetically altering *Metarhizium* to overexpress the gene encoding the natural cuticle-degrading protease Pr1. The resulting transformant killed insects 25% faster (St. Leger et al., 1996).

Additionally, the chitinase gene constitutively expressed in *B. bassiana* greatly improved the fungus's ability to kill insects (Fan et al., 2007). Moreover, fungal virulence is genetically enhanced by applying foreign genes that encode proteins toxic to insects. For instance, an insect-specific neurotoxic peptide gene (AaIT) was created and utilized to modify *M. anisopliae* to express it specifically in insect hemolymph. The fungal toxicity against mosquitoes, tobacco hornworms (*Manduca sexta*), and coffee berry borers (*Hypothenemus hampei*) increased by more than nine times as a result of this change (Pava-Ripoll et al., 2008; Wang and St. Leger, 2007).

The virulence of *B. bassiana* was not enhanced by the combined use of Pr1 and AaIT genetic modification, as Pr1's protease breaks down the toxic peptide AaIT (Lu et al., 2008). However, *M. acridum* modified with several neurotoxic genes showed a synergistic effect (Fang et al., 2014). To prevent malaria in mosquitoes, *M. anisopliae* was genetically modified to express genes that are either insecticidal or anti-malarial (Fang et al., 2011). A similar process effectively integrated an insecticidal protein from the insect pathogenic bacteria *Bacillus thuringiensis* into *B. bassiana*. The recipient strain demonstrated a marked increase in virulence and the capacity to infect insects (Qin et al., 2010). The pathogenicity of *B. bassiana* was boosted by expressing endogenous insect genes, such as those for a diuretic hormone, neuropeptide, and serpin, to disrupt or inhibit normal hormone levels, necrophoretic behavior, or phenoloxidase activation in insects (Fan et al., 2012a, 2012b; Yang et al., 2014).

5.4.2.3 Enhance Fungal Environmental Stability via Genetic Engineering

In insect pathogen formulations, UV protectants and heat stabilizers are added to promote fungal spore stability and counteract abiotic stress (Behle et al., 2009). Genetic engineering has also attempted to enhance fungal stability in agricultural settings. For example, *B. bassiana*'s genetic modification to overexpress a tyrosinase gene resulted in a notable enhancement of the fungus's resistance to UV light (Shang et al., 2012). Fungal pathogenicity was increased, and *M. anisopliae* gained UV radiation resistance by introducing a PKS gene cluster for melanin synthesis from a plant pathogen (Tseng et al., 2011). Fungal resistance to solar radiation was further enhanced by designing *Metarhizium* or *Beauveria* to overexpress a DNA repair photolyase (Fang and St. Leger, 2012). Heat shock proteins (HSPs) play various roles in many organisms' stress reactions. When an endogenous HSP-encoding gene (Hsp25) was genetically altered in *M. robertsii*, the resultant mutant was resistant to oxidative stress, osmotic agents, and severe temperatures (Liao et al., 2014). Therefore, increasing the stability of the fungal habitat and, consequently, the effectiveness of field applications can be achieved by genetic engineering. Creating these genetically modified strains will increase the effectiveness of mycoinsecticides, which is a cost-effective way to manage various insect pests and disease vectors, depending on the legislation in different countries.

5.5 FUTURE PROSPECTS

Precision Agriculture: precision farming, which utilizes molecular technologies to optimize crop management based on real-time data on plant health and disease status.

Microbiome Studies: investigating the function of the plant microbiome—the community of microbes that reside in and around plants—in health and illness.

Climate Change Adaptation: creating agricultural types resistant to diseases and stressors brought on by climate change.

Although transgenic strains are necessary for the efficient and environmentally responsible control of diseases, the commercial dissemination of these strains is unacceptable to society. It is required to prove to the general population that these genetically modified strains have no negative effects. Genomic analysis is crucial for characterizing advantageous microorganisms. Genetic engineering offers an unusually strong chance of producing superior formulations. This will promote sustainable agriculture by reducing the need for chemical pesticides.

Recent advances in transcriptomics technology have illuminated previously unidentified transcriptional "noise," "junk DNA," or experimental artifacts—the hidden intergenic regions between protein-coding genes—through the availability of genome and transcriptome sequences. Genes, transposable elements, and the general genome structure are all included in genome assemblies. Comprehensive bioinformatics studies, such genome-wide gene family analyses, are made feasible and reinforced by the availability of a species' reference genome sequences. Gene annotation creation and gene promotion are the two most challenging tasks following the creation of a genomic sequence. Including external cues like RNA-Seq reads is still beneficial, even though these methods have advanced dramatically in recent years. If a high-quality genome and annotation are available for a species, resequencing can assess accession diversity. This makes it easier to spot minor and major structural changes, such as inversions and transpositions, as well as insertions and deletions. Complex bioinformatics techniques—many of which are new—are required to detect these alterations. Sequence variants can be used in the genetic mapping of traits. Several mapping-by-sequencing methods were developed to determine the underlying genes corresponding to relevant agricultural traits. Transcriptome methods complement these genomic approaches, with RNA-Seq technology leading the way. To ascertain the transcript abundance, the corresponding cDNA molecules are sequenced. For example, RNA-Seq data can be used for transcriptome assembly or gene expression analysis to compare the abundance of particular transcripts in different tissues, conditions, or genotypes. The rapid and ardent adoption of single-cell RNA sequencing (scRNA-seq) has demonstrated that it is much more than another technique for conducting transcriptome research. To say that the details of whole-transcriptome images are shifting from a tissue level to a cell level due to the advancement of scRNA-seq is not hyperbole. This revolutionary technique can now mine heterogeneity inside individual cells and between different tissue types. This facilitates the faster identification of rare and novel cell types, the simultaneous characterization of many cell types and states, and a deeper understanding of their roles in biological processes. However, the full range of scRNA-seq applications has not yet been fully realized. This is particularly true for single-cell transcriptome profiling in the study of plants, which is still in its infancy and faces many fascinating challenges. The field of spatial transcriptomics is expanding quickly. While studies utilizing single-cell and spatial transcriptomics are becoming more prevalent in the animal kingdom, they are still relatively new in the plant domain. In situations where cell dissociation is not possible, a novel combination of scRNA-seq and snRNA-seq could be used to achieve spatial single-cell resolution in plant tissues and broaden the scope of spatial transcriptomics to include a variety of species, from crops to forest trees. Our knowledge of cellular localization and cell–cell contact has greatly expanded due to the separation of the geographic locations of the various cell types common in tissue. As a result of climate change and the growing demands of a growing population on food and energy resources, the scientific community studying plants may now be able to offer innovative solutions to urgent issues. This new knowledge clarifies the complex biological processes involved in plant development and responses to biotic and abiotic external stimuli (Giacomello, 2021).

5.6 CONCLUSION

Cutting edge molecular biology techniques are combined with real-world applications in the dynamic and quickly developing field of molecular plant pathology to study and control plant diseases, promoting sustainable agricultural practices and food security. New, developing, previously reported, and re-emerging fungal plant diseases are now more easily detected and diagnosed, thanks to recent developments in molecular biology techniques. The culturable and unculturable fungal pathogens that cause symptomatic and asymptomatic diseases in solitary and co-infections have been successfully discovered and diagnosed using molecular-based techniques. When combined with other emerging technical breakthroughs, molecular approaches should be used as a point-of-care testing (POCT) method for diagnosing fungal diseases. Scientists need help creating workable methods for the molecular diagnostics of crop diseases affecting field, horticultural, floricultural, ornamental, and forest plant species that are significant to agriculture.

REFERENCES

Aegerter, B. J. and T. R. Gordon. 2006. Rates of pitch canker induced seedling mortality among *Pinus radiata* families varying in levels of genetic resistance to *Gibberella circinata (anamorph Fusarium circinatum)*. *Eco Manag* 235:14–17.

Ahmar, S., R. A. Gill, K. H. Jung, A. Faheem, M. U. Qasim, M. Mubeen and W. Zhou. 2020. Conventional and molecular techniques from simple breeding to speed breeding in crop plants: Recent advances and future outlook. *Int J Molec Sci* 21:2590.

Alfonso, E., C. Chiara, R. Valentino and L. C. Maria. 2016. Bioinformatics for agriculture in the next-generation sequencing era. *Chem Biol Technol Agric* 3:9.

Andrew, F. B., W. Jinrong and D. F. Mark. 2002. Probing plant-pathogen interaction and downstream defense signaling using DNA microarray. *Func Integrat Gen* 259–273.

Andrews, G. K. 2000. Regulation of metallothionein gene expression by oxidative stress and metal ions. *Biochem Pharmacol* 59:95–104.

Askolin, S., M. Penttila, H. A. B. Wösten and T. Nakari-Setala. 2005. The *Trichoderma reeesei* hydrophobin genes hfb1 and hfb2 have diverse functions in fungal development. *FEMS Microbiol Lett* 253:281–288.

Bachem, C. W. B., R. S. van der Hoeven, S. M. de Bruijn, D. Vreugdenhil, M. Zabeau and G. F. Visser. 1996. Visualization of differential gene expression using a novel method of RNA fingerprinting based on AFLP. *Plant J* 9:745–753.

Bailey, M. J., S. Askolin, N. Horhammer, M. Tenkanen, M. Linder, M. Penttila and T. Nakari-Setala. 2002. Process technological effects of deletion and amplification of hydrophobins I and II in transformants of *Trichoderma reesei*. *Appl Microbiol Biotechnol* 58:721–727.

Behle, R. W., D. L. Compton, J. A. Laszlo and D. I. Shapiro-Ilan. 2009. Evaluation of soyscreen in an oil-based formulation for UV protection of *Beauveria bassiana* conidia. *J Econ Entomol* 102:1759–1766.

Bittner-Eddy, P. D., R. L. Allen, A. P. Rehmany, P. Birch and J. L. Beynon. 2003. Use of suppression subtractive hybridization to identify downy mildew genes expressed during infection of *Arabidopsis thaliana*. *Mol Plant Pathol* 4:501–507.

Brazma, A., P. Hingamp, J. Quacklenbush, G. Sherlock, P. Spellman, C. Stoeckert, J. Aach, W. Ansorge, C. A. Ball, H. C. Causton and T. Gaasterland. 2001. Minimum information about a microarray experiment (MIAME)-towards standards of microarray data. *Nat Genet* 29:365–71.

Breakspear, A. and M. Momany. 2007. The first fifty microarray studies in filamentous fungi. *Microbiology* 153:7–15.

Brigand, K. L., R. Russell, C. Moreilhon, J. M. Rouillard, B. Jost, F. Amiot, V. Magnone, C. Bole-Feysot, P. Rostagno, V. Zirolle and V. Defamie. 2006. An open-access long oligonucleotide microarray resource for analysis of the human and mouse transcriptomes. *Nucl Acids Res* 34:e87.

Brown, P. O. and D. Botstein. 1999. Exploring the new world of genome with DNA microarrays. *Nat Genet* 21: 33–7.

Chen, Z. Z. and R. E. Gerszten. 2020. Metabolomics and proteomics in type 2 diabetes. *Circ Res* 126:1613–1627.

Collinge, D. B. 2018. Transgenic crops and beyond: How can new breeding technologies contribute to the sustainable control of plant diseases. *Eur J Plant Pathol* 152:977–986.

D'Souza, C. A. and J. Heitman. 2001. Conserved cAMP signaling cascades regulate fungal development and virulence. *FEMS Microbiol Rev* 25:349–364.

Datson, N. A., J. van der Perk-de Jong, M. P. van den Berg, E. R. de Kloet and E. Vreugdenhil. 1999. MicroSAGE: A modified procedure for serial analysis of gene expression in limited amounts of tissue. *Nucl Acids Res* 27:1300–1307.

De Wit, P., M. H. Pespeni, J. T. Ladner, D. J. Barshis, F. Seneca, H. Jaris, N. O. Therkildsen, M. Morikawa and S. R. Palumbi. 2012. The simple fool's guide to population genomics via RNA-seq: An introduction to high-throughput sequencing data analysis. *Molec Ecol Res* 12:1058–1067.

Devi, E. L., C. P. Devi, S. Kumar, S. K. Sharma, A. Beemrote, S. K. Chongtham, C. H. Singh, C. Tania, T. B. Singh, A. Ningombam and R. Akoijam. 2017. Marker assisted selection (MAS) towards generating stress tolerant crop plants. *Plant Gene* 11:205–218.

Diatchenko, L., U. F. C. Lau, A. P. Campbell, A. Chenchik, F. Moqadam, B. Huang, S. Lukyanov, K. Lukyanov, N. Gurskaya, E. D. Sverdlov and P. D. Siebert. 1996. Suppression subtractive hybridization: A method for generating differentially regulated or tissue-specific cDNA probes and libraries. *Proc Natl Acad Sci USA* 93:6025–6030.

Dodds, P. N. and J. P. Rathjen. 2010. Plant immunity: Towards an integrated view of plant-pathogen interactions. *Nat Rev Gen* 539–548.

Dong, X. 2001. Genetic dissection of systemic acquired resistance. *Cur Opin Plant Biol* 309–314.

Dong, Z. and Y. Chen. 2013. Transcriptomics: Advances and approaches. *Sci China Life Sci* 56:960–967.

Donoso, A. and S. Valenzuela. 2018. In-field molecular diagnosis of plant pathogens: Recent trends and future perspectives. *Plant Pathol* 67:1451–1461.

Erdoğan, İ., B. Cevher-Keskin, Ö. Bilir, Y. Hong and M. Tör. 2023. Recent developments in CRISPR/Cas9 genome-editing technology related to plant disease resistance and abiotic stress tolerance. *Biology* 12:1037.

Eyiwumi, O. F., K. N. Phuong and M. Hofte. 2015. Recent advances in Pseudomonas biocontrol. In J. Murillo, B. Vinatzer, R. Jackson and D. Arnold (eds.), *Bacteria-Plant Interactions: Advanced Research and Future Trends*. Caister Academic Press, Wymondham, 167–198.

Fan, Y., D. Borovsky, C. Hawkings, A. Ortiz-Urquiza and N. O. Keyhani. 2012a. Exploiting host molecules to augment mycoinsecticide virulence. *Nat Biotechnol* 30:35–37.

Fan, Y., W. Fang, S. Guo, X. Pei, Y. Zhang, Y. Xiao, D. Li, K. Jin, M. J. Bidochka and Y. Pei. 2007. Increased insect virulence in *Beauveria bassiana* strains overexpressing an engineered chitinase. *Appl Environ Microbiol* 73:295–302.

Fan, Y., R. M. Pereira, E. Kilic, G. Casella and N. O. Keyhani. 2012b. Pyrokinin β-neuropeptide affects necrophoretic behavior in fire ants (*S. invicta*), and expression of β-NP in a mycoinsecticide increases its virulence. *PLoS One* 7:e26924.

Fang, W. and R. J. St. Leger. 2012. Enhanced UV resistance and improved killing of malaria mosquitoes by photolyase transgenic entomopathogenic fungi. *PLoS One* 7:e43069.

Fang, W., H. L. Lu, G. F. King and R. J. St. Leger. 2014. Construction of a hypervirulent and specific mycoinsecticide for locust control. *Sci Rep* 4:7345.

Fang, W., J. Vega-Rodriguez, A. K. Ghosh, M. Jacobs-Lorena, A. Kang and R. J. St. Leger. 2011. Development of transgenic fungi that kill human malaria parasites in mosquitoes. *Science* 331:1074–1077.

Giacomello, S. 2021. A new era for plant science: Spatial single-cell transcriptomics. *Curr Opin Plant Biol* 60:102041.

González-Fernández. R., E. Prats and J. V. Jorrin-Novo. 2010. Proteomics of plant pathogenic fungi. *BioMed Res Int*. https://doi.org/10.1155/2010/932527.

Gopal, M., A. Gupta and G. V. Thomas. 2013. Bespoke microbiome therapy to manage plant diseases. *Front Microbiol* 4:355.

Gossen, B. D. and M. R. McDonald. 2020. New technologies could enhance natural biological control and disease management and reduce reliance on synthetic pesticides. *Can J Plant Pathol* 42:30–40.

Goswami, M., K. Attri and I. Goswami. 2022. Applications of molecular markers in fruit crops: a review. *Int J Econ Plants* 9:121–126.

Hammond-Kosack, K. E. and J. D. G. Jones. 2000. Responses to plant pathogens. In B. B. Buchanan, W. Gruissem and R. L. Jones (eds.), *Biochemistry and Molecular Biology of Plants*. American Society of Plant Physiology, Rockville, 1102–1156.

Hammond-Kosack, K. E. and K. Kanyuka. 2007. Resistance genes (R genes) in plants. *Encycl Life Sci* 1–21.

Handley, D., N. Serban, D. Peters, R. O'Doherty, M. Field, L. Wasserman, P. Spirtes, R. Scheines and C. Glymour. 2004. Evidence of systematic expressed sequence tag IMAGE clone cross-hybridization on cDNA microarrays. *Genomics* 83:1169–1175.

Harman, G. E., C. R. Howell, A. Viterbo, I. Chet and M. Lorito. 2004. Trichoderma spp.—Opportunistic avirulent plant symbionts. *Nat Microbiol Rev* 2:43–56.

He, X., T. Lu and X. Zhou. 2021. Whole genome sequencing and comparative genomics analysis of *Pectobacterium carotovorum* identifies key pathogenic genes. *Mol Phylog Evol* 162:107114.

Herrera-Estrella, A. 2014. Genome-wide approaches toward understanding mycotrophic Trichoderma species. In V. Gupta, M. Schmoll, A. Herrera-Estrella, R. S. Upadhyay and M. G. Tuohy (eds.), *Biotechnology and Biology of Trichoderma*. Elsevier, Oxford, 455–464.

Hughes, T. R., M. Mao, A. R. Jones, J. Burchard, M. J. Marton, K. W. Shannon, S. M. Lefkowitz, M. Ziman, J. M. Schelter, M. R. Meyer and S. Kobayashi. 2001. Expression profiling using microarrays fabricated by an ink-jet oligonucleotide synthesizer. *Nat Biotechnol* 19:342–347.

Hwang, H. H., M. Yu and E. M. Lai. 2017. Agrobacterium-mediated plant transformation: Biology and applications. In *The Arabidopsis Book*. American Society of Plant Biologists, 15.

Irie, T., H. Matsumura, R. Terauchi and H. Saitoh. 2003. Serial analysis of gene expression (SAGE) of *Magnaporthe grisea*: Genes involved in appressorium formation. *Mol Gen Genom* 270:181–189.

Kamoun, S., H. Lindqvist and F. Govers. 1997. A novel class of elicitin-like genes from *Phytophthora infestans*. *Mol Plant Microbe Interact* 10:1028–1030.

Kamoun, S., P. van West, V. G. Vleeshouwers, K. E. de Groot and F. Govers. 1998. Resistance of *Nicotiana benthamiana* to *Phytophthora infestans* is mediated by the recognition of the elicitor protein INF1. *Plant Cell* 10:1413–1426.

Kane, M. D., T. A. Jatkoe, C. R. Stumpf, J. Lu, J. D. Thomas and S. J. Madore. 2000. Assessment of the sensitivity and specificity of oligonucleotide (50 mer) microarrays. *Nucl Acids Res* 28:4552–4557.

Kanehisa, M., S. Goto, S. Kawashima and A. Nakaya. 2002. The KEGG databases at genome net. *Nucl Acids Res* 30:42–46.

Kang, X., L. Wang, Y. Guo, M. Z. ulArifeen, X. Cai, Y. Xue, Y. Bu, G. Wang and C. Liu. 2019. A comparative transcriptomic and proteomic analysis of hexaploid wheat's responses to colonization by *Bacillus velezensis* and *Gaeumannomyces graminis*, both separately and combined. *Mol Plant Microbe Interact* 32:1336–1347.

Kasuga, T., J. P. Townsend, C. Tian, L. B. Gilbert, G. Mannhaupt, J. W. Taylor and N. L. Glass. 2005. Long-oligomer microarray profiling in *Neurospora crassa* reveals the transcriptional program underlying biochemical and physiological events of conidial germination. *Nucl Acids Res* 33:6469–6485.

Kazan, K., P. M. Schenk, J and M. Manners. 2001. DNA microarrays: New tools in the analysis of plant defense responses. *Mol Plant Pathol* 2:177–185.

Kloet, R., T. B. de Cock and J. Bunders. 2011. An introduction to and a reflection on the eco genomics promise. *J Integr Environ Sci* 8:23–38.

Klosterman, S. J., J. R. Rollins, M. R. Sudarshana and B. A. Vinatzer. 2016. Disease management in the genomics era: Summaries of focus issue papers. *Phytopathology* 106:1068–1070.

Knief, C. 2014. Analysis of plant microbe interactions in the era of next generation sequencing technologies. *Front Plant Sci* 5:216.

Kreps, J. A., Y. Wu, H. S. Chang, T. Zhu, X. Wang and J. F. Harper. 2002. Transcriptome changes for arabidopsis in response to salt, osmotic, and cold stress. *Plant Physiol* 2129–2141.

Kumar, K., G. Gambhir, A. Dass, A. K. Tripathi, A. Singh, A. K. Jha, P. Yadava, M. Choudhary and S. Rakshit. 2020. Genetically modified crops: Current status and future prospects. *Planta* 251:91.

Lamesch, P., K. Dreher, D. Swarbreck, R. Sasidharan, L. Reiser and E. Huala. 2010. Using the Arabidopsis information resource (TAIR) to find information about Arabidopsis genes. *Curr Protoc Bioinf* 30:1–11.

Larraya, L. M., K. J. Boyce, A. So, B. R. Steen, S. Jones, M. Marra and J. W. Kronstad. 2005. Serial analysis of gene expression reveals conserved links between protein kinase A, ribosome biogenesis, and phosphate metabolism in Ustilago maydis. *Eukaryot Cell* 4:2029–2043.

Lee, N., C. A. D'Souza and J. W. Kronstad. 2003. Of smuts, blasts, mildews, and blights: cAMP signaling in phytopathogenic fungi. *Annu Rev Phytopathol* 41:399–427.

Liang, P. and A. B. Pardee. 1992. Differential display of eukaryotic messenger RNA by means of the polymerase chain reaction. *Science* 257:971–976.

Liao, X., H. L. Lu, W. Fang and R. J. St. Leger. 2014. Overexpression of a *Metarhizium robertsii* HSP25 gene increases thermotolerance and survival in soil. *Appl Microbiol Biotechnol* 98:777–783.

Lockhart, D. and E. Winzeler. 2000. Genomics, gene expression and DNA arrays. *Nature* 405:827–836.

Lodha, T. D. and J. Basak. 2012. Plant-pathogen interaction: What microarray tells about it? *Molec Biotechnol* 87–97.

Loper, J. E., K. A. Hassan, D. V. Mavrodi, E. W. Davis, C. K. Lim, B. T. Shafer, L. D. Elbourne, V. O. Stockwell, S. L. Hartney, K. Breakwell, M. D. Henkels, S. G. Tetu, L. I. Rangel, T. A. Kidarsa, N. L. Wilson, J. E. vandeMortel, C. Song, R. Blumhagen, D. Radune, J. B. Hostetler, L. M. Brinkac, A. S. Durkin, D. A. Kluepfel, W. P. Wechter, A. J. Anderson, Y. C. Kim, L. S. Pierson, E. A. Pierson, S. E. Lindow, D. Y. Kobayashi, J. M. Raaijmakers, D. M. Weller, L. S. Thomashow, A. E. Allen and I. T. Paulsen. 2012. Comparative genomics of plant-associated Pseudomonas spp.: Insights into diversity and inheritance of traits involved in multitrophic interactions. *PLoS Genet* 8:e1002784.

Lora, J. M., J. A. Pintor-Toro, T. Benitez and L. C Romero. 1995. Qid3 protein links plant bimodular proteins with fungal hydrophobins. *Mol Microbiol* 18:380–382.

Low, L. Y., S. K. Yang, D. X. A. Kok, J. Ong-Abdullah, N. P. Tan and K. S. Lai. 2018. Transgenic plants: Gene constructs, vector and transformation method. *New Vis Plant Sci* 41.

Lu, D., M. Pava-Ripoll, Z. Li and C. Wang. 2008. Insecticidal evaluation of *Beauveria bassiana* engineered to express a scorpion neurotoxin and a cuticle degrading protease. *Appl Microbiol Biotechnol* 81:515–522.

Lu, J. P., T. B. Liu and F. C. Lin. 2005. Identification of mature appressorium-enriched transcripts in *Magnaporthe grisea*, the rice blast fungus, using suppression subtractive hybridization. *FEMS Microbiol Lett* 245:131–137.

Lucas. J. A. 1998. *Plant Pathology and Plant Pathogens*. Blackwell Science, Oxford.

Lysee, E., M. W. Dees and M. B. Brurberg. 2017. A three-way transcriptomic interaction study of a biocontrol agent (*Clonostachys rosea*), a fungal pathogen (*Helminthosporium solani*) and a potato host (*Solanum tuberosum*). *Mol Plant Microbe Interact* 30:644–655.

Machado, D., R. S. Costa, M. Rocha, E. C. Ferreira, B. Tidor and I. Rocha. 2011. Modeling formalisms in systems biology. *AMB Express* 1:1–14.

Maleck, K. and R. A. Dietrich. 1999. Defense on multiple fronts: How do plants cope with diverse enemies? *Trends Plant Sci* 4:215–219.

Mann, M., R. C. Hendrickson and A. Pandey. 2001. Analysis of proteins and proteomes by mass spectrometry. *Ann Rev Biochem* 70:437–473.

Martin, F. 2013. *The Ecological Genomics of Fungi*. Wiley, Hoboken, 385.

Martinez-Espinoza, A. D., M. D. Garcia-Pedrajas and S. E. Gold. 2002. The Ustilaginales as plant pests and model systems. *Fungal Genet Biol* 35:1–20.

Massart, S. and H. M. Jijakli. 2007. Use of molecular techniques to elucidate the mechanisms of action of fungal biocontrol agents: A review. *Microbiol Methods* 69:229–241.

Massart, S., M. Martinez-Medina and M. H. Jijakli. 2015a. Biological control in the microbiome era: Challenges and opportunities. *Bio Control* 89:98–108.

Massart, S., M. Perazzolli, M. Hofte, I. Pertot and M. H. Jijakli. 2015b. Impact of the omic technologies for understanding the modes of action of biological control agents against plant pathogens. *Bio Control* 60:725–746.

Matsumura, H., S. Reich, A. Ito, H. Saitoh, S. Kamoun, P. Winter P, G. Kahl, M. Reuter, D. H. Krüger and R. Terauchi. 2003. Gene expression analysis of plant host-pathogen interactions by SuperSAGE. *Proc Natl Acad Sci USA* 100:15718–15723.

Matsumura, M., A. Ito, H. Saitoh, P. Winter, G. Kahl, M. Reuter, D. H. Krüger and R. Terauchi. 2005. SuperSAGE. *Cell Microbiol* 7:11–18.

Mazzola, M. and S. Freilich. 2017. Prospects for biological soilborne disease control: Application of indigenous versus synthetic microbiomes. *Phytopathology* 107:256–263.

McCafferty, H. R. K. and N. J. Talbot. 1998. Identification of three ubiquitin genes of the rice blast fungus *Magnaporthe grisea*, one of which is highly expressed during initial stages of plant colonization. *Curr Genet* 33:352–361.

Meteignier, L. V., E. Oirdi, M. Cohen, M. Barf, T. Matteau, D. J. F. Lucier, S. Rodrigue, P. E. Jacques, K. Yoshioka and P. Moffett. 2017. Translatome analysis of an NB-LRR immune response identifies important contributors to plant immunity in Arabidopsis. *J Exp Bot* 68:2333–2344.

Mhlongo, M. I., L. A. Piater, N. E. Madala, N. Labuschagne and I. A. Dubery. 2018. The chemistry of plant-microbe interactions in the rhizosphere and the potential for metabolomics to reveal signaling related to defense priming and induced systemic resistance. *Front Plant Sci* 9:112.

Mukherjee, P. K., B. A. Horwitz, A. Herrera-Estrella, M. Schmoll and C. M. Kenerley. 2013. Trichoderma research in the genome era. *Ann Rev Phytopathol* 51:105–129.

Munoz, G., T. Nakari-Setala, E. Agosin and M. Penttila. 1997. Hydrophobin gene srh1, expressed during sporulation of the biocontrol agent *Trichoderma harzianum*. *Curr Genet* 32:225–230.

Mur, L. A., P. Kenton, A. J. Lloyd, H. Ougham and E. Prats. 2008. The hypersensitive response; The centenary is upon us but how much do we know? *J Exp Bot* 501–520.

Nakanishi, T., T. Oka and T. Akagi. 2001. Recent advances in DNA microarrays. *Acta Med Okayama* 55(6):319–328.

Ortiz-Urquiza, A., Z. B. Luo and N. O. Keyhani. 2015. Improving mycoinsecticides for insect biological control. *Appl Microbiol Biotechnol* 99:1057–1068.

Otto, A., J. Bernhardt, M. Hecker and D. Becher. 2012. Global relative and absolute quantitation in microbial proteomics. *Curr Opin Microbiol* 15:364–372.

Pava-Ripoll, M., F. Posada, B. Momen, C. Wang and R. St. Leger. 2008. Increased pathogenicity against coffee berry borer, *Hypothenemus hampei* (Coleoptera: *Curculionidae*) by *Metarhizium anisopliae* expressing the scorpion toxin (AaIT) gene. *J Invertebr Pathol* 99:220–226.

Peters, D. G., A. B. Kassam, H. Yonas, E. H. O'Hare, R. E. Ferrell and A. M. Brufsky. 1999. Comprehensive transcript analysis in small quantities of mRNA by SAGE-Lite. *Nucl Acids Res* 27:e39.

Pieterse, C. M. J. and L. C. Van Loon. 2004. NPR1: The spider in the web of induced resistance signaling pathways. *Curr Opin Plant Biol* 456–464.

Qin, Y., S. H. Ying, Y. Chen, Z. C. Shen and M. G. Feng. 2010. Integration of insecticidal protein Vip3Aa1 into *Beauveria bassiana* enhances fungal virulence to *Spodoptera litura* larvae by cuticle and per os infection. *Appl Environ Microbiol* 76:4611–4618.

Rathore, J. S. and C. Ghosh. 2018. Pathogen-associated molecular patterns and their perception in plants. In *Molecular Aspects of Plant-Pathogen Interaction*, Springer; 79–113.

Rebrikov, D. V., S. M. Desai, P. D. Siebert and S. A. Lukyanov. 2004. Suppression subtractive hybridization. *Methods Mol Biol* 258:107–134.

Reithner, B., E. Ibarra-Laclette, R. L. Mach and A. Herrera-Estrella. 2011. Identification of mycoparasitism-related genes in *Trichoderma atroviride. Appl Environ Microbiol* 77:4361–4370.

Sabat, A. J., A. Budimir, D. Nashev, R. Sá-Leão, J. M. Van Dijl, F. Laurent, H. Grundmann, A. W. Friedrich and ESCMID Study Group of Epidemiological Markers (ESGEM). (2013). Overview of molecular typing methods for outbreak detection and epidemiological surveillance. *Eurosurveillance* 18:20380.

Sackton, T. B. 2019. Comparative genomics and Transcriptomics of host-pathogen interactions in insects: Evolutionary insights and future directions. *Curr Opin Insect Sci* 31:106–113.

Saha, S., A. B. Sparks, C. Rago, V. Akmaev, C. J. Wang, B. Vogelstein, K. W. Kinzler and V. E. Velculescu. 2002. Using the transcriptome to annotate the genome. *Nat Biotechnol* 20:508–512.

Schenk, P. M., J. H. Choo and C. L. Wong. 2009. Microarray analyses to study plant defense and rhizosphere microbe interaction. In *CAB Review: Perspective in Agriculture, Veterinary, Science, Nutrition, and Natural Resources*, CABI Reviews, 1–14.

Schenk, P. M., K. Kazan, I. Wilson, J. P. Anderson, T. Richmond, S. C. Somerville and J. M. Manners. 2000. Coordinated plant defense responses in Arabidopsis revealed by microarray analysis. *Proc Natl Acad Sci USA* 97:11655–11660.

Schmidt, J. M., A. Acebes-Doria, B. Blaauw, A. Kheirodin, S. Pandey, K. Lennon, A. D. Kaldor, P. E. Toledo and E. E. Grabarczyk. 2021. Identifying molecular-based trophic interactions as a resource for advanced integrated pest management. *Insects* 12:358.

Shang, Y., Z. Duan, W. Huang, Q. Gao and C. Wang. 2012. Improving UV resistance and virulence of *Beauveria bassiana* by genetic engineering with an exogenous tyrosinase gene. *J Invertebr Pathol* 109:105–109.

Sharma, V., R. Salwan, P. N. Sharma and A. Gulati. 2017. Integrated translatome and proteome: Approach for accurate portraying of widespread multifunctional aspects of Trichoderma. *Front Microbiol*. https://doi.org/10.3389/fmicb.2017.01602.

Sloots, T. P., M. D. Nissen, A. N. Ginn and J. R. Iredell. 2015. Rapid identification of pathogens using molecular techniques. *Pathol J RCPA* 47: 191–198.

Srivastava, M., M. Shahid, S. Pandey, A. Singh, V. Kumar, S. Gupta and M. Maurya. 2014. Trichoderma genome to genomics: A review. *J Data Min Genom Proteom* 5:162.

St. Leger, R. J. and C. S. Wang. 2010. Genetic engineering of fungal biocontrol agents to achieve greater efficacy against insect pests. *Appl Microbiol Biotechnol* 85:901–907.

St. Leger, R. J., L. Joshi, M. J. Bidochka and D. W. Roberts. 1996. Construction of an improved mycoinsecticide overexpressing a toxic protease. *PNAS* 93:6349–6354.

Stein, S. and P. Liang. 2002. Differential display analysis of gene expression in mammals: A p53 story. *Cell Mol Life Sci* 59:1274–1279.

Syed Ab Rahman, S. F., E. Singh, C. M. J. Pieterse and P. M. Schenk. 2018. Emerging microbial biocontrol strategies for plant pathogens. *Plant Sci* 267:102–111.

Talakayala, A., S. Ankanagari and M. Garladinne. 2022. CRISPR-Cas genome editing system: A versatile tool for developing disease-resistant crops. *Plant Stress* 3:100056.

Thomas, S. W., M. A. Glaring, S. W. Rasmussen, J. T. Kinane and R. P. Oliver. 2002. Transcript profiling in the barley mildew pathogen *Blumeria graminis* by serial analysis of gene expression (SAGE). *Mol Plant Microbe Interact* 15:847–856.

Tippmann, H. F. 2004. Analysis for free: Comparing programs for sequence analysis. *Briefings Bioinf* 5:82–87.

Tomczynska, I., M. Stumpe, T. G. Doan and F. Mauch. 2020. A Phytophthora effector protein promotes symplastic cell-to-cell trafficking by physical interaction with plasmodesmata-localised callose synthases. *New Phytol* 227:1467–1478.

Torres, M. A., D. G. Jonathan and J. L. Dangl. 2006. Reactive oxygen species signaling in response to pathogen. *Plant Physiol* 373–378.

Tseng, M. N., P. C. Chung and S. S. Tzean. 2011. Enhancing the stress tolerance and virulence of an entomopathogen by metabolic engineering of dihydroxynaphthalene melanin biosynthesis genes. *Appl Environ Microbiol* 77:4508–4519.

Vallad G. E. and R. M. Goodman. 2004. Systemic acquired resistance and induced systemic resistance in conventional agriculture. *Crop Sci* 1920–1934.

Velculescu, V. E., L. Zhang, B. Vogelstein and K. W. Kinzler. 1995. Serial analysis of gene expression. *Science* 270:484–487.

Venkatesh, B., U. Hettwer, B. Koopmann and P. Karlovsky. 2005. Conversion of cDNA differential display results (DDRT-PCR) into quantitative transcription profiles. *BMC Genom* 6:51–62.

Virlon, B., L. Cheval, J. M. Buhler, E. Billon, A. Doucet and J. M. Elalouf. 1999. Serial microanalysis of renal transcriptomes. *Proc Natl Acad Sci USA* 96:15286–15291.

Wang, C. S. and M. G. Feng. 2014. Advances in fundamental and applied studies in China of fungal biocontrol agents for use against arthropod pests. *Biol Control* 68:129–135.

Wang, C. S and R. J. St. Leger. 2007. A scorpion neurotoxin increases the potency of a fungal insecticide. *Nat Biotechnol* 25:1455–1456.

Wu, D. Q., J. Ye, H. Y. Ou, X. Wei, X. Huang, Y. W. He and Y. Xu. 2011. Genomic analysis and temperature-dependent transcriptome profiles of the rhizosphere originating strain Pseudomonas aeruginosa M18. *BMC Genome* 12:438.

Wu, S., L. Shan and P. He. 2014. Microbial signature-triggered plant defense responses and early signaling mechanisms. *Plant Sci* 228:118–126.

Xiong, L. Z., M. W. Lee, M. Qi and Y. N. Yang. 2001. Identification of defense-related rice genes by suppression subtractive hybridization and differential screening. *Mol Plant Microbe Interact* 14:685–692.

Yang, J., G. Duan, C. Li, L. Liu, G. Han, Y. Zhang and C. Wang. 2019. The crosstalks between jasmonic acid and other plant hormone signaling highlight the involvement of jasmonic acid as a core component in plant response to biotic and abiotic stresses. *Front Plant Sci* 10:1349.

Yang, L., N. O. Keyhani, G. Tang, C. Tian, R. Lu, X. Wang, Y. Pei and Y. Fan. 2014. Expression of a Toll signaling regulator serpin in a mycoinsecticide for increased virulence. *Appl Environ Microbiol* 80:4531–4539.

Yedidia, I., N. Benhamou and I. Chet. 1999. Induction of defence responses in cucumber plants (*Cucumis sativus L.*) by the biocontrol agent *Trichoderma harzianum*. *Appl Environ Microbiol* 65:1061–1070.

Young, R. 2000. Biomedical discovery with DNA microarrays. *Cell* 102:9–16.

Zeki, Ö. C., C. C. Eylem, T. Reçber, S. Kır and E. Nemutlu. 2020. Integration of GC–MS and LC–MS for untargeted metabolomics profiling. *J Pharm Biomed Anal* 190:113509.

Zhao, H., T. Zhou, J. Xie, J. Cheng, T. Chen, D. Jiang and Y. Fu. 2020. Mycoparasitism illuminated by genome and transcriptome sequencing of *Coniothyrium minitans*, an important biocontrol fungus of the plant pathogen *Sclerotinia sclerotiorum*. *Microb Genom* 6:6.

Zipfel, C. 2008. Pattern-recognition receptors in plant innate immunity. *Curr Opin Immunol* 10–16.

Zotti, M. J. and G. Smagghe. 2015. RNAi technology for insect management and protection of beneficial insects from diseases: Lessons, challenges and risk assessments. *Neotr Entomol* 44:197–213.

6 Fungal Metabolomics and Phenomics

Barbaros Çetinel, İhsan Barışcan Tunçkol, Güliz Tepedelen Ağaner, and Ghalia Nadeem

6.1 FUNGAL METABOLOMICS

6.1.1 Introduction

How long fungi have been on Earth is still being determined, and it is impossible to pinpoint their place of origin. Fungal cells are typically heterotrophs with saprophytic or parasitic patterns; they lack plastids and chloroplasts. According to a recent study, there may be up to 6 million species of fungus globally, more than 600 of which are thought to be closely linked to humans. They play a significant role as resident fungi or disease-causing pathogens and can contribute to developing the human micro-ecosystem. Fungi significantly impact biomedicine, agriculture, biodiversity, natural ecology, and human health (Li et al., 2022).

Metabolomics is defined as a "complete and quantitative identification of all small molecules or metabolites in biological samples." It focuses on the study of metabolites with a molecular weight less than 1000. Metabolomics involves strategies for quantitatively detecting cellular metabolites and understanding how the movement of these biochemical messengers across the metabolic network affects phenotype. Metabolomics, using advanced analytical techniques with high resolution and sensitivity and chemometric approaches, examines fluctuations following interference or simulation in biological systems (Chen et al., 2019). Fungal metabolomics is a subfield of metabolomics that aims to comprehensively analyze all metabolites produced by fungi. It is an advanced field that can assist in providing better insight into fungal metabolism and understanding, characterizing, and exploiting fungi. Fungi have been actively explored as a target for metabolomics due to their widespread usage in the manufacturing of chemicals; they are well-recognized for having a diversified metabolic landscape and usefulness as great model organisms (eukaryotic) for the study of systems biology and metabolism. Metabolomics captures the distinctive features of fungi, particularly the remarkable chemical variety in their metabolites and the use of eukaryotic cellular characteristics, to ascertain the proper operation of specific pathways in distinct compartments. By providing a window into key attributes that underlie various phenotypes, metabolite analysis utilizes the information derived from secondary and primary metabolism to determine the basic principles that illustrate the link between phenotype and genotype. Metabolite studies have proved useful in systems biology applications, the discovery of bioactive compounds

DOI: 10.1201/9781003452362-6

with potential for therapeutic use, and the classification of fungi (Jewett et al., 2006). Metabolomics is most often used to study plants infected by fungi to gain a better insight into plants' defense mechanisms (Chen et al., 2019).

The interaction between fungi and different hosts, for example, insects, plants, humans, or animals, constitutes a captivating study area. The interactions between fungal pathogens and plants are crucial in agricultural production. The current state of metabolomics research in fungi and plants has been well-reviewed, but it has yet to emphasize the interactions between fungal pathogens and plants. Researchers may fully use the potential of fungi by using metabolomics for several fungus-related research topics, such as the identification and classification of fungi, their metabolic pathways, the discovery of natural products produced by fungi, and interactions between plants and fungi (Chen et al., 2019).

6.1.2 History

The term "metabolomics" is derived from metabolite profiling. The concept of metabolic profiling was first published by Baylor College of Pharmacy researchers in the 1970s. Various chemists have started metabolomics research with the advances in other omics fields such as genomics. Stephen Oliver first suggested the "metabolome" in 1998 (Oliver et al., 1998). The term "metabolomics" was proposed by Fiehn in 2001 as a "complete and quantitative investigation of nearly all metabolites in biological samples" (Fiehn, 2001). Metabolomics research emphasizes all small molecules or metabolites and different cell types or cell variations (Shulaev, 2006).

The systematic investigation of secondary metabolites produced by fungi started in 1922 under the guidance of Harold Raistrick, who successfully identified over 200 mold metabolites. Penicillin, discovered in 1929, began the development of fungus-specific metabolites such as griseofulvins, cyclosporins, and cephalosporins. However, only the development and discovery of penicillin led to significant attention to fungal metabolites. Large-scale screening programs were started by pharmaceutical companies, and by 1950, extensive microbial compounds with medicinal uses had been identified. In the last 60 years, terrestrial fungi have provided enormous opportunities for identifying new natural compounds, many of which have the potential for development in biomedicine (Anke and Erkel, 2002; Keller et al., 2005).

This search for secondary metabolites with bioactive properties has continued, identifying numerous compounds that impede the growth of fungi, bacteria, parasites, protozoa, viruses, and insects. Various other compounds that exhibit mutagenic, teratogenic, cytotoxic, carcinogenic, allelopathic, enzyme-inhibitory, immunosuppressive, and other biological properties have also been discovered. A current literature review of fungal metabolites, which surveyed about 1500 compounds identified and characterized from 1993 to 2001, revealed that most of these compounds had anti-fungal, anti-tumor, and antibacterial activity (Pelaez, 2005).

According to estimates, about 3000–4000 fungus-specific metabolites have been identified, whereas 5000–7000 taxonomic groups have been investigated. *Aspergillus*, *Acremonium*, *Fusarium*, and *Penicillium* are fungi that are particularly capable of creating various secondary metabolites. Secondary metabolites from fungi encompass a wide range of compounds that hold significant value in industrial, agricultural, and pharmaceutical contexts or can harm human health, animals, and agriculture due to their toxicity. The complex nature of these compounds and the genes that regulate their synthesis have been extensively studied by pharmacologists and microbiologists due to their beneficial and detrimental attributes (Zain et al., 2009; Awaad et al., 2012).

6.1.3 Classes of Fungal Metabolites

The kingdom of fungi represents a highly significant class of microorganisms. Though more than 1 million fungal species are thought to exist, only 70,000 species are currently known. Most of these organisms belong to filamentous fungi, which differ from yeasts in their higher metabolic complexity and complicated morphology and developmental processes (i.e., sexual and asexual structures). Particularly, they are widely recognized for producing secondary metabolites and secreted enzymes; many of these compounds have been utilized by humans. Fungi continue to be a significant source of novel metabolites due to the diversity of their species and environments. Fungi can survive in nearly every ecological niche worldwide and produce small molecules of natural substances or secondary metabolites. The kingdom of fungi includes several species with unusual and unique biochemical pathways. During the last decade, genetic studies regarding secondary metabolic pathways have uncovered certain common themes, providing novel approaches for utilizing natural substances. These pathways yield important drugs like statins, cyclosporine, and penicillin; several toxins like trichothecenes and aflatoxins; and dual-natured metabolites like ergot alkaloids, which have both pharmaceutical value and toxicity. These natural compounds and various other low-weight fungal metabolites are categorized as secondary metabolites. The production of secondary metabolites occurs through common biosynthetic routes, but the ensuing intermediate compounds undergo various enzyme-catalyzed reactions, giving rise to a very broad range of chemical structures (Zain et al., 2014). Thus, most secondary metabolites may be easily classified according to their biosynthetic origins, including alkaloids, polyketides, terpenoids, cyto-chalasins, and nonribosomal peptides. Penicillin, an antibacterial agent, is the most well-known fungal metabolite, discovered by Alexander Fleming in 1928. In addition to common fungal isolates, marine-derived and endophytic fungi are gaining popularity. The extensive use of plants in traditional medicine has paved the way for identifying endophytic fungi, a few of which have greatly benefitted the pharmaceutical sector. This contribution aligns with secondary metabolites found within these fungi that are isolated, modified, and used for research purposes. The endophytic characteristics of fungi are beneficial and harmful, as they can provide plants with essential nutrients while posing detrimental impacts by exploiting the plant's machinery to generate secondary metabolites (Daley et al., 2017).

Recent research shows that marine fungi are a rich source of novel natural compounds. In contrast to terrestrial fungi, they might have evolved secondary metabolic processes due to their distinctive habitat conditions, nutrition, salinity, variations in temperature, higher pressure, and competition with viruses, bacteria, and fungi. To date, various antibiotics have been isolated from the culture media of filamentous fungi. Recent studies on marine filamentous fungi searching for bioactive secondary compounds illustrate their potential to produce novel medications (Zain et al., 2014). Secondary metabolites produced by fungi exhibit a range of biological activities and have gained interest in various pharmaceutical, agrochemical, and food industries. In recent years, novel fungal metabolites have been discovered due to new screening techniques utilizing inventive genetic, biological, and chemical methods (Schueffler and Anke, 2014; Gholami-Shabani et al., 2019).

Primary metabolites, compounds commonly found in intermediary metabolism, usually differ from secondary metabolites. Primary metabolism has traditionally been the realm of biochemists, while secondary metabolism has mostly been explored by organic chemists. Primary metabolites are substances synthesized due to primary metabolism, for example, monosaccharides, amino acids, vitamins, nucleotides, fatty acids, etc. (Li et al., 2022). Secondary metabolites are generally bioactive compounds with low molecular weights. They are often produced as groups of related compounds restricted to certain parts of an organism's lifecycle, with production typically linked to a particular stage of morphological differentiation. Most of these fungal secondary metabolites are produced from just a few essential precursors, following pathways with a relatively limited number of reactions that diverge from primary metabolic processes at specific points. Acetyl-CoA has been regarded as the most common precursor of fungal secondary compounds, leading to the production of steroids, terpenes, polyketides, and metabolites from fatty acids. In addition, some secondary metabolites are formed using intermediary substances from the tricarboxylic acid cycle, shikimic acid pathway, and amino acids (Zain et al., 2014).

Since many of these secondary metabolites do not directly play a role in the normal development, growth, or reproduction of the host fungi, they may be crucial for ecological relationships with other species and stress tolerance. Fungi are great producers of natural compounds and thus are essential to human health and welfare on a global scale. In the 20th century, fungal metabolites greatly influenced the quality of human existence. Over the last five decades, fungal compounds have been utilized in the clinic to produce antifungals, antibiotics, cholesterol-lowering medications, and immunosuppressants. These advancements have greatly improved human welfare and contributed to the remarkable increase in life expectancy during the latter half of the century (Pela, 2004).

6.1.4 Metabolomic Approaches for Fungal Research

Techniques for metabolomics research are currently being developed. This leads to the constant evolution of methodologies as the scope of application expands (Li et al., 2016). The three primary steps in the technological approach to metabolomic research on fungus are preparing samples, collecting and processing data, and conducting analysis.

6.1.4.1 Experimental Design

Metabolomics research involves proper experimental design, in which factors such as groups, nature, and timing for collecting samples should be carefully evaluated. It is essential to identify the specific issues, and then optimal research parameters and accurate experimental methodologies must be chosen to obtain high-quality data.

Numerous approaches for designing experiments rely on statistics, including regression, orthogonal and central combination, and multiple- and single-factor designs. It is crucial to consider the optimal number of samples and the detection of the appropriate quantity of metabolites to statistically represent the accuracy of the experimental findings (Roessner et al., 2000).

Sample size is essential when designing an experiment for metabolomics research. Based on statistical principles, a minimum of 30 cases must be counted, and at least 20 cases must have statistical significance. Except for particular conditions, for example, valuable classic instances where it is challenging to obtain samples, a sample size of fewer than five will result in imprecise statistical results (Werf et al., 2005). The sampling error will increase with decreasing sample size. Suppose the size of the sample is very low. In that case, there is a possibility of repeating differences, leading to reduced testing efficiency, and the impact of accidental factors cannot be excluded, leading to poor scientific validity and authenticity. In contrast, if the sample size is too big, properly regulating the test conditions will be challenging, wasting time, labor, and material resources (Chen et al., 2019).

6.1.4.2 Sample Preparation

Metabolomics aims to examine all the data within a metabolome following simulation or disruption in a biological system. Although there is a possibility of analyzing all metabolite levels, several variables must be considered, such as the sample source, genetic information, growth parameters, time of sampling, interval between sampling, and control parameters. To guarantee consistent microbial growth under stable culture conditions, the chosen samples should accurately represent the study subject while not disrupting the research goal (Chen et al., 2019).

Small molecules or metabolites within the organism will change quickly with variation in the external environment. Sample preparation may influence the measured metabolite content and the biological interpretation of the obtained data. Therefore, the steps involving the collection and preparation of samples should be performed carefully to avoid disrupting the metabolomic analysis. While comprehensive studies have been conducted on sample preparation in tissues, biological fluids, plants, and mammalian cells, there is limited research regarding sample preparation techniques in fungal metabolomics studies (Ceglarek et al., 2009; Li et al., 2016). The ideal method for preparing samples in metabolomics should be quick, simple, highly selective, reproducible, and have the ability of quenching to ascertain the accurate metabolite composition at the time of sampling. General steps in sample preparation include the rapid collection, quenching, and extraction of samples (Vuckovic, 2012). Rapid freezing of samples is typically required after collecting them, followed by their storage at temperatures below −60 °C to maintain the integrity of all the metabolites within the organism.

Rapid sampling from shake flasks or fermentation tanks is the initial step in the preparation of fungal samples. Harrison et al. initially performed rapid sampling from lab bioreactors (small scale) for the first time in 1969. This research was evolved by Theobald et al. and Iversen, who conducted the refining of sampling systems to possess minimal dead volume, thereby facilitating manual feeding (Iversen, 1981; Theobald et al., 1993). Nowadays, rapid sampling systems are distinguished by high frequency, motorized sampling, efficient inactivation, and negligible dead zones (Schadel and Franco-Lara, 2009; Lameiras et al., 2015). Adenine nucleotides were utilized as an indicator in the method proposed by Van Gulik to examine how they dynamically responds to varying glucose concentrations and rapidly sample fungal metabolites. The method has a high sampling rate while also ensuring long-term sterility. Another sampling technique devised by Hannes Link et al. involved directly injecting fungi into mass spectrometers with high resolution for metabolomic analysis in real time (Link et al., 2015). However, this method is impractical and expensive for many researchers.

6.1.4.3 Metabolite Quenching

Metabolic quenching is a crucial stage to obtain biological specimens. The enzyme activity in the body leads to the degradation of metabolites more rapidly and easily than that of proteins and mRNAs. It is necessary to take into account certain steps to impede degrading enzyme activities for the purpose of reducing the degradation of metabolites. The primary objective of quenching is to rapidly inhibit numerous metabolic processes occurring within cells, maintain a constant level of metabolites, inactivate enzymes, and decrease the rate of metabolite degradation.

6.1.4.3.1 Cold Methanol Quenching

A quenching technique using cold methanol was a common approach utilized in the past, and investigation into cell quenching techniques largely relied on quenching temperature and varying methanol concentrations. One of the common techniques for the analysis of microbial metabolites is quenching using 60% (v/w) methanol at 40 °C. However, this technique leads to the escape of cellular metabolites (Link et al., 2008). So far, the use of cold methanol for quenching has been a controversial issue. Some studies have consistently refined the initial quenching technique utilizing cold methanol because of the potential advantage of extracting target intracellular metabolites efficiently. By considering factors such as temperature and exposure time, utilizing rapid sampling equipment, adjusting the ratio of sample to quencher liquid to 1:5 or less during quenching, and using pure methanol at temperature of ≤−40 °C help to prevent the leakage of metabolites (Canelas et al., 2008).

6.1.4.3.2 Rapid Filtration Utilizing Cold Methanol

Rapid filtration with cold methanol was shown to be effective for quenching the cellular metabolic processes of filamentous fungi, allowing researchers to distinguish between extracellular and intracellular metabolites (Smart et al., 2010). On the other hand, rapid filtration takes about 45 seconds per sample, which could cause significant fluctuations in the assessment of intracellular metabolites with fast turnover rates.

6.1.4.3.3 Rapid Filtration with Liquid Nitrogen

In addition, new quenching techniques are also being developed. Rapid filtration with liquid nitrogen has been considered a more optimal technique as compared to cold methanol quenching, causing less damage to the integrity of the cells and promoting the extraction of target intracellular metabolites (Zheng et al., 2019). However, the formation of ice crystals during liquid nitrogen freezing can cause potential damage to the cell membrane, leading to metabolite leakage within the cell.

Some quenching techniques using strong acids (i.e., perchloric acid) can result in significant metabolite reduction and the degradation of certain compounds in a highly acidic environment (Chassagnole et al., 2002). This implies that finding the ideal quenching technique is difficult. However, it is believed that a quenching technique utilizing methanol solution (60%, −40 °C) results in minimal leakage of metabolites. It was determined that metabolite analysis using this approach is insufficient and still unable to achieve complete leakage avoidance for all screening indicators (Meinert et al., 2013). Moreover, the majority of quenching techniques are involved in the study of yeast cells, Gram (–) bacteria, and Gram (+) bacteria, with limited research conducted on filamentous fungi (Moritz et al., 2000; Hans et al., 2001; Spura et al., 2009).

6.1.4.4 Extraction of Metabolites

Metabolite extraction is a crucial stage in the process of sample preparation. An ideal extraction technique is that in which the chemical structure and physical properties of the metabolites remains unchanged while simultaneously enhancing the capture of the target metabolites. Current extraction techniques can be divided into chemical and physical categories. Physical approaches commonly include equipment such as ultrasound instruments, homogenizers, microwave devices, etc. On the other hand, chemical approaches involve water, base, acid, ethanol, methanol, and mixed solvents, as well as different ratios of organic solvents and water. The choice of different solvents is based on various samples. Strong bases and acids might result in significant leakage of metabolites, leading to a relatively low rate of recovery of resulting metabolites. Though it is a traditional extraction technique, it is still not advised to use strong bases and acids like perchloric acid. Optimal solvents with high efficiency of extraction can be chosen by experimental means. Gas chromatography-coupled mass spectrometry was used to assess five solvent combinations and three different sample preparation techniques for a fungus named *Mortierella alpina* (Lu et al., 2019). The results demonstrated better recovery and reproducibility of the lyophilized material. A combination of water and methanol with a 1:1 ratio was more useful in the extraction of metabolites from *Mortierella alpina*. In contrast to the methanol extraction, which is a quick, simple, and ideal technique for extracting metabolites from *Phanerochaete chrysosporium*, biphasic extraction at varying pH levels is more preferred for specific analysis (Madla et al., 2012).

6.1.4.4.1 Methods Used for the Extraction of Metabolites

There are several methods used for the extraction of target metabolites from microbial cells (e.g., fungi). These methods include chemical lysis, boiling ethanol, cold methanol, buffered methanol–chloroform–water, acidic extraction, alkaline extraction, supercritical fluid extraction (SFE), and pressurized liquid extraction (PLE).

6.1.4.4.1.1 Chemical Lysis The most widely used techniques for the extraction of intracellular metabolites involve the utilization of chemical agents. Metabolites are categorized into two phases on the basis of their relative volume, solubility, partitioning coefficients, and solvent temperature. In this case, the goal is concentrating the metabolites into a single phase by utilizing chemical agents. It is essential to take into account the extraction rates of particular metabolites while choosing a chemical agent. The rate of extraction can be influenced by factors such as temperature and rate of diffusion between the two phases. This will cause the solvent to enter the cell membrane, allowing for the extraction of intracellular metabolites. Consequently, there is a direct correlation between the extraction rate of a chemical substance and the level of cell permeabilization. Therefore, the extraction conditions and the selection of chemical substances is determined by specific microbial cells being targeted and particular groups of metabolites that are of the interest. The primary aim of the majority of chemical extraction techniques is to extract specific metabolite species such as amino acids and fatty acids. However, the optimal technique for metabolite analysis intends to extract a maximum number of species and classes of metabolites. For this reason, the metabolomics industry has been highlighting the need to employ various extraction techniques to obtain a complete metabolite profile (Duportet et al., 2012; Kim et al., 2013).

Both non-polar (chloroform, hexane, and ethyl acetate) and polar (ethanol or methanol) solvents are often used to extract microbial cellular metabolites. The organic solvents used may degrade the cell wall, as well as cell membrane lipids and proteins. This leads to the formation of pores in the cell envelope, resulting in the release of intracellular metabolites via the pore, which are then collected in the organic solvent. The ideal methods for solvent extraction, however, necessitate a minimal quantity of solvent. Additionally, critical factors such as the extraction time, extent of metabolite coverage, and volume of sample are essential to consider when choosing an optimal extraction protocol (Villas-Boas et al., 2005).

6.1.4.4.1.2 Boiling Ethanol Utilizing buffered solution of boiling ethanol (75% v/v) is a quick and easy method for extracting cellular metabolites. This particular extraction method was initially described by Entian et al. for the purpose of extracting polar metabolites or compounds from yeast cells. This approach deactivates proteins and enzymes by subjecting the quenched microbial cells to a buffered solution of boiling ethanol (80 °C) for a few minutes. Additionally, the heat increases cell disruption, which makes it possible to extract intracellular (water-soluble) metabolites. The combination of water and ethanol is then allowed to evaporate, and the resulting pellets are resuspended in water before analysis. The key feature of this approach is its remarkable reproducibility. Nevertheless, some researchers claimed that this method results in minimal recovery of several metabolites classes such as tricarboxylic acids, phosphorylated metabolites, and nucleotides. Additionally, this protocol is not appropriate for heat-sensitive metabolites and poses a risk of oxidation for reduced metabolites. Despite this, the technique of boiling ethanol extraction has stood out as the most well-liked approach for extracting intracellular metabolites for several years (Maharjan and Ferenci, 2003; Villas-Boas et al., 2005).

6.1.4.4.1.3 Cold Methanol Cold methanol extraction is a widely employed technique and is an easy and quick way to extract intracellular metabolites from a variety of microbial organisms. For example, metabolites from filamentous fungi have been extracted using cold methanol (Hajjaj et al., 1998). This technique is very powerful since it utilizes a single organic solvent, which is readily extracted from the samples by evaporating them. Moreover, the extraction procedure is often carried out at temperatures below −20 °C, making it appropriate for metabolites that are thermo-labile. The primary drawback of this approach is that it does not completely inactivate the enzyme, which increases the possibility of altering intracellular metabolite pools. Additionally, this approach demonstrated good reproducibility and remarkable recovery for mid-polar and polar metabolites. Non-polar metabolites, however, do not recover as well as polar metabolites (Villas-Boas et al., 2005). In some cases, the technique of cold methanol extraction is used in combination with sonication or freeze–thaw cycles to increase the cellular permeability (Duportet et al., 2012).

6.1.4.4.1.4 Buffered Methanol–Chloroform–Water This technique was first described by de Koning and colleagues for extracting intracellular metabolites (de Koning and van Dam, 1992). They utilized a mixture of methanol–chloroform–water in a buffered solution at temperature ranging from −40 to −20 °C. During the extraction process, the mixture was shaken vigorously for about 45 minutes to isolate polar metabolites from fungal (yeast) cells. This technique is highly effective for extracting non-polar and polar metabolites from filamentous fungi. Additionally, this technique can be used for extracting heat-sensitive compounds, as it works well at low temperatures. While chloroform has been recognized as a carcinogenic and toxic agent, it assists in the denaturation of all microbial enzymes and prevents other chemical reactions. However, proper safety measures must be taken when employing this technique to prevent the adverse effects of chloroform. Furthermore, the approach is time-consuming and laborious, and the buffers may cause issues for certain analytical methods. However, this technique results in the excellent recovery of thermo-labile and phosphorylated compounds (Pinu et al., 2017).

6.1.4.4.1.5 Acidic Extraction One of the traditional techniques that has been in use for a long time is the utilization of acids such as hydrochloric acid, trichloroacetic acid, and perchloric acid for extracting intracellular metabolites. This technique has been shown to be effective for acid-stable and polar metabolites and has been utilized for extracting intracellular metabolites from filamentous fungi (Hajjaj et al., 1998; Villas-Boas et al., 2005). This extraction technique is typically carried out at temperatures ranging from 0 to 4 °C and generally employs a freeze–thaw cycle to improve the extraction process. However, once the process is completed, it is essential to adjust the sample's pH to neutral levels. The technique of acidic extraction has been found useful for recovery of polyamines and amines; however, low recovery of other metabolites was observed. Moreover, there is a chance that polymers and proteins will hydrolyze, alongside the oxidation of reduced metabolites during the extraction process (Pinu et al., 2017).

6.1.4.4.1.6 Alkaline Extraction Alkaline extraction is a popular technique from extracting intracellular metabolites from filamentous fungi and yeast cells. This extraction technique is performed at temperatures ranging from 0 to 4 °C, alongside a freeze–thaw cycle. This method proves to be highly effective in breaking down the microbial cell wall, which otherwise rapidly degrades enzymes at extremely high pH. Nevertheless, this approach has several limitations, such as the hydrolysis of polymers and proteins, lipid saponification, and low recovery of metabolites (Maharjan and Ferenci, 2003). To counter these issues, an additional step of neutralization is needed for adjusting the pH, resulting in the formation of salts, which are then extracted from the sample using centrifugation (Pinu et al., 2017).

6.1.4.4.1.7 Pressurized Liquid Extraction (PLE) Pressurized liquid extraction, also known as accelerated solvent extraction, has been used to extract secondary metabolites from fungal organisms. The method of pressurized liquid extraction was initially studied and published in the late 1990s (Azmir et al., 2013). Many metabolomics researchers, however, have not employed this technique. Conversely, this technique yields highly concentrated metabolites and is appropriate for high-throughput screening (Gomez-Ariza et al., 2004). However, PLE can only be used to extract heat-labile metabolites (Alvarez-Rivera et al., 2020).

6.1.4.4.1.8 Supercritical Fluid Extraction (SFE) Supercritical fluid extraction (SFE) is a widely used method for extracting mid-polarity and non-polar compounds from filamentous fungi (Cocks et al., 1995). This extraction technique typically relies on the properties of supercritical fluid, that is critical temperature and pressure. When the temperature and pressure of a substance exceed their critical values, it reaches a supercritical state. Supercritical fluids were originally discovered in 1822 by Charles Cagniard. Supercritical fluids provide several benefits over liquid solvents since they possess properties of both gases and liquids. Supercritical fluids as solvents exhibit physiochemical characteristics such as diffusivity, dielectric constant, density, and viscosity. The density of supercritical solvents determines their dissolving capacity, which may be changed by varying the pressure and temperature. In addition, the low viscosity and high diffusivity of supercritical solvents facilitate their diffusion through solid materials, thus resulting in higher extraction rates (Larson and King, 1986; Munshi and Bhaduri, 2009).

Typically, this technique utilizes CO_2 as a supercritical solvent to extract intracellular metabolites. CO_2 is a preferred solvent for this technique because it is non-flammable, non-toxic, and affordable. Occasionally, xenon and nitrous oxide are also utilized. Moreover, ethanol or methanol is further added as a modifier, along with carbon dioxide, to facilitate the extraction of polar compounds from microbial cells. This is a quick approach that uses minimal solvents and sample amounts. Furthermore, this technique has the potential for automation and integration with analytical instruments like gas chromatographs and mass spectrometers. However, there is a possibility of the degradation of labile metabolites because SFE operates at high pressure (Larson and King, 1986).

6.1.5 Analytical Techniques in Fungal Metabolomics

The primary goals in metabolomics studies are to separate, detect, and identify the target metabolites. The main analytical techniques in the detection of fungal metabolites are liquid or gas chromatography-coupled mass spectrometry (LC-MS or

GC-MS), nuclear magnetic resonance (NMR) spectrometry, Fourier transform infrared spectroscopy (FTIR), MALDI-MS, and capillary electrophoresis/mass spectrometry (CE–MS). Each of these analytical techniques has its own benefits and drawbacks. It is impossible to directly and accurately quantify or characterize all small molecules or metabolites produced by fungal organisms. A proper set of tools is usually required, relying on particular experimental conditions to enhance the analysis of target fungal metabolites (Li et al., 2022).

6.1.5.1 Gas Chromatography–Mass Spectrometry (GC-MS)

The main idea behind MS analysis is the ionization of an individual compound to identify the presence of particular ions, thereby facilitating the qualitative evaluation of various properties of compounds. This technique allows for the simultaneous measurement of several metabolites with distinct chemical properties such as high volatility and thermal stability, including amino acids, organic acids, sugar alcohols, sugars, fatty acids, and aromatic amines. It has several advantages such as excellent robustness, high selectivity and separation capacity, effective sensitivity, and a reliable retention time. The primary benefit of GC-MS lies in its ability to utilize established libraries for structural analysis, enabling the retrieval of a vast array of libraries. However, this technique has certain drawbacks as well. The majority of compounds with polar functional groups require a derivatization step to achieve excellent separation (Chen et al., 2019, Li et al., 2022).

6.1.5.2 Liquid Chromatography–Mass Spectrometry

LC-MS is a highly efficient analytical instrument that combines the separation capability of LC with the high resolution of mass spectrometry to measure metabolites on the basis of their mass/charge (m/z) ratio. Once the sample is introduced into the instrument, ionization takes place, resulting in the formation of distinctive peak signals for the detection of metabolites. Electrospray ionization (ESI) is the most commonly used ionization technique employed in LC-MS-based metabolomics. However, the high resolution of mass spectrometry allows researchers to separate and differentiate metabolites that share the same nominal mass but differing elemental compositions (Lei et al., 2011; Alonso et al., 2015).

Studies using LC-MS to analyze metabolites in endophytic fungi have grown significantly in the past few years, representing about 80% of the literature that has been published. The popularity has demonstrated the unique benefits of LC-MS-based metabolomics. In general, the LC-MS technique is highly selective and sensitive. It is capable of detecting samples at concentrations as low as nanomolar (nM), enabling the analysis of a wide range of metabolites. It is an excellent tool for metabolomics investigations, both non-targeted and targeted. Additionally, LC-MS-based metabolomics studies have the advantage of the wide availability of spectral databases, which is essential for the identification of metabolites (Sands et al., 2021).

6.1.5.3 Nuclear Magnetic Resonance (NMR) Spectrometry

NMR is a high-throughput and non-destructive detection method that relies on the magnetic characteristics of nuclei possessing spin properties. These nuclei capture radio frequency (RF) emission, and when subjected to an external magnetic field, they emit it at a specific radio frequency. NMR spectroscopy may be used to measure the emission's frequency and provide information on the structure and composition of target metabolites in the sample. The NMR spectrum provides detailed knowledge on all metabolites or small molecules within the organism. Statistical procedures such as pattern recognition processing and multivariate analysis are used to construct a comprehensive metabolic map of related organisms (Chen et al., 2019).

NMR-based metabolomics provides various benefits over MS techniques. These include accurate quantification, high reproducibility, easy sample preparation, identification of undefined metabolites, detectable analytes using different solvents, and comprehensive detection of metabolites. Nevertheless, the low sensitivity of NMR makes it challenging to simultaneously identify metabolites within biological systems with significant concentration variations, which limits its use in fungal metabolomics (Markley et al., 2017).

6.1.5.4 Fourier Transform Infrared Spectroscopy (FTIR)

FTIR spectroscopy is a safe and rapid technique that allows for the identification of biochemical characteristics by creating a unique fingerprint for each cell component. Though it has been used for structural chemical analysis for many years, FTIR spectroscopy gained popularity in the 1990s as a method for identifying and characterizing biological materials. FTIR spectroscopy helps in the identification of individual compounds, primarily concentrating on discerning the functional groups of all components present within the sample, as well as the vibrations occurring within the polar bonds. This technique has been widely utilized in biotechnology and applied microbiology to analyze a wide range of microorganisms such as yeasts and fungi. Extracellular metabolites and growth media are also measured using FTIR spectroscopy. It has been demonstrated that FTIR spectroscopy is a reliable and precise technique for identifying and analyzing polyphosphates, chitosan/chitin, and microbial lipids. FTIR can function as a rapid monitoring instrument for highly efficient research, which includes determining which fungal strains produce low- and high-value lipids. However, there are several drawbacks to FTIR, including the incapability to assess ion suppression quantitatively, the effects of complex ions and component fragments on the analysis, and the inability to identify isomers (Johnson et al., 2003; Dzurendova et al., 2020).

6.1.5.5 MALDI Mass Imaging Spectrometry

MALDI-MSI is a rapid, simple, and reliable method for identifying, differentiating, and classifying microorganisms. The development of MALDI-coupled mass spectrometry to analyze fungal microorganisms started in the year 2000 by initial trials involving *Scytalidium dimidiatum*, *Aspergillus niger*, *Penicillium* spp., and *Trichophyton rubrum* (Chalupova et al., 2014). It is commonly employed in combination with various mass analyzer types, including ion-trap, QqToF, and ToF (Goodwin et al., 2008). The primary factors contributing to the successful use of MALDI-MSI are minimal operational costs, rapid speed, and high specificity, along with the flexibility and universality of detection (Posteraro et al., 2013).

6.1.5.6 Capillary Electrophoresis/Mass Spectrometry (CE–MS)

CE–MS has proven to be an effective approach to analyze charged and highly polar metabolites within biological samples. It enables the high-resolution and rapid analysis of a variety of chemical compounds, including amino acids, organic acids, carboxylic acids, nucleosides and nucleotides, carbohydrates, vitamins, thiols, peptides, and inorganic ions. It has several advantages, including low cost, minimal required samples, rapid analysis, and less consumption of reagents. However, this method is not suitable for the detection of complex metabolites due to its low sensitivity and reproducibility as well as poor migration time (Ramautar et al., 2013; Chen et al., 2019).

6.1.6 Data Processing and Analysis

The extensive range of metabolomics data poses significant issues for researchers; therefore, software tools are necessary for its analysis and processing. Numerous software tools are available for the computational analysis of metabolite spectrum data generated by various analytical tools. These *in silico* tools enable researchers to gain significant insight from the complex and large datasets obtained through analytical methods. The most commonly used software tools are XCMS, MZmine, MBRole, MetaboAnalyst, 3Omics, MetAlign, MAVEN, MetaCoreTM, and InCroMA (Li et al., 2022).

XCMS is an online software tool that is widely used in the processing as well as analysis of metabolites. Using this platform, users may quickly upload and analyze data. This tool provides a comprehensive solution for the entire untargeted metabolomic process, including alignment, feature detection, statistical analysis, and visualization of data. Results are available in the form of a customizable, interactive table showing chromatograms and statistics for each metabolite (Tautenhahn et al., 2012). MZmine is another web-based software that is easy to use on a variety of computer platforms. It was developed as a modular system with a particular emphasis on its robust visualization models, which differentiate it from other bioinformatics tools (Pluskal et al., 2020). MetaboAnalyst is an efficient software tool for the analysis and interpretation of metabolomics data. It deals with most common types of metabolomics data from various metabolomics platforms (Xia and Wishart, 2016).

6.1.7 Applications of Metabolomics in Fungal Research

Metabolomics is an advanced technique that has gained popularity in recent years due to its wide range of applications in the field of fungal research. Some of these applications include chemotaxonomic fungal classification, the evaluation of fungal response under stress conditions, the detection of important fungal metabolites, fungal metabolic engineering, and understanding fungal–plant interactions.

6.1.7.1 Taxonomic Classification of Fungi Using Metabolomics

Traditional methods for the classification of fungal organisms have some drawbacks. The accuracy of classification is impacted by factors such as similar morphologies, fungal growth environments, and other factors. Metabolomics is an efficient technique that has become popular in recent years for the chemotaxonomic identification of fungi (Li et al., 2022). A study was performed by Kang et al. in which secondary metabolites were examined in *Trichoderma* species using a chemotaxonomic comparison based on sequencing and metabolome data. It was discovered that the secondary metabolite-based chemical taxonomy was more precise than its sequence (Kang et al., 2011). In another study, it was found that an NMR metabolomics approach was faster and less time-consuming in identifying and differentiating *Ganoderma lucidum* (Wen et al., 2010).

6.1.7.2 Understanding Fungal Responses to Stress

The application of metabolomics for fungal research can help to gain insight into the ways that fungi react to stress during growth or when they infect their hosts. Additionally, it proves valuable for enhancing the utilization of fungi in biotechnology, improving the environment, and preventing fungal diseases (Li et al., 2022).

6.1.7.3 Detection of Unknown Fungal Metabolites

Various metabolites for example, secondary and primary metabolites, are produced in fungi. Metabolomics is an advanced approach that helps to discover and identify new metabolites in various fungal organisms. The use of metabolomics, along with advanced analytical platforms, enables the screening and extraction of thousands of natural compounds with bioactive properties and valuable fungal secondary metabolites (Li et al., 2022).

6.1.7.4 Metabolic Engineering in Fungi

Metabolic engineering is defined as the enhancement of cellular activity by utilizing recombinant DNA technology to manipulate the regulatory, transport, and enzymatic processes within cells (Bailey, 1991). Fungi demonstrate a range of capabilities and are utilized in industrial applications such as secondary metabolism, the fermentation of organic acids, and protein production. The use of DNA recombination can result in alterations in various fungal metabolites that can be examined using metabolomics techniques.

6.1.7.5 Understanding Fungus–Plant Interactions

It is crucial to understand the interactions between plants and fungi for the purpose of controlling and preventing plant diseases. In the past ten years, metabolomics approaches have been found useful in various study disciplines on plant–fungus interactions, including the determination of mode of infection and the identification of fungi. In addition, metabolomics technology assists in understanding the pathogenesis and mechanisms of defending plants against pathogenic fungi, as well as developing efficient treatment and preventive measures for fungal infections (Li et al., 2022).

6.2 FUNGAL PHENOMICS

6.2.1 Introduction

A critical step in identifying the biological function of unidentified genes is to analyze the effects of gene loss. Model species like *Caenorhabditis elegans* and *Saccharomyces cerevisiae* have recently been used to develop systematic techniques for the targeted silencing of each gene in a genome. These projects have paved the way for the development of phenomics, a cutting-edge subject that aims to provide a comprehensive analysis of phenotypic traits across the entire genome. The qualitative nature of traditional phenotypic analysis should be a major concern, since most of the phenotypes are anticipated to have marginal characteristics. Furthermore, it is anticipated that a number of genes are associated with various physiologic functions of cells. The comprehensive illustration of the cellular function of a gene would therefore depend on the capability to differentiate diverse physiological characteristics such as the timeframe required to adjust to environmental conditions and growth speed or efficiency (Warringer et al., 2003).

Phenomics is a broad term that refers to the collection of high-dimensional phenotypic information on an organismal level. Although the genome serves as an organism's "blueprint," the behavior of that organism is displayed through its phenotype, which includes cell differentiation, growth rates, the synthesis of enzymes and secondary metabolites, and environmental responses (Smedsgaard and Nielsen, 2005). Geneticists have suggested the concepts of phenomics and phenotypes in an effort to understand the link between desired traits and genes. The main objective of these phenomics projects is to elucidate the intricate relationship between the genotype and phenotype of an organism, thereby facilitating the interpretation of the genome's function (Yeung et al., 2022). Fungal phenomics is a sub-branch of phenomics that refers to the study of phenotypic traits of fungal organisms on a large scale. Through the use of advanced analytical techniques, phenomics aims to analyze various phenotypic characteristics of fungi, including morphology, growth rates, spore formation, metabolic activities, cell cycle, and interactions with different hosts such as plants. The data obtained can thus be used to understand fungal biology as well as the applications of fungi in biotechnology and other fields (Houle, 2010).

6.2.2 History

The term "phenotype" is derived from the Greek words "typos" and "phainein," which mean "type" and "show," respectively. Wilhelm Johannsen characterized the term "phenotype" in 1911, stating that "all organisms can be differentiated through direct observation or with more refined techniques of measurement or depiction." The term phenotype is defined as "the observable characteristics and traits of an organism." Dr. Steven A. Garan first used the term "phenomics" to describe the process of measuring phenomes in 1996 (Yang et al., 2020). In 2010, phenomics was described as the collection of high-dimensional and complex phenotypic information on an organismal level (Houle, 2010). Numerous authors have suggested that phenomics, which involves phenotyping on a large scale, is an ideal counterpart to the process of genome sequencing, serving as a means of accelerating biological processes over the last 15 years. A phenome refers to a set of quantifiable attributes comprising the chemical, biological, and physical characteristics of individuals as well as populations, resulting from the complex interplay of multiple factors such as genes and environmental exposures. With the growing popularity of phenomics research in a variety of domains, significant initiatives relying on advanced integrated phenotyping facilities and global collaborative ventures have been implemented to comprehensively explore phenotypes (Jin, 2021).

Most of the research has concentrated on genes that were previously well studied, a circumstance that has undergone minimal change in the last four decades of genetic, biochemical, and biological investigations. Even though whole-genome sequences are available, the biological functions of the majority of genes are still poorly understood or unknown. A significant bias in publications and citations that has not changed over the previous 20 years reflects this disparity. Yeast-like microbes are ideal for this kind of high-throughput phenotyping. In addition to providing clear functional clues about how certain genes impact biological processes, they also provide diverse phenotypic fingerprints that reflect systemic reactions to genetic changes under various conditions. These broad phenotypic data offer valuable functional information on particular genes and mechanisms that influence cellular fitness either positively or negatively in a variety of physiological and environmental contexts (Rodriguez-Lopez et al., 2023).

6.2.3 Phenotypic Traits of Fungi

It is estimated that there are more than 3 million fungal species, only about 150,000 of which have been described and categorized. The fungal kingdom represents a major eukaryotic lineage that rules over terrestrial ecosystems. Fungi are an integral part of ecosystems and play a variety of ecological roles, from epiphytes to endophytes and saprobes to pathogens. Fungi are a type of eukaryotic microorganisms that utilize an advanced inherent capacity to produce phenotypic variability. This effective approach enables them to quickly adapt to environmental changes. They shared a more recent common ancestor with plants than with animals, yet they differ from other living organisms in the context of distinct phenotypic traits, including nutritional, morphological, and physiological traits (Jain et al., 2008; Bhunjun et al., 2023). Some of these traits include the development of fruiting bodies, hyphae, mycelia, multicellular growth, mycorrhizae, and metabolic activities. These multicellular structures allow fungi to adjust to a variety of ecological niches. However, there are some other significant traits that contribute to the development of a remarkably diverse fungal kingdom in terms of both taxonomy and phenotype (Nagy et al., 2017).

6.2.3.1 Morphological Traits

Phenomics is an advanced field that allows researchers to visualize various morphological features or traits of fungi, including size, shape, color, and texture. Fruiting bodies are one of the striking morphological features of fungi. They exist in a variety of shapes, forms, and sizes and have a short life span ranging from a few hours to a few days. The emergence of fruiting bodies on vegetative mycelium marks a significant transition in the fungal life cycle. These are reproductive structures that assist in the production and dispersal of spores. Fruiting bodies have two main functions: to protect the maturing reproductive organs and to facilitate the dispersal of spores either by providing sporogenous tissue with a large supporting area and/or elevating it above the ground's surface (Nagy et al., 2017). The visualization of different morphological traits of fungal organisms through phenomics will help to understand fungal taxonomy and the diversity of fungal species.

6.2.3.2 Metabolic Traits

Small molecules, known as specialized or secondary metabolites, play key functions in environmental or ecological interactions. Secondary metabolites produced by fungi, for instance, may function as antibiotics, act as signals for other organisms, or even change the amount of nutrients available in the soil or host tissues. By analyzing metabolite expression profiles through phenomics approach, researchers can identify intricate regulatory networks and interactions that control metabolism (Nagy et al., 2017; Stengel et al., 2022).

6.2.3.3 Growth Patterns

Phenomics enables researchers to visualize the growth of fungi under different environmental conditions or when exposed to various stressors. It can also assist in understanding branching patterns and mycelial network formation. The fungal mycelium is made up of a mass of branching, thread-like structures called hyphae. Fungi take up nutrients from their surroundings through their mycelia. This information can be valuable to understand the development of fungi and their growth rate, as well as fungal responses to environmental changes, nutrient availability, etc. (Islam et al., 2017; Watts-Williams et al., 2019).

6.2.3.4 Interactions with Multiple Hosts

Fungi have a number of key characteristics, including the ability to form interactions with different hosts such as plants, bacteria, etc. By visualizing these interactions through a phenomics approach, researchers can gain insight into associated traits or characteristics such as the development of mycorrhizal networks in the roots of plants (Nagy et al., 2017).

6.2.4 Phenotyping Techniques

The characterization of fungi has traditionally relied on their phenotypic traits or molecular studies. However, about 70% of the identified fungal species have not yet been sequenced. These issues undermine the validity of the data and the reliability of the analysis. Additionally, the time needed for analysis is another issue, as analysis is often highly expensive and time-consuming (Puliga et al., 2022). Analytical methods refer to the techniques or tools that are used to measure and analyze the phenotypic traits of an organism. These phenotyping techniques include microscopy, inflorescence microscopy, high-content screening, spectroscopy, non-invasive imaging, microfluidics, image analysis, robotics, and high-speed computing techniques. These techniques are often used in combination to achieve the accurate and rapid analysis of phenotypic traits.

6.2.4.1 Microscopy

Microscopy is a primary technique for analyzing gene function and cellular phenotypes. Even though advanced microscopes can capture several thousand high-quality images in a single day, phenotypic interpretation and image analysis continue to be major obstacles. Expert biologists have used fission yeast to perform large-scale screening that involves visual inspection to enable qualitative phenotypic categorization (Rallis and Bahler, 2016).

6.2.4.2 Immunofluorescence Microscopy

This technique was first developed in the 1950s by Coons and Kaplan. Since the development of immunofluorescence microscopy, this technique has been extensively used in cell biology studies. This technique allows researchers to explore morphological phenotypes, thereby facilitating the quantitative and systematic analyses of mutant traits in yeast (fungal) species. The comprehensive phenotypic analysis of yeast strains helps to understand the function of individual genes within the yeast (fungal) genome (Ohya et al., 2005; Vizeacoumar et al., 2009).

6.2.4.3 High-Content Screening

The number of phenotypes suitable for genome-scale image analysis has greatly increased as a result of developments in highly efficient computational image analysis and microscopy techniques, also known as high-content screening. The term "high-content screening" was originally used for the first time during late 1990s (Vizeacoumar et al., 2009). These advanced screening techniques along with automated methods are particularly useful for evaluating genetic influences on subcellular morphology. Improvements in computational and imaging technologies will enhance the complexity and throughput of phenomics projects, ranging from single-cell analyses to multi-gene interactions (Yeung et al., 2022).

6.2.4.4 Raman Spectroscopy

The identification of fungi typically relies on measurable phenotypic features with subjective analysis and interpretation. In recent years, Raman spectroscopy has acquired widespread popularity as a useful method for the rapid and non-invasive identification and characterization of individual or multiple analytes or molecules. It features easy operability, high efficiency, great convenience, and extensive information on the molecular structure, chemical composition, and interactions between molecules. Raman spectra display unique signatures of analyzed compounds, characterized by their spectral patterns. This technique is often combined with digital microscopic imaging to enable high-quality visualization of morphological features, chemical composition, and spatial distribution in microscopic images (Cai et al., 2023).

6.2.4.5 Microfluidics

The incorporation of cutting-edge techniques like microfluidics has made it possible to perform high-throughput temporal single-cell analysis. These techniques enable the investigation of environmental and genetic factors that govern lifespan regulation as well as the tracking of the morphological variations that yeast (fungal) cells experience throughout their entire replicative lifespan, encompassing the aging process (Lee et al., 2012). The microfluidic chip is expected to assist in phenotypic (genome-wide) investigations with the help of other microfluidics elements. Using this technology, the morphology of yeast cells was studied in an experiment to uncover the molecular principles underlying cell morphogenesis and to discover crucial elements of various processes, including cell cycle progression. Morphological variations that take place throughout the cell cycle process were effectively assessed and characterized. However, because of its low throughput, this technique is not ideal for morphological analysis of a large number of yeast mutants to accurately predict gene functions, particularly during the manual image acquisition phase (Ohnuki et al., 2009).

6.2.4.6 Phenotypic Array Analysis

Phenotypic array analysis is a technique that is based on measuring growth rates by cultivating on agar medium. It is based on spotting a few microliter-sized droplets of diluted strains onto the surface of agar. After image acquisition at various time intervals, the exponential growth rate can be determined over 4–5 generations (Warringer and Blomberg, 2014). Such an assay provides quantitative phenotypic data; however, it is laborious and time-consuming. These limitations make the process unsuitable for phenotypic analysis on a large scale. Furthermore, the information collected on solid media cannot be used to infer the precise effects of a specific mutation and/or environment on various growth parameters of fungi, including lag phase, growth rate, and yield (Joubert et al., 2010).

6.2.4.7 Laser Nephelometry

Nephelometry is a technique of directly calculating the amount of light dispersed by suspended particles. Since the intensity of scattered light is directly correlated to the number of suspended particles, nephelometry is an effective approach for monitoring microbial growth, particularly for investigating filamentous fungi (Joubert et al., 2010).

6.2.5 Bioinformatics Tools and Databases

Over the past 300 years, the phenotypic classification idea has been used to identify, categorize, and predict fungal relationships using morphological characteristics that are quite simple to detect and record. The main objective of phenomics is to establish a link between the phenotype and genotype of an organism. There is an urgent need for an accessible phenotype database, as it can provide valuable information about fungal traits, allowing researchers to understand the biology and ecology of fungal organisms. To create a comprehensive database of phenotypic traits for a model species, the initial prerequisite is to have data that are objectively described, particularly in a searchable, mathematical, and easily digitized format. Faces of Fungi is a popular fungal database that enables the collection of phenotypic details, taxonomic data, and other valuable information that will improve our grasp of current taxonomy and eventually assist mycologists in acquiring better and more recent insights into the present fungal classification system (Jayasiri et al., 2015). FungalTraits is another database that offers a complementary option to currently available trait databases and aims to exchange information with other phenotypic databases to serve as a robust platform for the advancement of fungal biology (Polme et al., 2020).

These database systems can be used to store and retrieve data; however, data mining and statistical analysis techniques are still in their early stages of development. The technique of machine learning is currently being utilized in image segmentation to identify particular patterns and to identify the underlying characteristics of imaging data, distinguishing mutant phenotypes from wild-type ones and enabling functional predictions. Some software packages are commercially available for the efficient and rapid analysis of data. These include SpotFire, CellmineT, GeneData Screener, and Accelrys. These programs not only provide quick data analysis but also give valuable information about the quality and reliability of the data (Vizeacoumar et al., 2009).

6.2.6 Applications of Fungal Phenomics

Fungal phenomics is an advanced field that has a wide range of applications in various fields. Some of its applications are presented herein.

6.2.6.1 Studying the Link between Genotype and Phenotype

Phenomics approaches allow researchers to identify the link between genotype, phenotype, and environmental variables. In recent times, phenomics initiatives have begun in several species that link genomic data with information regarding quantitative variation in phenotypes in an attempt to comprehend the G–P map. Phenomics helps to anticipate and modify the traits of organisms, which enables a wide range of applications and advances in various domains.

6.2.6.2 Determining the Genetic Basis of Complex Traits

Fungi display diverse phenotypic variation, which is a reflection of their ability to adapt to a variety of environmental conditions. Fungal phenomics helps to determine the underlying genetic mechanisms of complex traits and can help to study phenotypic variation, thereby facilitating research in different fields of biotechnology.

6.2.6.3 Drug Discovery

The phenomics approach helps researchers to uncover compounds with distinctive phenotypic profiles, possibly resulting in the identification of new drug candidates.

6.2.7 Conclusion

Metabolomics is an effective and powerful tool which is commonly employed in fungal research. Fungal metabolomics has advanced significantly over the past several years because of the improvement and development of analytical methods and sample preparation techniques. However, there are a few issues that are required to be solved. These include the lack of a standardized method for quenching and extracting metabolites from fungal organisms, the limitation of automatic platform technology for data processing, and the complexity of data processing. A significant barrier to the advancement of fungal metabolomics is the lack of a thorough and comprehensive fungal metabolomics database. However, even though over a thousand metabolites have been found, this is still just the beginning for fungal metabolites. More significantly, the majority of researchers are only able to conduct secondary and primary metabolite investigations on fungi and have insufficient knowledge of the metabolic pathways in fungi. Fungal metabolomics is still in its early developmental stages compared to the use of metabolomics in the development of new drugs and disease diagnosis. It is anticipated that with the advancement in the field of technology and science, fungal metabolomics will also advance more rapidly.

Fungal phenomics has transformed fungal research by providing a comprehensive understanding of fungi's ecological and biological functions. It has a lot of potential for tackling crucial issues across various domains and promises to be an important technique in uncovering the unknown facts about the fungal kingdom. The systematic and comprehensive collection of precisely measured phenotypes is one strategy for reducing the gap between organismal phenotypes and genetic variation. Phenomics research has so far typically investigated a small number of phenotypes. The limited availability of phenotypic assays can be another major challenge in the analysis of diverse fungal phenotypes. However, improvements in equipment, computational biology, and functional genomics technologies is providing researchers with the means to thoroughly examine a larger amount of phenotypes.

REFERENCES

Alonso, A., Marsal, S., & Julia, A. (2015). Analytical methods in untargeted metabolomics: State of the art in 2015. *Frontiers in Bioengineering and Biotechnology*, *3*, 23–43.

Alvarez-Rivera, G., Bueno, M., Ballesteros-Vivas, D., Mendiola, J. A., & Ibañez, E. (2020). Pressurized liquid extraction. In *Liquid-phase extraction* (pp. 375–398). Elsevier.

Anke, T., & Erkel, G. (2002). Non-β-lactam antibiotics. In M. Hofrichter (Ed.), *Industrial Applications* (pp. 93–108). Berlin: Springer.

Awaad, A. S., Nabilah, A. A., & Zain, M. E. (2012). New antifungal compounds from *Aspergillus terreus* isolated from desert soil. *Phytotherapy Research*, *26*, 1872–1877.

Azmir, J., Zaidul, I. S., & Rahman, M. M. (2013). Techniques for extraction of bioactive compounds from plant materials: A review. *Journal of Food Engineering*, *117*, 426–436.

Bailey, J. E. (1991). Toward a science of metabolic engineering. *Science*, *252*, 1668–1675.

Bhunjun, C. S., Phukhamsakda, C., & Hyde, K. D. (2023). Do all fungi have ancestors with endophytic lifestyles? *Fungal Divers*, *16*, 1–26.

Cai, J., Zou, C., & Yin, L. (2023). Characterization and recognition of citrus fruit spoilage fungi using Raman scattering spectroscopic imaging. *Vibrational Spectroscopy*, *124*, 103474–103485.

Canelas, A. B., Ras, C., & Ten Pierick, A. (2008). Leakage-free rapid quenching technique for yeast metabolomics. *Metabolomics*, *4*, 226–239.

Ceglarek, U., Leichtle, A., & Bruegel, M. (2009). Challenges and developments in tandem mass spectrometry based clinical metabolomics. *Molecular and Celularl Endocrinology*, *301*, 266–271.

Chalupova, J., Raus, M., Sedlarova, M., & Sebela, M. (2014). Identification of fungal microorganisms by MALDI-TOF mass spectrometry. *Biotechnology Advances*, *32*, 230–241.

Chassagnole, C., Noisommit-Rizzi, N., & Schmid, J. W. (2002). Dynamic modeling of the central carbon metabolism of *Escherichia coli*. *Biotechnology and Bioengineering*, *79*, 53–73.

Chen, F., Ma, R., & Chen, X. L. (2019). Advances of metabolomics in fungal pathogen–plant interactions. *Metabolites*, *9*, 169–188.

Cocks, S., Wrigley, S. K., Chicarelli-Robinson, M. I., & Smith, R. M. (1995). High-performance liquid chromatography comparison of supercritical-fluid extraction and solvent extraction of microbial fermentation products. *Journal of Chromatography*, *697*, 115–122.

Daley, D. K., Brown, K. J., & Badal, S. (2017). Fungal metabolites. In S. Badal & R. Delgoda (Eds.), *Pharmacognosy* (pp. 413–421). Amsterdam: Elsevier.

de Koning, W., & van Dam, K. (1992). A method for the determination of changes of glycolytic metabolites in yeast on a subsecond time scale using extraction at neutral pH. *Analytical Biochemistry*, *204*, 118–123.

Duportet, X., Aggio, R. B. M., Carneiro, S., & Villas-Boas, S. G. (2012). The biological interpretation of metabolomic data can be misled by the extraction method used. *Metabolomics*, *8*, 410–421.

Dzurendova, S., Zimmermann, B., & Kohler, A. (2020). Microcultivation and FTIR spectroscopy-based screening revealed a nutrient-induced co-production of high-value metabolites in oleaginous *Mucoromycota* fungi. *PLoS One*, *15*, 1–34.

Fiehn, O. (2001). Combining genomics, metabolome analysis, and biochemical modelling to understand metabolic networks. *Comparative and Functional Genomics*, *2*, 155–168.

Gholami-Shabani, M., Shams-Ghahfarokhi, M., & Razzaghi-Abyaneh, M. (2019). Natural product synthesis by fungi: Recent trends and future prospects. In A. N. Yadav., S. Singh., S. Mishra & A. Gupta (Eds.), *Recent Advances in White Biotechnology Through Fungi, Vol. 2: Perspective for Value-Added Products and Environments* (pp. 195–228). Berlin: Springer.

Gomez-Ariza, J. L., de la Torre, M. A. C., Giraldez, I., & Morales, E. (2004) Speciation analysis of selenium compounds in yeasts using pressurised liquid extraction and liquid chromatography–microwave-assisted digestion–hydride generation–atomic fluorescence spectrometry. *Analytica Chimica Acta*, *524*, 305–314.

Goodwin, R. J. A., Pennington, S. R., & Pitt, A. R. (2008). Protein and peptides in pictures: Imaging with MALDI mass spectrometry. *Proteomics, 8,* 3785–3800.

Hajjaj, H., Blanc, P. J., Goma, G., & Francois, J. (1998). Sampling techniques and comparative extraction procedures for quantitative determination of intra-and extracellular metabolites in filamentous fungi. *FEMS Microbiology Letters, 164,* 195–200.

Hans, M., Heinzle, E., & Wittmann, C. (2001). Quantification of intracellular amino acids in batch cultures of *Saccharomyces cerevisiae. Applied Microbiology and Biotechnology*, *56*, 776–779.

Houle, D. (2010). Numbering the hairs on our heads: The shared challenge and promise of phenomics. *PNAS*, *107*, 1793–1799.

Islam, M. R., Tudryn, G., & Bucinell, R. (2017). Morphology and mechanics of fungal mycelium. *Scientific Reports*, *7*, 13070–13082.

Iversen, J. J. (1981). Rapid sampling valve with minimal dead space for laboratory scale fermenters. *Biotechnology and Bioengineering*, *23*, 437–440.

Jain, N., Hasan, F., & Fries, B. C. (2008). Phenotypic switching in fungi. *Current Fungal Infection Reports*, *2*, 180–188.

Jayasiri, S. C., Hyde, K. D., & Ariyawansa H. A. (2015). The Faces of Fungi database: Fungal names linked with morphology, phylogeny and human impacts. *Fungal Divers*, *74*, 3–18.

Jewett, M. C., Hofmann, G., & Nielsen, J. (2006). Fungal metabolite analysis in genomics and phenomics. *Current Opinion in Biotechnology*, *17*, 191–197.

Jin, L. (2021). Welcome to the phenomics journal. *Phenomics*, *1*, 1–2.

Johnson, H. E., Broadhurst, D., Goodacre, R., & Smith, A. R. (2003). Metabolic fingerprinting of salt-stressed tomatoes. *Phytochemistry*, *62*, 19–928.

Joubert, A., Calmes, B., & Berruyer, R. (2010). Laser nephelometry applied in an automated microplate system to study filamentous fungus growth. *Biotechniques*, *48*, 399–404.

Kang, D. J., Kim, J. Y., & Choi J. N. (2011). Chemotaxonomy of *Trichoderma* spp. using mass spectrometry-based metabolite profiling. *Journal of Microbiology and Biotechnology*, *21*, 5–13.

Keller, N. P., Turner, G., & Bennett, J. W. (2005) Fungal secondary metabolism—From biochemistry to genomics. *Nature Reviews Microbiology*, *3*, 937–947.

Kim, S., Lee, D. Y., & Wohlgemuth, G. (2013). Evaluation and optimization of metabolome sample preparation methods for *Saccharomyces cerevisiae. Analytical Chemistry*, *85*, 2169–2176.

Lameiras, F., Heijnen, J. J., & van Gulik, W. M. (2015). Development of tools for quantitative intracellular metabolomics of *Aspergillus niger* chemostat cultures. *Metabolomics*, *11*, 1253–1264.

Larson, K. A., & King, M. L. (1986). Evaluation of supercritical fluid extraction in the pharmaceutical industry. *Biotechnology Progress*, *2*, 73–82.

Lee, S. S., Vizcarra, I. A., & Huberts, D. H. E. W. (2012). Whole lifespan microscopic observation of budding yeast aging through a microfluidic dissection platform. *PNAS*, *109*, 4916–4920.

Lei, Z., Huhman, D. V., & Sumner, L. W. (2011). Mass spectrometry strategies in metabolomics. *Journal of Biological Chemistry*, *286*, 25435–25442.

Li, G., Jian, T., & Liu, X. (2022). Application of metabolomics in fungal research. *Molecules*, *27*, 7365–7390.

Li, N., peng Song, Y., Tang, H., & Wang, Y. (2016). Recent developments in sample preparation and data pre-treatment in metabonomics research. *Archives of Biochemistry and Biophysics*, *589*, 4–9.

Link, H., Anselment, B., & Weuster-Botz, D. (2008). Leakage of adenylates during cold methanol/glycerol quenching of *Escherichia coli*. *Metabolomics*, *4*, 240–247.

Link, H., Fuhrer, T., & Gerosa, L. (2015). Real-time metabolome profiling of the metabolic switch between starvation and growth. *Nature Methods*, *12*, 1091–1097.

Lu, H., Chen, H., & Tang, X. (2019). Evaluation of metabolome sample preparation and extraction methodologies for oleaginous filamentous fungi *Mortierella alpina*. *Metabolomics*, *15*, 1–10.

Madla, S., Miura, D., & Wariishi, H. (2012). Optimization of extraction method for GC-MS based metabolomics for filamentous fungi. *Journal of Microbial and Biochemical Technology*, *4*, 5–9.

Maharjan, R. P., & Ferenci, T. (2003). Global metabolite analysis: The influence of extraction methodology on metabolome profiles of *Escherichia coli*. *Analytical Biochemistry*, *31*, 45–154.

Markley, J. L., Bruschweiler, R., & Edison, A. S. (2017). The future of NMR-based metabolomics. *Current Opinion in Biotechnology*, *43*, 34–40.

Meinert, S., Rapp, S., & Schmitz, K. (2013). Quantitative quenching evaluation and direct intracellular metabolite analysis in *Penicillium chrysogenum*. *Analytical Biochemistry*, *438*, 47–52.

Moritz, B., Striegel, K., de Graaf, A. A., & Sahm, H. (2000). Kinetic properties of the glucose-6-phosphate and 6 phosphogluconate dehydrogenases from *Corynebacterium glutamicum* and their application for predicting pentose phosphate pathway flux *in vivo*. *European Journal of Biochemistry*, *267*, 3442–3452.

Munshi, P., & Bhaduri, S. (2009). Supercritical CO2: A twenty-first century solvent for the chemical industry. *Current Science*, *97*, 63–72.

Nagy, L. G., Toth, R., & Kiss, E. (2017). Six key traits of fungi: Their evolutionary origins and genetic bases. *Microbiology Spectrum*, *5*, 10–1128.

Ohnuki, S., Nogami, S., & Ohya, Y. (2009). A microfluidic device to acquire high-magnification microphotographs of yeast cells. *Cell Division*, *4*, 1–8.

Ohya, Y., Sese, J., & Yukawa, M. (2005). High-dimensional and large-scale phenotyping of yeast mutants. *PNAS*, *102*, 19015–19020.

Oliver, S. G., Winson, M. K., Kell, D. B., & Baganz, F. (1998). Systematic functional analysis of the yeast genome. *Trends in Biotechnology*, *16*, 373–378.

Pela, F. (2004). Biological activities of fungal metabolites. In Z. An (Ed.), *Handbook of Industrial Mycology* (pp. 68–111). US: CRC Press.

Pelaez, F. (2005). Biological activities of fungal metabolites. *Handbook of Industrial Mycology*, *22*, 49–92.

Pinu, F. R., Villas-Boas, S. G., & Aggio, R. (2017). Analysis of intracellular metabolites from microorganisms: Quenching and extraction protocols. *Metabolites*, *7*, 53–73.

Pluskal, T., Korf, A., & Smirnov, A. (2020). Metabolomics data analysis using MZmine. In R. Winkler (Ed.), *Processing Metabolomics and Proteomics Data with Open Software: A Practical Guide* (pp. 232–254). UK: Royal Society of Chemistry.

Polme, S., Abarenkov, K., & Henrik Nilsson, R. (2020). FungalTraits: A user-friendly traits database of fungi and fungus-like stramenopiles. *Fungal Divers*, *105*, 1–16.

Posteraro, B., De Carolis, E., Vella, A., & Sanguinetti, M. (2013). MALDI-TOF mass spectrometry in the clinical mycology laboratory: Identification of fungi and beyond. *Expert Review of Proteomics*, *10*, 151–164.

Puliga, F., Zuffi, V., & Zambonelli, A. (2022). Spectroscopic analysis of mushrooms by surface-enhanced Raman scattering (SERS). *Chemical and Biological Technologies in Agriculture*, *9*, 1–10.

Rallis, C., & Bahler, J. (2016). Cell-based screens and phenomics with fission yeast. *Critical Reviews in Biochemistry and Molecular Biology*, *51*, 86–95.

Ramautar, R., Somsen, G. W., & de Jong, G. J. (2013). CE-MS for metabolomics: Developments and applications in the period 2010–2012. *Electrophoresis*, *34*, 86–98.

Rodriguez-Lopez, M., Bordin, N., & Lees, J. (2023). Broad functional profiling of fission yeast proteins using phenomics and machine learning. *BioRxiv*, *12*, 1–47.

Roessner, U., Wagner, C., & Kopka, J. (2000). Simultaneous analysis of metabolites in potato tuber by gas chromatography–mass spectrometry. *Plant Journal*, *23*, 131–142.

Sands, C. J., Gomez-Romero, M., & Correia, G. (2021). Representing the metabolome with high fidelity: Range and response as quality control factors in LC-MS-based global profiling. *Analytical Chemistry*, *93*, 1924–1933.

Schadel, F., & Franco-Lara, E. (2009) Rapid sampling devices for metabolic engineering applications. *Applied Microbiology and Biotechnology*, *83*, 199–208.

Schueffler, A., & Anke, T. (2014). Fungal natural products in research and development. Natural *Products Reports*, *31*, 1425–1448.

Shulaev, V. (2006). Metabolomics technology and bioinformatics. *Brief Bioinformatics*, *7*, 128–139.

Smart, K. F., Aggio, R. B. M., Van Houtte, J. R., & Villas-Boas, S. G. (2010). Analytical platform for metabolome analysis of microbial cells using methyl chloroformate derivatization followed by gas chromatography–mass spectrometry. *Nature Protocols*, *5*, 1709–1729.

Smedsgaard, J., & Nielsen, J. (2005). Metabolite profiling of fungi and yeast: From phenotype to metabolome by MS and informatics. *Journal of Experimental Botany*, *56*, 273–286.

Spura, J., Reimer, L. C., & Wieloch, P. (2009). A method for enzyme quenching in microbial metabolome analysis successfully applied to gram-positive and gram-negative bacteria and yeast. *Analytical Biochemistry*, *394*, 192–201.

Stengel, A., Stanke, K. M., Quattrone, A. C, & Herr, J. R. (2022). Improving taxonomic delimitation of fungal species in the age of genomics and Phenomics. *Frontiers in Microbiology*, *13*, 847067–847077.

Tautenhahn, R., Patti, G. J., Rinehart, D., & Siuzdak, G. (2012). XCMS Online: A web-based platform to process untargeted metabolomic data. *Analytical Chemistry*, *84*, 5035–5039.

Theobald, U., Mailinger, W., Reuss, M., & Rizzi, M. (1993). *In vivo* analysis of glucose-induced fast changes in yeast adenine nucleotide pool applying a rapid sampling technique. *Analytical Biochemistry*, *214*, 31–37.

Villas-Boas, S. G., Hojer-Pedersen, J., & Akesson, M. (2005). Global metabolite analysis of yeast: Evaluation of sample preparation methods. *Yeast*, *22*, 1155–1169.

Vizeacoumar, F. J., Chong, Y., Boone, C., & Andrews, B. J. (2009). A picture is worth a thousand words: Genomics to phenomics in the yeast *Saccharomyces cerevisiae*. *FEBS Letters*, *583*, 1656–1661.

Vuckovic, D. (2012). Current trends and challenges in sample preparation for global metabolomics using liquid chromatography–mass spectrometry. *Analytical and Bioanalytical Chemistry*, *403*, 1523–1548.

Warringer, J., & Blomberg, A. (2014). *Yeast Phenomics-Large-Scale Mapping of the Genetic Basis for Organismal Traits*. Boca Raton: CRC Press.

Warringer, J., Ericson, E., & Fernandez, L. (2003). High-resolution yeast phenomics resolves different physiological features in the saline response. *PNAS*, *100*, 15724–15729.

Watts-Williams, S. J., Jewell, N., & Brien, C. (2019). Using high-throughput phenotyping to explore growth responses to mycorrhizal fungi and zinc in three plant species. *Plant Phenomics, 2019*, 1–12.

Wen, H., Kang, S., & Song, Y. (2010). Differentiation of cultivation sources of *Ganoderma lucidum* by NMR-based metabolomics approach. *Phytochemical Analysis: An International Journal of Plant Chemical and Biochemical Techniques*, *21*, 73–79.

Werf, M. J., Jellema, R. H., & Hankemeier, T. (2005). Microbial metabolomics: Replacing trial-and-error by the unbiased selection and ranking of targets. *Journal of Industrial Microbiology and Biotechnology*, *32*, 234–252.

Xia, J., & Wishart, D. S. (2016). Using MetaboAnalyst 3.0 for comprehensive metabolomics data analysis. *Current Protocols in Bioinformatics*, *55*, 10–14.

Yang, W., Feng, H., & Zhang, X. (2020). Crop phenomics and high-throughput phenotyping: Past decades, current challenges, and future perspectives. *Molecular Plant*, *13*, 187–214.

Yeung, C. H. L., Sahin, N., & Andrews, B. (2022). Phenomics approaches to understand genetic networks and gene function in yeast. *Biochemical Society Transactions*, *50*, 713–721.

Zain, M. E., Awaad, A. S., & Al-Othman, M. R. (2014). Biological activity of fungal secondary metabolites. *International Journal of Chemistry and Applied Biological Sciences*, *1*, 14–22.

Zain, M. E., Awaad, A. S., & Razak, A. A. (2009). Secondary metabolites of *Aureobasidium pullulans* isolated from Egyptian soil and their biological activity. *Journal of Applied Sciences Research*, *5*, 1582–1591.

Zheng, X., Yu, J., & Cairns, T. C. (2019). Comprehensive improvement of sample preparation methodologies facilitates dynamic metabolomics of *Aspergillus niger*. *Biotechnology Journal*, *14*, 1800315–1800324.

7 Fungal Glycomics and Lipidomics

Nasir Ahmed Rajput, Muhammad Atiq, Muhammad Hamza Shahid, Muniza Baig, Nuzhat Jamal, Muhammad Wahab, Babar Khan, Faizan Ali, and Yonghao Ye

7.1 INTRODUCTION

A fascinating wave of new scientific disciplines, collectively referred to as *omics*, emerged in the last decade of the 20th century (Yadav, 2007). The term "omic" is derived from the Latin suffix "ome," which mean mass or many. It involves mass measurements per endpoint rather than a few measurements (Lay et al., 2006). Omics fields are dedicated to the in-depth examination of complex biological systems at the molecular level, employing extensive data-driven approaches to unravel intricate molecular interactions (Zhang et al., 2010). Currently, life sciences-related omics techniques top the list, particularly genomics and proteomics, whereas other scientific disciplines such as metabolomics, glycomics, lipidomics, neuromics, transcriptomics, interactomics, and degradomics retain a close association with proteomics and genomics (Figure 7.0) (Zaikin and Borisov, 2021).

Glycomics and lipidomics are interdisciplinary fields of study that focus on the comprehensive analysis of carbohydrates and lipids, respectively, within biological systems. These molecules play crucial roles in various biological processes and are of paramount importance in understanding the biology of organisms. These interdisciplinary domains offer profound insights into the molecular structures, functions, and dynamics of carbohydrates and lipids within biological systems (Aoki-Kinoshita, 2008; Wenk, 2010). Glycomics and lipidomics stand at the nexus of chemistry, biology, and bioinformatics. They are branches of omics sciences, akin to genomics, proteomics, and metabolomics, that aim to decode the molecular intricacies underpinning life processes. While genomics delves into the genetic code, proteomics examines proteins, and metabolomics explores metabolites, glycomics and lipidomics are employed to decode the complex carbohydrates and lipids that adorn the cellular architecture of organisms.

The significance of these fields reverberates through the annals of biology and medicine. Carbohydrates and lipids, once relegated to the roles of mere energy storage and structural components, have emerged as critical players in numerous cellular functions. Their roles encompass energy storage, cell signaling, structural integrity, host–pathogen interactions, and a plethora of other cellular activities. Understanding

DOI: 10.1201/9781003452362-7

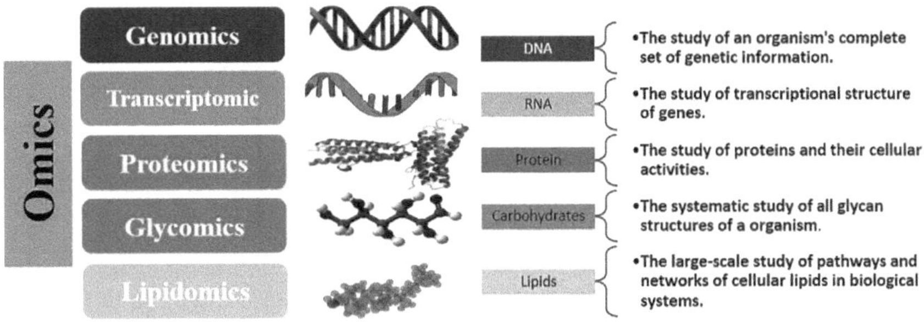

FIGURE 7.0 Omics constitutes five sub-branches, i.e., genomics, transcriptomics, proteomics, glycomics, and lipidomics.

the nuances of these molecules is not merely a scientific pursuit but a gateway to unraveling the mysteries of life itself (Bohm et al., 2019; Mehta et al., 2021).

Carbohydrates, known as sugars or saccharides, are organic compounds comprising carbon, hydrogen, and oxygen atoms. They serve as the primary energy source for most living organisms, facilitating the conversion of chemical energy into a form that cells can utilize—adenosine triphosphate (ATP). Yet carbohydrates are far from being mere energy sources; their functions transcend the boundaries of metabolism (von Ballmoos et al., 2009).

In fungi, carbohydrates manifest in a rich diversity of forms. Glucose, the quintessential sugar, reigns supreme as a source of quick energy. Mannose and galactose add to the repertoire, each with its unique metabolic significance. Chitin, a polymer of N-acetylglucosamine, emerges as a structural cornerstone, forming the robust cell walls that distinguishes fungi from other organisms (Gooday, 1995).

The roles carbohydrates play in fungal life are multifaceted. Beyond energy provision, they contribute substantially to fungal cell wall synthesis, a pivotal aspect of fungal biology. Carbohydrates also act as signaling molecules, orchestrating the growth and development of fungal organisms. Notably, the obscure role of carbohydrates in fungi extends beyond mere sustenance, shaping the fungus–host interactions that underlie both mutualistic and pathogenic relationships (Zeilinger et al., 2016).

Lipids, on the other hand, are a diverse class of hydrophobic molecules that include fats, phospholipids, sterols, and glycolipids (Fahy et al., 2005). While carbohydrates serve as the energy currency, lipids are the architects of cellular membranes, meticulously crafting the boundaries that demarcate the cellular world from its surroundings. In fungi, the significance of lipids extends well beyond mere membrane structure (Manna et al., 2019).

Fungal lipidomics reveals a captivating array of lipids at play. Phospholipids, with their amphipathic nature, serve as the foundation of cell membranes, providing both structural support and a platform for dynamic processes (Suetsugu et al., 2014). Sterols, like ergosterol, maintain membrane fluidity and integrity, analogous to cholesterol in animal cells.

Lipids not only serve as the backbone of cell membranes but also act as storage depots for energy in the form of triglycerides. Lipids participate in various cellular processes, including signal transduction, intracellular trafficking, and membrane remodeling. Moreover, they are integral to the fungal response to changing environmental conditions, enabling these organisms to adapt and thrive in diverse niches (He et al., 2018).

The study of glycomics and lipidomics in fungi transcends the realm of basic science. It has profound implications in several key areas.

Fungal infections pose a significant global health challenge, affecting both humans and plants. Understanding the glycobiology and lipidomics of fungal pathogens allows for the development of more precise diagnostic tools and targeted therapeutic interventions. By targeting unique fungal carbohydrate and lipid structures, researchers can design strategies to combat these infections effectively (Reece et al., 2018).

The intricate interplay of carbohydrates and lipids in fungi is central to their biology. Glycomics and lipidomics provide a lens through which we can decode the molecular mechanisms governing fungal growth, development, and virulence. A deepened understanding of these aspects aids in the elucidation of fungal pathogenesis and the exploration of potential vulnerabilities in fungal systems (Ball et al., 2020).

Fungi are a source of diverse bioactive compounds with applications in biotechnology and the pharmaceutical industry. Exploring fungal glycomics and lipidomics offers opportunities to discover novel compounds, biosynthetic pathways, and metabolic processes that can be harnessed for the development of bioproducts such as antibiotics, enzymes, and biofuels (Chen et al., 2018).

As we delve deeper into the fascinating realms of glycomics and lipidomics within the fungal context, we embark on a journey of discovery that not only enriches our scientific understanding but also holds promise for transformative applications. In the following sections, we will explore the intricate details of glycomics and lipidomics, shedding light on their roles, methods of study, and the exciting applications they bring to the world of fungi.

7.2 FUNGAL GLYCOMICS

Fungal glycomics refers to the application of glycomics techniques and approaches specifically to the study of fungi. Fungal glycans play crucial roles in various biological processes, such as cell adhesion, host–pathogen interactions, and immune responses. Research in fungal glycomics may involve the analysis of fungal glycans (the glycome), which includes the identification and characterization of the different types of glycans present in fungi, as well as studying their functions and roles in fungal biology. Fungal glycans are composed of monosaccharides linked through glycosidic bonds. Their high molecular weight and structural complexity make them essential components of fungal biology (Yang et al., 2019). The glycans of fungi encompass a diverse array of structures, ranging from protein-linked compounds to extensive polymers integral to cell walls. These glycans also serve as crucial molecules for energy storage and contribute to the development of intricate surface structures. These surface structures play a vital role in aiding pathogens in evading the host immune response. The significance of these molecules extends beyond mere structural components; they actively participate in processes essential for the survival of the organism (Reilly and Doering, 2010). This area of study could have implications for understanding fungal pathogenesis, developing antifungal therapies, and exploring biotechnological applications of fungal glycans.

7.2.1 Types of Glycans in Fungi

Complex polysaccharides predominantly make up the extracellular matrix of fungi, which includes chitin, mannans, glucans, and galactans in particular. Mannose-rich glycans, historically called yeast gum, have been recognized since the 1890s (Routier et al., 2022).

7.2.1.1 β-(1,3)-Glucan

It is constituted by repeating glucose (Glc) residues linked through β-(1,3) glycosidic linkages (Figure 7.1A). The glucan synthase complex located in the cell membrane synthesizes β-(1,3)-glucan (Bernard and Latgé, 2001; Latgé, 2007) from nucleotide sugar, that is, UDP-Glc, to generate a growing linear chain of glucan polysaccharide (Gow et al., 2017). It is then ejected into periplasmic space, where it binds to the other cell wall constituents such as chitin through glucanosyltransferases or through hydrogen bonds (Bernard and Latgé, 2001; Gow et al., 2017). Approximately 50 % of the cell wall is constituted by β-(1,3)-glucan (Campoy and Adrio, 2017).

7.2.1.2 β-(1,6)-Glucan

β-(1,6)-Glucan interacts with and interconnects other components of cell wall such as β-(1,3)-glucan, chitin, and cell wall proteins. Though it is an important and integral part of the cell wall, its biosynthetic pathway remains obscure (Wang et al., 2018; Cortés et al., 2019). It is branched and varies among species (Figure 7.1B). It accounts for 10–15 % of the fungal cell wall, with each chain comprising 140 to 350 glucose residues (Lesage and Bussey, 2006).

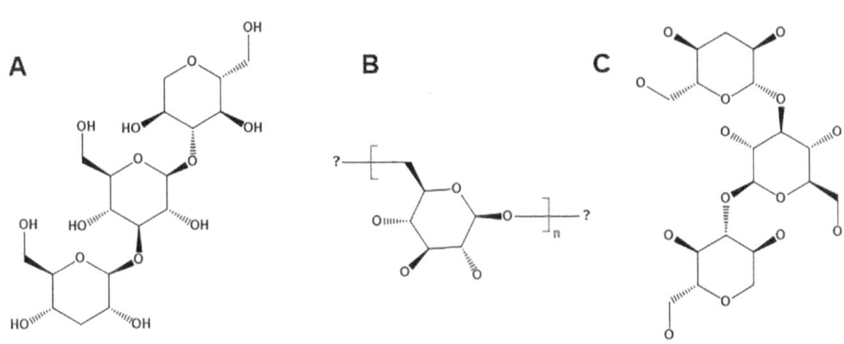

FIGURE 7.1 Structures of (A) β-(1,3)-glucan, (B) β-(1,6)-glucan, and (C) α-(1,3)-glucan.

7.2.1.3 α-(1,3)-Glucan

α-(1,3)-Glucan is present in minute quantities in most fungal cell walls (Figure 7.1C). It is completely absent in *Candida* spp., but a relatively high quantity can be found in *Cryptococcus* spp. (Gow et al., 2017). It is synthesized from UDP-Glc by transmembrane α-(1,3)-glucan synthase enzyme (Doering, 2009). Except for its role as a virulence factor in animal and plant pathogenic fungi, its biological functions remain debated (Yoshimi et al., 2017).

7.2.1.4 Chitin

Chitin is a linear polysaccharide (β-(1,4)-*N*-acetylglucosamine) (Figure 7.2A) (Gastebois et al., 2009) that is enzymatically synthesized by the transfer of a glycosyl of N-GlcNAc from the donor UDP-GlcNAc (Latgé et al., 2005). It confers cell wall strength and structure through covalent bonding with other constituents of the cell wall (Wang et al., 2018). In ascomycetes, basidiomycetes, and zygomycetes, a deacetylated form of chitin known as chitosan is found, which is a more soluble and more flexible polysaccharide (Figure 7.2B) (Latgé, 2007).

7.2.1.5 Galactomannan

Galactomannans (GMs) in fungi comprise β-(1,5)-/β-(1,6)-galactofuranosyl residues and α-(1,2)-/α-(1,6)-mannosyl (Oka, 2018). Two kinds of GMs are found in fungi, namely, *O*-mannose type galactomannan (OMGM) and fungal-type galactomannan (FTGM). In FTGM, β-(1,5)-galactofuranotetraose contains a β-(1,2)-, β-(1,3)- or β-(1,6)-group attached to the α-core-mannan chain. Nine to ten sets of α-(1,2)-mannotetraose units are linked by α-(1,6)-linkages in α-core-mannan chains. In OMGM, the non-reducing terminal side of the *O*-mannose-type glycan is linked

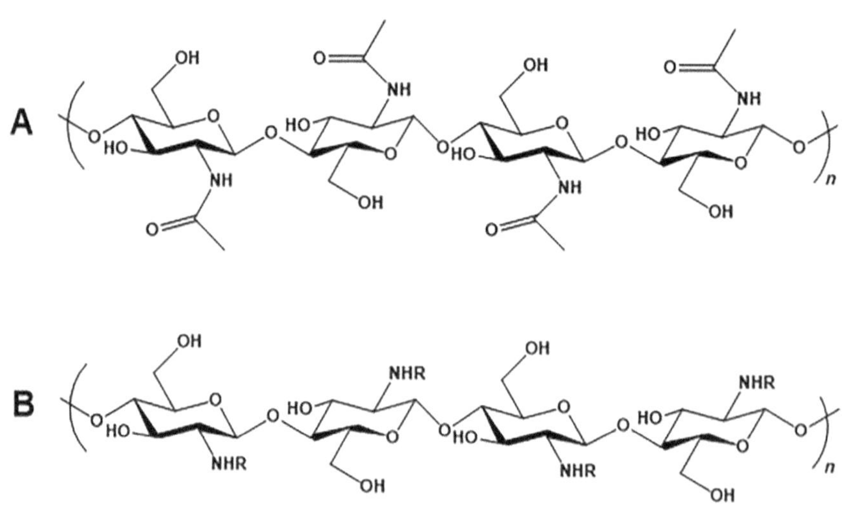

FIGURE 7.2 Chemical structures of (A) chitin and (B) chitosan.

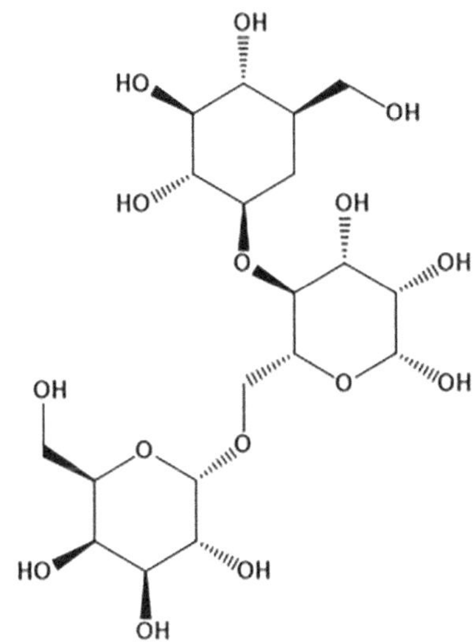

FIGURE 7.3 Chemical structure of galactomannan.

to β-(1,5)-galactofuranotetraose, while mannose (Man) is attached to the protein skeleton through a serine or threonine residue (Figure 7.3) (Latge et al., 1994; Kudoh et al., 2014). GMs are covalently linked to the cell wall and cell membrane through α-(1,3)-glucan and glycosylphosphatidylinositols (GPI) anchors, respectively (Tefsen et al., 2011; Engel et al., 2012).

7.2.1.6 Glycoproteins

Glycoproteins are also called "mannans" or "mannoproteins" (Masuoka, 2004). The glycosylation of fungal proteins results in three types of glycans, namely, *N*-linked glycans, *O*-linked glycans, and glycosylphosphatidylinositols (GPIs).

7.2.1.6.1 N-Linked Glycans

N-linked glycans are attached through as *N*-glycosidic covalent bond to asparagine (Asn) (Figure 7.4) (Imperiali and O'Connor, 1999). They are formed by dolichol-linked oligosaccharides, which are transformed into Dol-PP-GlcNAc with the help of glycosyl-phosphotransferase. After the addition of a second molecule of GlcNAc, mannosyltransferases further modify the structure by adding mannose residues derived from GDP-mannose. The resulting molecule, that is, Dol-PPGlcNAc2Man5, is translocated across the luminal leaflet of the ER. Four additional Man residues derived from Dol-P-Man are supplemented to the oligosaccharide core by mannosyltransferases. Three glucose residues derived from Dol-P-Glc cap Dol-PPGlcNAc2Man5, resulting in a Dol-PP-GlcNAc2Man9Glc3 molecule, which can be attached to a polypeptide. The glycan core, that is, GlcNAc2Man9Glc3, undergoes necessary structural modifications. In the case of *Saccharomyces cerevisiae*, after arriving in Golgi apparatus, chains of Man residues (α-(1,6)-Man) form the

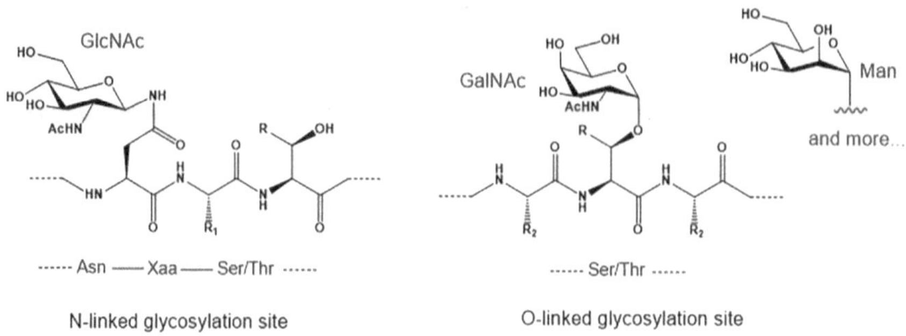

FIGURE 7.4 Chemical structures of *N*-linked and *O*-linked glycans.

backbone, and α-(1,2)- and α-(1,3)- make the side chains. This can take up to 200 Man residues (Cutler, 2001; Reilly and Doering, 2010; Hall and Gow, 2013).

7.2.1.6.2 O-Linked Glycans

In *O*-linked glycans, an *O*-linked mannose core is attached to serine (Ser)/threonine (Thr) residues through an *O*-glycosidic covalent bond (Figure 7.4). In contrast to *N*-linked mannans, *O*-linked mannans are synthesized by linear chains of α-linked Man residues and tend to be short; however, in *Aspergillus fumigatus* and other molds, long chains of core polysaccharides have been found (Gow et al., 2017; Latgé, 2007).

7.2.1.6.3 Glycosylphosphatidylinositols (GPIs)

These are complex glycolipids, which are covalently attached to the C-termini of proteins during posttranslational modifications. *N*-acetyl glucosamine, phosphatidylinositol, and mannosyl are the key components of GPIs (Oliveira-Garcia and Deising, 2016). GPIs help anchor cell wall proteins and have been extensively studied in *Saccharomyces cerevisiae* (Pittet and Conzelmann, 2007). GPI anchors play a central role in cell wall rigidity, as well as the functional reprogramming of the cell wall during infection processes (Oliveira-Garcia and Deising, 2016).

7.2.1.7 Galactosaminogalactan (GAG)

GAG is a heteropolysaccharide resulting from the combination of galactosamine (GalN), acetyl galactosamine (GalNAc), and α-1,4 linked galactose. It is secreted by growing hyphae and forms the extracellular matrix between hyphae and also envelops the organism. GAG is associated with infection processes, as its presence in *Aspergillus* species is essential for severe virulence in aspergillosis (Speth et al., 2019).

7.2.1.8 Sialic Acids

Sialic acids are derivatives of neuraminic acid and have a backbone of nine carbon atoms. They are present at the terminal residues of glycoconjugates. Among 50 reported sialic acids, *N*-acetylneuraminic acid (Neu5Ac) is the most abundant (Tiralongo et al., 2013). In fungi, sialic acids (5,9-*N,O*-acetylneuraminic acids, *N*-glycolyl-, *N*-acetyl-) have been reported in *Candida albicans*, *Paracoccidioides brasiliensis*, *Cryptococcus neoformans*, *Aspergillus fumigatus*, and *Fonsacaea pedrosoi* (Lee et al., 2019). They are found to be involved in cellular communication, survival of the pathogen, and infection (Cavalcante et al., 2021).

7.2.2 Constituents of Glycans

Glycans in both eukaryotes and prokaryotes originate during secretary pathways catalyzed by glycosyltransferases utilizing sugar donor molecules such as dolichol phosphate sugars and nucleotide-linked sugars (Parker and Newstead, 2017).

7.2.2.1 Nucleotide-Linked Sugars

As many as 70 nucleotide-linked sugars (Bar-Peled and O'Neill, 2011) have been identified, the majority of which are limited to prokaryotes. Nucleotide-linked sugars comprise a nucleoside and two phosphate groups attached to a hexose (UDP-glucose, UDP-mannose) or amino sugar (UDP-N-acetylglucosamine). However, in fungi, three nucleotide-linked sugars, namely, UDP-glucose (UDP-Glc), UDP-*N*-acetylglucosamine (UDP-GlcNAc), and GDP-mannose (GDP-Man), are principally involved in the biosynthesis of glycoproteins and polysaccharides (Figure 7.5A, 7.5B, and 7.5C) (Schwerdt et al., 2021). UDP-Glc is utilized by glucan synthases such as FKS, which is a building block of β-(1,3)-glucans. FKS is a primary target for numerous antifungal drugs (Hu et al., 2023). UDP-GlcNAc is a raw material for chitin (Moussian, 2008), and GDP-Man is essential for glycoprotein, galactomannan, and glycosylphosphatidylinositol (GPI) anchors (Jiang et al., 2008).

7.2.2.2 Dolichol Phosphate sugars

Mannose (Man) is the most abundant sugar in fungi (Cottrell et al., 2007) and actively participates in the composition of mannose-containing polysaccharides. Dolichol phosphate mannose (Dol-P-Man) is formed by the transfer of Man from cytosolic GDP-Man to membrane-associated dolichol monophosphate (Figure 7.5D), which is catalyzed by dolichol phosphate mannose synthase. An unidentified flippase translocates the active sugar donor to the leaflet of the lumen of the endoplasmic

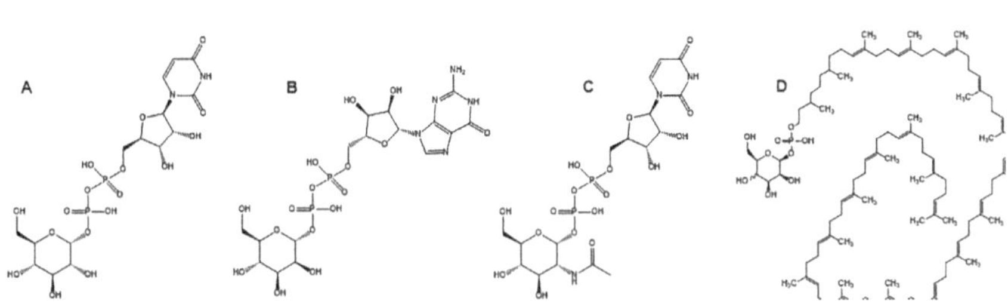

FIGURE 7.5 Chemical structures of (A) UDP-glucose, (B) GDP-mannose, (C) UDP-GlcNAc, and (D) Dol-P-Man.

Source: Retrieved from NCBI.

reticulum (ER) membrane, an appropriate place for glycan synthesis (Reilly and Doering, 2010).

7.2.3 Functional Role of Glycans

7.2.3.1 Cell Wall Development

Polysaccharides, which constitute 90 % of the fungal cell wall composition (Latgé, 2007), are responsible for its dynamic and complex nature. It protects fungal organisms from extraneous factors and allows for fungus–host interactions (Hopke et al., 2018). Its flexibility allows it to adapt to external triggers (Arana et al., 2009), and its rigidity helps to penetrate the host tissue. The cell wall can be conveniently divided into two sub-layers, that is, the inner layer and outer layer (Gow et al., 2017). The inner layer is usually conserved among different fungal species and comprises chitins and β-(1,3)-glucan, which are covalently bonded to other polysaccharides like β-(1,6)-glucan or galactomannan. However, the outer layer is heterogenous as well as variable among fungal species and is composed of α-(1,3)-glucan, mannan, melanin, galactomannan, galactosaminogalactan, and glycosylated proteins (Figure 7.6) (Latgé et al., 2005; Arana et al., 2009; Wang et al., 2018; Hopke et al., 2018). A third, loosely attached, gel-like layer of glycans known as the "extracellular polysaccharide matrix" (EPS) is present on the outer layer.

7.2.3.1.1 Inner Cell Wall Layer

The inner layer is composed of densely packed hydrophobic chitin, resulting in a rigid skeletal layer next to the cell membrane. Microfibrils of chitin consist of linear chains of β-1,4-linked GlcNAc interconnected through hydrogen bonding. The chitin layer of penetrating structures such as appressoria is impregnated with melanin and enhances resilience and strength. β-glucan chains are linked to the chitin microfibrils via covalent linkages (Ludwig et al., 2014; Geoghegan et al., 2017).

7.2.3.1.2 Outer Cell Wall Layer

β-glucan generally makes up the prime component (~65–90 % of polysaccharide content) of this layer, which is linked through β-1,3- linkages. After every 2–5 glucose units, β-1,6-side branches occur (Zeković et al., 2005; Bowman and Free, 2006).

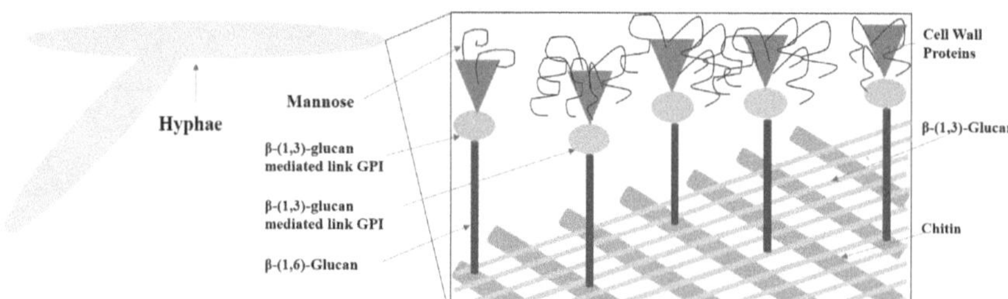

FIGURE 7.6 Schematic diagram of a fungal cell wall.

Mixed-linkage β-1,3/1,4-glucose polymers and β-1,4-linked moieties have been found in some fungi (Fontaine et al., 2000; Pettolino et al., 2009; Kang et al., 2018). β-glucans can make coils or helices and spread throughout the cell wall, forming a non-crystalline branched network. Mixed-linkage mannose polymers or interlinked α-1,3-glucans are present on the top of the β-glucan layer and are attached to the mannoproteins (Geoghegan et al., 2017; Gow et al., 2017; Mélida et al., 2018).

7.2.3.1.3 Extracellular Polysaccharide Matrix (EPS)

Various plant and human pathogenic fungi have a third, loosely attached, gel-like layer encompassing the hyphae, which is called the "extracellular polysaccharide matrix" (EPS) (Gow et al., 2017; Wawra et al., 2019). Its ascribed role is to act as a frame for macromolecules such as lipids, proteins, and extracellular DNA, and it can be easily washed away (Martins et al., 2010; Lee and Sheppard, 2016; Wawra et al., 2019). It is composed of glucose, galactose, xylose, rhamnose, mannose, and fucose (Mahapatra and Banerjee, 2013). Moreover mannose and fucose chains linked with a linear α-1,6-galactose backbone have been found (Scarpari et al., 2017). EPS matrices of plant-colonizing fungi such as *Botrytis cinerea* and *Serendipita indica* contain high-molecular-weight β-1,3/1,6-glucans, whereas EPS matrices of human fungal pathogens such as *Aspergillus fumigatus* and *Candida albicans* comprise galactans and mannans in addition to galactosaminogalactan (GAG) (El Oirdi et al., 2011; Gravelat et al., 2013; Gow et al., 2017; Wawra et al., 2019). *Cryptococcus neoformans*, a human pathogen, is surrounded by a highly organized EPS matrix in the form of capsule comprising glucuronoxylomanogalactan (GalXM) and glucuronoxylomannan (GXM) (Denham et al., 2018; Decote-Ricardo et al., 2019).

7.2.3.2 Adhesion to the Host

Adhesion is the most important phase of fungal life cycle and is consequential for pathogenesis (Lipke, 2018). Most fungal adhesins are mannoproteins, and adhesins of human fungal pathogens are well characterized as compared to plant fungal pathogens (Epstein and Nicholson, 2016). Cell wall proteins of *Candida* spp. comprise 80 to 90 % mannose residues linked by glycosylphosphatidylinositol (GPI) anchoring, *N*-glycosylation, and *O*-glycosylation. The majority of proteins have been characterized as GPI proteins found on the outer cell wall, which govern host–pathogen interactions (de Groot et al., 2013).

7.2.3.3 Fungus–Plant Interactions

Complex carbohydrates such as mannoproteins, phospholipomannan, beta-glucan, and chitin give rise to PAMPs (pathogen-associated molecular patterns) of fungi, which bind to pattern recognition receptors (PRRs) of host organisms (Sorrell and Chen, 2009). Carbohydrates on the surface of fungi play vital role in the recognition of fungus–plant interactions. Particular patterns of carbohydrates at the infection site will determine the success or failure of infection (Callow, 1987). The cells of fungi and bacteria are enclosed in the envelope of glycans. These glycans play roles in protection from the surrounding environment (Reid et al., 2010). The glycans of fungi are of particular interest, as the cell walls of these pathogens are recognized by the immune system of the host and initiate the immune response. To avoid recognition, fungi and bacteria are capable of modifying the glycan chains in such a way that host becomes unable to recognize the pathogen (Sukhithasri et al., 2013). Moreover, the plant cells are also surrounded by a cell wall made up of carbohydrates, but most of the diverse plant carbohydrates are parts of small biologically active molecules produced by plant pathogens (Augustin et al., 2011). The cell wall modification of fungi occurs to initiate host recognition.

Studies have revealed that the exchange of carbon nutrients during the pathogenesis of *Sclerotinia sclerotiorum* and *B. cinerea* showed the sink capacity of carbohydrates of the fungi linked to the multigenic hexose transport system and invertase expression during infection. Studies showed that the reserve sugars and polyol pool revealed that glycerol and mannitol were produced *in planta* by the pathogen (Jobic et al., 2007). During the interaction of a necrotrophic pathogen and s plant, carbon nutrition and metabolism have been studied. Fungal invertase was detected in grapes in great concentrations after infection with *B. cinerea*. It has also been revealed that in *B. cinerea*, glucokinase is needed for the growth of the fungus and the establishment of infection in the presence of hexoses (Rui and Hahn, 2007). The vital role of fructose as an inducer of germination and the role of trehalose in the survival of conidia are very important (Doehlemann et al., 2006).

Fungal microbes and plants use glycoconjugates as structural energy reserve entities for carrying out cellular processes, and these also act as components for cellular recognition between the pathogen and plant. The heterogeneity of carbohydrates is the outcome of the carbohydrate biosynthesis enzyme for the reorientation of sugar monomers in a variety of forms. During the interaction of a pathogen with the plant, the surface glycans, cell wall-derived glycans, and glycoproteins of the pathogen initiate the plant immune signaling cascade response through a number of events. The glycans of the fungal pathogen initiate the plant immune response and also lead to the modification of host plant glycan structures (Chaliha et al., 2018).

The study investigated glycan-mediated immunity in plant, focusing on fungal cell surface glycans (structural components of microbial cells) and their role as regulators of plant defense signaling during host–pathogen interactions (Trouvelot et al., 2014). Among important sugar signals modulating defense responses through complex sugar-sensing pathways, trehalose and trehalose-6 phosphate are considered of great importance. During the infection of wheat with *Blumeria graminis*, which causes powdery mildew disease, trehalose led to the activation of the defense genes phenylalanine ammonia lyase and peroxidase (Muchembled et al., 2006). Trehalose-6 phosphate from plant cells is supposed to act as molecular switch that regulates the activity of the sucrose non-fermenting-related protein negatively. Sucrose non-fermenting-related protein kinase is a regulator that is capable of controlling metabolism during biotic stress (Hulsmans et al., 2016). Sensitive cultivars of rice to *Magnaporthe grisea* were found to have fewer metabolites involved in the metabolism of sugar as compared to those varieties that are resistant (Jones et al., 2011).

7.3 FUNGAL LIPIDOMICS

Lipidomics is a nascent field that primarily involves the lipid profiles of biological systems, including reaction pathways in lipid metabolism (Sousa et al., 2021).

Cellular lipids are highly complex and play essential roles as cellular barriers, signaling molecules, energy reserves, and membrane matrices. Lipoproteins are crucial to retain cell membrane integrity (Yetukuri et al., 2007; Yang et al., 2009; Yang and Han, 2016). Fungal lipidomics can be ascribed as the study of lipid profiles in fungi.

7.3.1 Types of Lipids in Fungi and Their Roles

Fungi contain various types of lipids such as fatty acids, phospholipids, sterols, glycolipids, oxylipins, and sphingolipids. These lipids play a crucial role in membrane systems, cell walls, energy storage, and signaling systems (Beccaccioli et al., 2019).

7.3.1.1 Fatty Acids

Fatty acids (FAs) constitute phospholipids; they are the building blocks of cellular membranes and act as energy reserves such as triacylglycerols and steryl esters. FA synthesis initiates in the cytosol with the development of acyl-CoA, having an acyl chain length of 16 to 18 atoms. The elongation of the acyl chain takes place in the endoplasmic reticulum (ER). Acetyl-CoA carboxylase and the cytosolic multi-enzyme fatty acid synthase complex catalyze the formation of acyl-CoA (Beccaccioli et al., 2019).

7.3.1.2 Oxylipins

The term oxylipins means oxidized fatty acids (Funk, 2001). Oxylipins are formed by the peroxidation of polyunsaturated fatty acids (PUFAs), catalyzed by lipoxygenases (LOXs) via LOX pathways (Andreou and Feussner, 2009). These hydroperoxy fatty acids are further modified by certain enzymes (divinyl ether synthase (DES), allene oxide synthase (AOS), hydroperoxide lyase (HPL), and peroxygenase), thus generating structurally different products (Brodhun and Feussner, 2011). There is scarce knowledge about the physiological role and biosynthetic pathways of oxylipins in fungi. Oxylipins act as modulators in cell-to-cell communication and signaling pathways during host–pathogen interactions. Linolenic, linoleic, and oleic acid-mediated oxylipins play an important role in plant–pathogen interactions (Beccaccioli et al., 2022).

7.3.1.3 Sphingolipids

Sphingolipids can be found in cell walls and the extracellular environment. Their role can be traced to fungal virulence, cell polarity, growth, and hyphal development (Martin and Konopka, 2004; Takeshita et al., 2008; Farnoud et al., 2015). Moreover, they also act as signaling molecules (Singh and Del Poeta, 2010). Sphingolipids comprise a long-chain base amide, also known as a sphingoid base, attached to a fatty acid, forming a ceramide backbone. This ceramide backbone is synthesized in the endoplasmic reticulum (ER), from where these ceramides are transported to the Golgi complex, and complex sphingolipids are formed (Dickson and Lester, 2002; Ren and Hannun, 2019).

7.3.1.4 Phospholipids

The fungal cell membrane generally contains several key phospholipids, including phosphatidylcholines (PCs), phosphatidylethanolamines (PEs), phosphatidylserine

(PS), and phosphatidylinositol (PI) (de Kroon et al., 2013). Both PE and PC make up 80 % of phospholipids, and the remaining 20 % are anionic phospholipids (Löffler, 2000). The synthesis of PE and PC has been thoroughly characterized in *Saccharomyces cerevisiae*, which includes two pathways, that is, the endogenous pathway (de novo pathway) and the exogenous pathway (the Kennedy pathway). PC and PE are essential for fungal vegetative growth; additionally, their role has been traced to environmental adaptability, signal transduction, and energy reserves (Wang et al., 2019).

7.3.1.5 Sterols

Sterols are necessary for cell viability and maintaining cell membrane fluidity and permeability, as well as regulating membrane-bound enzymes. Fungal membranes are composed of ergosterols, which are essentially similar to cholesterol, and their biosynthetic pathways are strikingly similar (Joffrion and Cushion, 2010).

REFERENCES

Andreou, A., & Feussner, I. (2009). Lipoxygenases–structure and reaction mechanism. *Phytochemistry*, 70, 1504–1510.

Aoki-Kinoshita, K. F. (2008). An introduction to bioinformatics for glycomics research. *PLoS Computational Biology*, 4(5), e1000075. https://doi.org/10.1371/journal.pcbi.1000075

Arana, D. M., Prieto, D., Román, E., Nombela, C., Alonso-Monge, R., & Pla, J. (2009). The role of the cell wall in fungal pathogenesis. *Microbial Biotechnology*, 2(3), 308–320. https://doi.org/10.1111/j.1751–7915.2008.00070.x

Augustin, J. M., Kuzina, V., Andersen, S. B., & Bak, S. (2011). Molecular activities, biosynthesis and evolution of triterpenoid saponins. *Phytochemistry*, 72(6), 435–457.

Ball, B., Langille, M., & Geddes-McAlister, J. (2020). Fun (gi) omics: Advanced and diverse technologies to explore emerging fungal pathogens and define mechanisms of antifungal resistance. *MBio*, 11(5), 10–1128.

Bar-Peled, M., & O'Neill, M. A. (2011). Plant nucleotide sugar formation, interconversion, and salvage by sugar recycling. *Annual Review of Plant Biology*, 62(1), 127–155. https://doi.org/10.1146/annurev-arplant-042110–103918

Beccaccioli, M., Pucci, N., Salustri, M., Scortichini, M., Zaccaria, M., Momeni, B., Loreti, S., Reverberi, M., & Scala, V. (2022). Fungal and bacterial oxylipins are signals for intra— and inter-cellular communication within plant disease. *Frontiers in Plant Science*, 13. https://doi.org/10.3389/fpls.2022.823233

Beccaccioli, M., Reverberi, M., & Scala, V. (2019). Fungal lipids: Biosynthesis and signalling during plant-pathogen interaction. *Frontiers in Bioscience (Landmark Edition)*, 24(1), 172–185. https://doi.org/10.2741/4712

Bernard, M., & Latgé, J.-P. (2001). Aspergillus fumigatuscell wall: Composition and biosynthesis. *Medical Mycology*, 39(1), 9–17. https://doi.org/10.1080/mmy.39.1.9.17

Bohm, M., Bohne-Lang, A., Frank, M., Loss, A., Rojas-Macias, M. A., & Lütteke, T. (2019). Glycosciences. DB: An annotated data collection linking glycomics and proteomics data (2018 update). *Nucleic Acids Research*, 47(D1), D1195–D1201.

Bowman, S. M., & Free, S. J. (2006). The structure and synthesis of the fungal cell wall. *Bioessays*, 28, 799–808.

Brodhun, F., & Feussner, I. (2011). Oxylipins in fungi. *FEBS Journal*, 278(7), 1047–1063. https://doi.org/10.1111/j.1742–4658.2011.08027.x

Callow, J. A. (1987). Models for host-pathogen interactions. In: *Genetics and Planl Pathogenesis* (Ed. by P. R. Day & G. J. Jellis), pp. 283–295. Blackwell Scientific Publications, Oxford, UK.

Campoy, S., & Adrio, J. L. (2017). Antifungals. *Biochemical Pharmacology*, 133, 86–96. https://doi.org/10.1016/j.bcp.2016.11.019

Cavalcante, T., Medeiros, M. M., Mule, S. N., Palmisano, G., & Stolf, B. S. (2021). The role of sialic acids in the establishment of infections by pathogens, with special focus on leishmania. *Frontiers in Cellular and Infection Microbiology*, 11. https://doi.org/10.3389/fcimb.2021.671913

Chaliha, C., Rugen, M. D., Field, R. A., & Kalita, E. (2018). Glycans as modulators of plant defense against filamentous pathogens. *Frontiers in Plant Science*, 928.

Chen, X., Gao, C., Guo, L., Hu, G., Luo, Q., Liu, J., . . . & Liu, L. (2018). DCEO biotechnology: Tools to design, construct, evaluate, and optimize the metabolic pathway for biosynthesis of chemicals. *Chemical Reviews*, 118(1), 4–72.

Cortés, J. C. G., Curto, M.-Á., Carvalho, V. S. D., Pérez, P., & Ribas, J. C. (2019). The fungal cell wall as a target for the development of new antifungal therapies. *Biotechnology Advances*, 37(6), 107352. https://doi.org/10.1016/j.biotechadv.2019.02.008

Cottrell, T. R., Griffith, C. L., Liu, H., Nenninger, A. A., & Doering, T. L. (2007). The pathogenic fungus cryptococcus neoformans expresses two functional GDP-mannose transporters with distinct expression patterns and roles in capsule synthesis. *Eukaryotic Cell*, 6(5), 776–785. https://doi.org/10.1128/ec.00015–07

Cutler, J. E. (2001). N-Glycosylation of yeast, with emphasis on Candida albicans. *Medical Mycology*, 39(1), 75–86. https://doi.org/10.1080/mmy.39.1.75.86

de Groot, P. W. J., Bader, O., de Boer, A. D., Weig, M., & Chauhan, N. (2013). Adhesins in human fungal pathogens: Glue with plenty of stick. *Eukaryotic Cell*, 12(4), 470–481. https://doi.org/10.1128/ec.00364–12

de Kroon, A. I. P. M., Rijken, P. J., & De Smet, C. H. (2013). Checks and balances in membrane phospholipid class and acyl chain homeostasis, the yeast perspective. *Progress in Lipid Research*, 52(4), 374–394. https://doi.org/10.1016/j.plipres.2013.04.006

Decote-Ricardo, D., LaRocque-de-Freitas, I. F., Rocha, J. D. B., Nascimento, D. O., Nunes, M. P., Morrot, A., Freire-de-Lima, L., Previato, J. O., Mendonça-Previato, L., & Freire-de-Lima, C. G. (2019). Immunomodulatory role of capsular polysaccharides constituents of Cryptococcus neoformans. *Frontiers in Medicine*, 6, 129.

Denham, S. T., Verma, S., Reynolds, R. C., Worne, C. L., Daugherty, J. M., Lane, T. E., & Brown, J. C. S. (2018). Regulated release of cryptococcal polysaccharide drives virulence and suppresses immune cell infiltration into the central nervous system. *Infection and Immunity*, 86.

Dickson, R. C., & Lester, R. L. (2002). Sphingolipid functions in Saccharomyces cerevisiae. *Biochimica et Biophysica Acta (BBA)—Molecular and Cell Biology of Lipids*, 1583(1), 13–25. https://doi.org/10.1016/s1388–1981(02)00210-x

Doehlemann, G., Berndt, P., & Hahn, M. (2006). Trehalose metabolism is important for heat stress tolerance and spore germination of Botrytis cinerea. *Microbiology*, 152(9), 2625–2634.

Doering, T. L. (2009). How sweet it is! Cell wall biogenesis and polysaccharide capsule formation incryptococcus neoformans. *Annual Review of Microbiology*, 63(1), 223–247. https://doi.org/10.1146/annurev.micro.62.081307.162753

El Oirdi, M., El Rahman, T. A., Rigano, L., El Hadrami, A., Rodriguez, M. C., Daayf, F., Vojnov, A., & Bouarab, K. (2011). Botrytis cinerea manipulates the antagonistic effects between immune pathways to promote disease development in tomato. *The Plant Cell*, 23, 2405–2421.

Engel, J., Schmalhorst, P. S., & Routier, F. H. (2012). Biosynthesis of the fungal cell wall polysaccharide galactomannan requires intraluminal GDP-mannose. *The Journal of Biological Chemistry*, 287(53), 44418–44424. https://doi.org/10.1074/jbc.M112.398321

Epstein, L., & Nicholson, R. (2016). Adhesion and adhesives of fungi and oomycetes. *Biological Adhesives*, 25–55. https://doi.org/10.1007/978-3-319-46082-6_2

Fahy, E., Subramaniam, S., Brown, H. A., Glass, C. K., Merrill, A. H., Murphy, R. C., Raetz, C. R., Russell, D. W., Seyama, Y., Shaw, W., & Shimizu, T. (2005). A comprehensive classification system for lipids1. *Journal of Lipid Research*, 46(5), 839–861.

Farnoud, A. M., Toledo, A., Konopka, J. B., Poeta, M. D., & London, E. (2015). Raft-like membrane domains in pathogenic microorganisms. *Current Topics in Membranes*, 233–268. https://doi.org/10.1016/bs.ctm.2015.03.005

Fontaine, T., Simenel, C., Dubreucq, G., Adam, O., Delepierre, M., Lemoine, J., Vorgias, C. E., Diaquin, M., & Latge, J. P. (2000). Molecular organization of the alkali-insoluble fraction of Aspergillus fumigatus cell wall. *Journal of Biological Chemistry*, 275, 41528.

Funk, C. D. (2001). Prostaglandins and leukotrienes: Advances in eicosanoid biology. *Science*, 294, 1871–1875.

Gastebois, A., Clavaud, C., Aimanianda, V., & Latgé, J.-P. (2009). Aspergillus fumigatus: Cell wall polysaccharides, their biosynthesis and organization. *Future Microbiology*, 4(5), 583–595. https://doi.org/10.2217/fmb.09.29

Geoghegan, I., Steinberg, G., & Gurr, S. (2017). The role of the fungal cell wall in the infection of plants. *Trends in Microbiology*, 25, 957–967.

Gooday, G. W. (1995). Cell walls. In: *The growing fungus*, pp. 43–62. Springer Netherlands, Dordrecht.

Gow, N. A. R., Latge, J. P., & Munro, C. A. (2017). The fungal cell wall: Structure, biosynthesis, and function. *Microbiology Spectrum*, 5. https://doi.org/10.1128/microbiolspec.FUNK-0035–2016

Gravelat, F. N., Beauvais, A., Liu, H., et al. (2013). Aspergillus galactosaminogalactan mediates adherence to host constituents and conceals hyphal β-glucan from the immune system. *PLoS Pathogens*, 9, e1003575.

Hall, R. A., & Gow, N. A. R. (2013). Mannosylation in Candida albicans: Role in cell wall function and immune recognition. *Molecular Microbiology*, 90(6), 1147–1161. https://doi.org/10.1111/mmi.12426

He, J., Ford, H. C., Carroll, J., Douglas, C., Gonzales, E., Ding, S., . . . & Walker, J. E. (2018). Assembly of the membrane domain of ATP synthase in human mitochondria. *Proceedings of the National Academy of Sciences*, 115(12), 2988–2993. https://doi.org/10.1073/pnas.1722086115

Hopke, A., Brown, A. J. P., Hall, R. A., & Wheeler, R. T. (2018). Dynamic fungal cell wall architecture in stress adaptation and immune evasion. *Trends in Microbiology*, 26(4), 284–295. https://doi.org/10.1016/j.tim.2018.01.007

Hu, X., Yang, P., Chai, C., Liu, J., Sun, H., Wu, Y., . . . & Yu, H. (2023). Structural and mechanistic insights into fungal β-1,3-glucan synthase FKS1. *Nature*, 616(7955), 190–198. https://doi.org/10.1038/s41586-023-05856-5

Hulsmans, S., Rodriguez, M., De Coninck, B., & Rolland, F., 2016. The SnRK1 energy sensor in plant biotic interactions. *Trends in Plant Science*, 21(8), 648–661.

Imperiali, B., & O'Connor, S. E. (1999). Effect of N-linked glycosylation on glycopeptide and glycoprotein structure. *Current Opinion in Chemical Biology*, 3(6), 643–649. https://doi.org/10.1016/s1367–5931(99)00021–6

Latgé J., Mouyna, I., Fredj Tekaia, Beauvais, A., Debeaupuis, J. P., & Nierman, W. C. (2005). Specific molecular features in the organization and biosynthesis of the cell wall of Aspergillus fumigatus. *Medical Mycology*, 43(Suppl. 1), S15–S22. https://doi.org/10.1080/13693780400029155

Jiang, H., Ouyang, H., Zhou, H., & Jin, C. (2008). GDP-mannose pyrophosphorylase is essential for cell wall integrity, morphogenesis and viability of Aspergillus fumigatus. *Microbiology*, 154(9), 2730–2739. https://doi.org/10.1099/mic.0.2008/019240–0

Jobic, C., Boisson, A. M., Gout, E., Rascle, C., Fèvre, M., Cotton, P., & Bligny, R. (2007). Metabolic processes and carbon nutrient exchanges between host and pathogen sustain the disease development during sunflower infection by Sclerotinia sclerotiorum. *Planta*, 226, 251–265.

Joffrion, T. M., & Cushion, M. T. (2010). Sterol biosynthesis and sterol uptake in the fungal pathogen Pneumocystis carinii. *FEMS Microbiology Letters*, 311(1), 1–9. https://doi.org/10.1111/j.1574–6968.2010.02007.x

Jones, O. A., Maguire, M. L., Griffin, J. L., Jung, Y. H., Shibato, J., Rakwal, R., Agrawal, G. K., & Jwa, N. S. (2011). Using metabolic profiling to assess plant-pathogen interactions: An example using rice (Oryza sativa) and the blast pathogen Magnaporthe grisea. *European Journal of Plant Pathology*, 129, 539–554.

Kang, X., Kirui, A., Muszyński, A., Widanage, M. C. D., Chen, A., Azadi, P., Wang, P., Mentink-Vigier, F., & Wang, T. (2018). Molecular architecture of fungal cell walls revealed by solid-state NMR. *Nature Communications*, 9, 2747.

Kudoh, A., Okawa, Y., & Shibata, N. (2014). Significant structural change in both O—and N-linked carbohydrate moieties of the antigenic galactomannan from Aspergillus fumigatus grown under different culture conditions. *Glycobiology*, 25(1), 74–87. https://doi.org/10.1093/glycob/cwu091

Latgé, J. (2007). The cell wall: A carbohydrate armour for the fungal cell. *Molecular Microbiology*, 66(2), 279–290. https://doi.org/10.1111/j.1365–2958.2007.05872.x

Latge, J. P., Kobayashi, H., Debeaupuis, J. P., Diaquin, M., Sarfati, J., Wieruszeski, J. M., . . . & Fournet, B. (1994). Chemical and immunological characterization of the extracellular galactomannan of Aspergillus fumigatus. *Infection and Immunity*, 62(12), 5424–5433.

Lay, Jr., J. O., Liyanage, R., Borgmann, S., & Wilkins, C. L. (2006). Problems with the "omics". *TrAC Trends in Analytical Chemistry*, 25(11), 1046–1056.

Lee, D. J., O'Donnell, H., Routier, F. H., Tiralongo, J., & Haselhorst, T. (2019). Glycobiology of human fungal pathogens: New avenues for drug development. *Cells*, 8(11), 1348. https://doi.org/10.3390/cells8111348

Lee, M. J., & Sheppard, D. C. (2016). Recent advances in the understanding of the Aspergillus fumigatus cell wall. *Journal of Microbiology*, 54, 232–242.

Lesage, G., & Bussey, H. (2006). Cell wall assembly in Saccharomyces cerevisiae. *Microbiology and Molecular Biology Reviews*, 70(2), 317–343. https://doi.org/10.1128/mmbr.00038–05

Lipke, P. (2018). What we do not know about fungal cell adhesion molecules. *Journal of Fungi*, 4(2), 59. https://doi.org/10.3390/jof4020059

Löffler, J. (2000). Phospholipid and sterol analysis of plasma membranes of azole-resistant Candida albicans strains. *FEMS Microbiology Letters*, 185(1), 59–63. https://doi.org/10.1016/s0378–1097(00)00071–9

Ludwig, N., Löhrer, M., Hempel, M., Mathea, S., Schliebner, I., Menzel, M., Kiesow, A., Schaffrath, U., Deising, H. B., & Horbach, R. (2014). Melanin is not required for turgor generation but enhances cell-wall rigidity in appressoria of the corn pathogen Colletotrichum graminicola. *Molecular Plant-Microbe Interactions*, 27, 315–327.

Mahapatra, S., & Banerjee, D. (2013). Fungal exopolysaccharide: Production, composition and applications. *Microbiology Insights*, 6, 1–16.

Manna, M., Nieminen, T., & Vattulainen, I. (2019). Understanding the role of lipids in signaling through atomistic and multiscale simulations of cell membranes. *Annual Review of Biophysics*, 48, 421–439.

Martin, S. W., & Konopka, J. B. (2004). Lipid raft polarization contributes to hyphal growth in candida albicans. *Eukaryotic Cell*, 3(3), 675–684. https://doi.org/10.1128/ec.3.3.675–684.2004

Martins, M., Uppuluri, P., Thomas, D. P., Cleary, I. A., Henriques, M., LopezRibot, J. L., & Oliveira, R. (2010). Presence of extracellular DNA in the Candida albicans biofilm matrix and its contribution to biofilms. *Mycopathologia*, 169, 323–331.

Masuoka, J. (2004). Surface glycans of candida albicans and other pathogenic fungi: Physiological roles, clinical uses, and experimental challenges. *Clinical Microbiology Reviews*, 17(2), 281–310. https://doi.org/10.1128/cmr.17.2.281–310.2004

Mehta, S., Chakraborty, A., Roy, A., Singh, I. K., & Singh, A. (2021). Fight hard or die trying: Current status of lipid signaling during plant–pathogen interaction. *Plants*, 10(6), 1098.

Mélida, H., Sopeña-Torres, S., Bacete, L., Garrido-Arandia, M., Jordá, L., López, G., Muñoz-Barrios, A., Pacios, L. F., & Molina, A. (2018). Non-branched β-1,3-glucan oligosaccharides trigger immune responses in Arabidopsis. *The Plant Journal*, 93, 34–49.

Moussian, B. (2008). The role of GlcNAc in formation and function of extracellular matrices. *Comparative Biochemistry and Physiology Part B: Biochemistry and Molecular Biology*, 149(2), 215–226. https://doi.org/10.1016/j.cbpb.2007.10.009

Muchembled, J., Sahraoui, A. L. H., Grandmougin-Ferjani, A., & Sancholle, M. (2006). Changes in lipid composition of Blumeria graminis f. sp. tritici conidia produced on wheat leaves treated with heptanoyl salicylic acid. *Phytochemistry*, 67(11), 1104–1109.

National Center for Biotechnology Information (2025). PubChem Compound Summary for CID 8629, Uridine Diphosphate Glucose. Retrieved April 13, 2025 from https://pubchem.ncbi.nlm.nih.gov/compound/Uridine-Diphosphate-Glucose

National Center for Biotechnology Information (2025). PubChem Compound Summary for CID 135398627, Guanosine diphosphate mannose. Retrieved April 13, 2025 from https://pubchem.ncbi.nlm.nih.gov/compound/Guanosine-diphosphate-mannose

National Center for Biotechnology Information (2025). PubChem Compound Summary for CID 445675, Uridine-Diphosphate-N-Acetylglucosamine. Retrieved April 13, 2025 from https://pubchem.ncbi.nlm.nih.gov/compound/UDP-N-acetylglucosamine

National Center for Biotechnology Information (2025). PubChem Substance Record for SID 56464075, dolichyl beta-D-mannosyl phosphate, Source: ChEBI. Retrieved April 13, 2025 from https://pubchem.ncbi.nlm.nih.gov/substance/56464075

Oka, T. (2018). Biosynthesis of galactomannans found in filamentous fungi belonging to Pezizomycotina. *Bioscience, Biotechnology, and Biochemistry*, 82(2), 183–191. https://doi.org/10.1080/09168451.2017.1422383

Oliveira-Garcia, E., & Deising, H. B. (2016). The glycosylphosphatidylinositol anchor biosynthesis genes GPI12, GAA1, and GPI8 are essential for cell-wall integrity and pathogenicity of the maize anthracnose fungus colletotrichum graminicola. *Molecular Plant-Microbe Interactions*, 29(11), 889–901. https://doi.org/10.1094/mpmi-09-16-0175-r

Parker, J. L., & Newstead, S. (2017). Structural basis of nucleotide sugar transport across the Golgi membrane. *Nature*, 551(7681), 521–524. https://doi.org/10.1038/nature24464

Pettolino, F., Sasaki, I., Turbic, A., Wilson, S. M., Bacic, A., Hrmova, M., & Fincher, G. B. (2009). Hyphal cell walls from the plant pathogen Rhynchosporium secalis contain (1,3/1,6)-beta-d-glucans, galacto—and rhamnomannans, (1,3;1,4)-beta-d-glucans and chitin. *The FEBS Journal*, 276, 3698–3709.

Pittet, M., & Conzelmann, A. (2007). Biosynthesis and function of GPI proteins in the yeast Saccharomyces cerevisiae. *Biochimica et Biophysica Acta (BBA)—Molecular and Cell Biology of Lipids*, 1771(3), 405–420. https://doi.org/10.1016/j.bbalip.2006.05.015

Reece, A. S., Wang, W., & Hulse, G. K. (2018). Pathways from epigenomics and glycobiology towards novel biomarkers of addiction and its radical cure. *Medical Hypotheses*, 116, 10–21.

Reid, C. W., Fulton, K. M., & Twine, S. M. (2010). Never take candy from a stranger: The role of the bacterial glycome in host–pathogen interactions. *Future Microbiology*, 5(2), 267–288.

Reilly, M. C., & Doering, T. L. (2010). Biosynthesis of fungal and yeast glycans. In *Microbial Glycobiology*, pp. 393–412. Academic Press.

Ren, J., & Hannun, Y. A. (2019). Metabolism and roles of sphingolipids in yeast Saccharomyces cerevisiae. Biogenesis of fatty acids, lipids and membranes. In *Handbook of Hydrocarbon and Lipid Microbiology*, pp. 341–361. Springer International Publishing, New York City, NY.

Routier, F. H., Doering, T. L., Cummings, R. D., et al. (2022). Fungi. In *Essentials of Glycobiology*, 4th edition (Ed. by A.Varki, R. D. Cummings, J. D. Esko, et al.), Chapter 23. Cold Spring Harbor Laboratory Press, Cold Spring Harbor, NY. https://www.ncbi.nlm.nih.gov/books/NBK579942/https://doi.org/10.1101/glycobiology.4e.23

Rui, O., & Hahn, M. (2007). The Botrytis cinerea hexokinase, Hxk1, but not the glucokinase, Glk1, is required for normal growth and sugar metabolism, and for pathogenicity on fruits. *Microbiology*, 153(8), 2791–2802.

Scarpari, M., Reverberi, M., Parroni, A., et al. (2017). Tramesan, a novel polysaccharide from Trametes versicolor. Structural characterization and biological effects. *PLoS One*, 12, e0171412.

Schwerdt, J. G., Qiu, H., Shirley, N. J., Little, A., & Bulone, V. (2021). Phylogenomic analyses of nucleotide-sugar biosynthetic and interconverting enzymes illuminate cell wall composition in fungi. *American Society for Microbiology*, 12(2). https://doi.org/10.1128/mbio.03540–20

Singh, A., & Del Poeta, M. (2010). Lipid signalling in pathogenic fungi. *Cellular Microbiology*, 13(2), 177–185. https://doi.org/10.1111/j.1462–5822.2010.01550.x

Sorrell, T. C., & Sharon C.-A. Chen. (2009). Fungal-derived immune modulating molecules. *Advances in Experimental Medicine and Biology*, 108–120. https://doi.org/10.1007/978-1-4419-1601-3_9

Sousa, B. C., Wakelam, M. J. O., & Lopez-Clavijo, A. F. (2021). Methods of lipid analysis. *Biochemistry of Lipids, Lipoproteins and Membranes*, 53–83. https://doi.org/10.1016/b978-0-12-824048-9.00010-9

Speth, C., Rambach, G., Lass-Flörl, C., Howell, P. L., & Sheppard, D. C. (2019). Galactosaminogalactan (GAG) and its multiple roles in Aspergillus pathogenesis. *Virulence*, 10(1), 976–983. https://doi.org/10.1080/21505594.2019.1568174

Suetsugu, S., Kurisu, S., & Takenawa, T. (2014). Dynamic shaping of cellular membranes by phospholipids and membrane-deforming proteins. *Physiological Reviews*, 94(4), 1219–1248.

Sukhithasri, V., Nisha, N., Biswas, L., Kumar, V. A., & Biswas, R. (2013). Innate immune recognition of microbial cell wall components and microbial strategies to evade such recognitions. *Microbiological Research*, 168(7), 396–406.

Takeshita, N., Higashitsuji, Y., Konzack, S., & Fischer, R. (2008). Apical sterol-rich membranes are essential for localizing cell end markers that determine growth directionality in the filamentous fungusaspergillus nidulans. *Molecular Biology of the Cell*, 19(1), 339–351. https://doi.org/10.1091/mbc.e07-06-0523

Tefsen, B., Ram, A. F., van Die, I., & Routier, F. H. (2011). Galactofuranose in eukaryotes: Aspects of biosynthesis and functional impact. *Glycobiology*, 22(4), 456–469. https://doi.org/10.1093/glycob/cwr144

Tiralongo, J., & Martinez-Duncker, I. (eds.). (2013). *Sialobiology: Structure, Biosynthesis and Function. Sialic Acid Glycoconjugates in Health and Disease*, Vol. 1. Bentham Science Publishers.

Trouvelot, S., Héloir, M. C., Poinssot, B., et al. (2014). Carbohydrates in plant immunity and plant protection: Roles and potential application as foliar sprays. *Frontiers in Plant Science*, 5, 592. https://doi.org/10.3389/fpls.2014.00592

von Ballmoos, C., Wiedenmann, A., & Dimroth, P. (2009). Essentials for ATP synthesis by F1F0 ATP synthases. *Annual Review of Biochemistry*, 78(1), 649–672. https://doi.org/10.1146/annurev.biochem.78.081307.104803

Wang, J., Wang, H., Zhang, C., Wu, T., Ma, Z., & Chen, Y. (2019). Phospholipid homeostasis plays an important role in fungal development, fungicide resistance and virulence in Fusarium graminearum. *Phytopathology Research*, 1(1). https://doi.org/10.1186/s42483-019-0023-9

Wang, Z. A., Li, L. X., & Doering, T. L. (2018). Unraveling synthesis of the cryptococcal cell wall and capsule. *Glycobiology*, 28(10), 719–730. https://doi.org/10.1093/glycob/cwy030

Wawra, S., Fesel, P., Widmer, H., Neumann, U., Lahrmann, U., Becker, S., Hehemann, J. H., Langen, G., & Zuccaro, A. (2019). FGB1 and WSC3 are in planta-induced β-glucan-binding fungal lectins with different functions. *New Phytologist*, 222, 1493–1506.

Wenk, M. R. (2010). Lipidomics: New tools and applications. *Cell*, 143(6), 888–895.

Yadav, S. P. (2007). The wholeness in suffix-omics,-omes, and the word om. *Journal of Biomolecular Techniques: JBT*, 18(5), 277.

Yang, D., Zhou, Z., & Zhang, L. (2019). An overview of fungal glycan-based therapeutics. *Progress in Molecular Biology and Translational Science*, 135–163. https://doi.org/10.1016/bs.pmbts.2019.02.001

Yang, K., Cheng, H., Gross, R. W., & Han, X. (2009). Automated lipid identification and quantification by multidimensional mass spectrometry-based shotgun lipidomics. *Analytical Chemistry*, 81(11), 4356–4368. https://doi.org/10.1021/ac900241u

Yang, K., & Han, X. (2016). Lipidomics: Techniques, applications, and outcomes related to biomedical sciences. *Trends in Biochemical Sciences*, 41(11), 954–969. https://doi.org/10.1016/j.tibs.2016.08.010

Yetukuri, L., Katajamaa, M., Medina-Gomez, G., Seppänen-Laakso, T., Vidal-Puig, A., & Orešič, M. (2007). Bioinformatics strategies for lipidomics analysis: Characterization of obesity related hepatic steatosis. *BMC Systems Biology*, 1(1). https://doi.org/10.1186/1752-0509-1-12

Yoshimi, A., Miyazawa, K., & Abe, K. (2017). Function and biosynthesis of cell wall α-1,3-glucan in fungi. *Journal of Fungi*, 3(4), 63. https://doi.org/10.3390/jof3040063

Zaikin, V. G., & Borisov, R. S. (2021). Options of the main derivatization approaches for analytical ESI and MALDI mass spectrometry. *Critical Reviews in Analytical Chemistry*, 52(6), 1287–1342. https://doi.org/10.1080/10408347.2021.1873100

Zeilinger, S., Gupta, V. K., Dahms, T. E., Silva, R. N., Singh, H. B., Upadhyay, R. S., Gomes, E. V., Tsui, C. K. M., & Nayak, S. C. (2016). Friends or foes? Emerging insights from fungal interactions with plants. *FEMS Microbiology Reviews*, 40(2), 182–207.

Zeković, D. B., Kwiatkowski, S., Vrvić, M. M., Jakovljević, D., & Moran, C. A. (2005). Natural and modified (1→3)-beta-d-glucans in health promotion and disease alleviation. *Critical Reviews in Biotechnology*, 25, 205–230.

Zhang, W., Li, F., & Nie, L. (2010). Integrating multiple 'omics' analysis for microbial biology: Application and methodologies. *Microbiology*, 156(2), 287–301.

8 Fungal Proteomics

Identification of Candidate Fungal Species by High-End Proteomics Approaches Leading Toward Beneficial Exploration

Amjad Abbas, Akhlas Shafique, Muhammad Usman, Effat Naz, Muhammad Huzaifa Tanveer, Kiaenat Nazir, Abdalmenem I. M. Hawamda, and Muhammad Amjad Ali

8.1 INTRODUCTION

8.1.1 CONTRIBUTIONS OF FUNGAL PROTEOME RESEARCH

Fungi are enormous in number and comprise an entire kingdom because of their unique features that differ from other eukaryotic organisms. With the advent of modern technology, information about the biology, genetics, nutritional uptake, and multiple roles of fungi in nature, as well as their exploitation in various industries, has evolved. Recently, research related to fungal genomes has become available for many of its species because, nowadays, high-end technologies like next-generation sequencing, liquid chromatography and mass spectrophotometry (LC-MS), matrix assisted laser desorption/ionization-time of flight mass spectrophotometry (MALDI-TOF-MS), and the upscaling of chromatography techniques have made it possible to characterize many fungal organisms. Additionally, the work and higher analyzing power of modern computational tools and available information is sufficient to identify many of fungal proteins, and this functional proteomics research has created a wealth of information to understand the biology, evolution, and adaptivity of fungi in nature, leads toward the extraction of maximum benefits that could be employed for multiple uses.

These uses include biofuel production and the industrial production of enzymes and pharmaceuticals. Proteomics research provides information on protein structure, crystalline structure, isoelectric behavior, gene expression, transcription factors

DOI: 10.1201/9781003452362-8

responsible for protein production, heterologous expression systems to produce proteins and enzymes, the use of inducible gene promotors, and post-translational modification (Dagenais and Keller, 2009; Schuster and Schmoll, 2010).

The extraction of secondary metabolites is one of the important uses of fungal organisms and is either understood by activating silenced genes that produce these metabolites or by high-end analytical approaches that are useful to detect these important metabolites. All fungal organisms are not friendly to lab conditions, and there are always issues with culturing such fungi under controlled conditions. So, to explore more useful secondary metabolites, it is important to have good hands-on information on the proteomics potential of a fungal organism after its successful cultivation in a controlled environment.

Proteomics biodiversity is a tool for exploring fungi and discovering new drugs. There is a challenge in the cultivation of fungi under optimum production conditions. The cultivability of important fungi could be increased by playing with growth conditions and the selection of suitable nutrient-rich media. Modern whole-genome databases are either linked with proteome output or there are offline tools to translate important genes into functional protein models. The information of the genome cannot be seen individually, as it is closely related to the second phase, which is protein translation. For more insight into the discovery of new enzymes and drugs, proteomic research is the end product, and it depends on the biology and cultivability of fungi under consideration. However, proteomics information is only useful if the fungal organism expresses its potential in the growth media. The co-culturing of a desirable fungal entity is also common when it is less responsive to producing novel proteins, enzymes, or metabolites. In such cases, desirable, slow-growing, responsive fungi are co-cultured with more adaptive fungi, but there is a problem with extracting the final product from such a mixture (Shaw and Hoch, 2007).

The production of enzymes from fungi at the industrial scale is a well-known fact, and many important biomolecules are prepared from various fungi. Fungi act like cell enzyme factories. Organic acids like citric acid are also key items prepared from fungus like *Aspergillus niger*. Similarly, proteolytic, cellulolytic, and catalase compounds are produced using these fungi on a large scale. But this production is not possible without a prior knowledge of the proteome of *A. niger* and the optimum conditions that help to generate these enzymes. *A. niger* is also a commonly used heterologous expression model that is efficient in the production of complex enzymes in animals and of certain lifesaving drugs. The implications of this important fungus are because of the genomic and proteomic information available and the research that guides us to grow this fungus to produce important enzymes and proteins at an industrial scale (De Vries and Visser, 2001; Berka et al., 1992).

The understanding of bioinformatics and omics has made it possible to advance in enzyme production. This saves time due to the fast characterization of the individual fungi, which helps researchers decide whether using a fungal strain to produce an industrial enzyme is suitable. For example, background proteomic information enables the use of strain CBS513.88 to produce enzymes, while the other strain, ATCC1015, is efficient in the production of citric acid (Pel et al., 2007; Andersen et al., 2011).

By unfolding the information available in the fungal proteome, it is possible to identify key proteins involved in metabolic pathways, different biotic stress responses, and environmental adaptation, essential for optimizing fungal strains for enhanced protein yield. Additionally, fungal proteomics enables the discovery of novel proteins with unique properties that can be harnessed for pharmaceutical, agricultural, and environmental applications. The insights gained from this research improve the efficiency of protein production processes and contribute to the development of sustainable and innovative biotechnologies (Pel et al., 2007).

The fungal proteome also provides information regarding the symbiosis of fungi with plants, which is beneficial for producing more food. Modern omics information is also used to monitor fungal genes that closely affect food decay, fruit spoilage, and the production of toxin molecules. The modern food industry heavily depends on the use of fungal products and processes; for example, fermentative yeasts and mushrooms are directly involved in the food system. Because fungi are fast-growing and consume resources efficiently, easy genetic manipulation makes them perfect candidates for the food industry (Bergquist et al., 2002). Preparing food and higher-quality industrial products using fungi is cheaper, faster, and easier, resulting in enhanced cultivation to meet customers' needs. Fungi serve as a source of food and are also used as value addition organisms in the food industry (Papagianni, 2004).

There is a long list of products derived from fungal enzymes and proteins. Further, proteases from fungal sources have been recognized for their significant contribution to brewing and preparing baked food products (Meyer et al., 2020). Biological wastes, such as leftover food, decayed food, and dead bodies, could be an alarming issue if fungi stop working to remove the heaps of dead organic wastes. The ability to biodegrade is enhanced based on proteome data and genetic manipulations about the species of fungi associated with such actions, and ultimately, the components are returned to the soil for the next growing seasons (Paul and Joshi, 2015). Mushrooms are raised industrially and are considered promoters of human health, as they provide advantages compared to other foods. Mushrooms are rich in vitamins, minerals, and dietary fibers and promote healthy portions. They also add probiotics to food, which are vital to producing many beneficial biomolecules in the gut. Due to these important constituents, mushrooms provide antioxidants, anti-inflammatory compounds, and antimicrobial drugs in the body for the long term. These advantages of mushrooms for health and consumption can lead to a well-balanced diet (Madhavan et al., 2022; Pel et al., 2007). Fungi have vital roles in every stage of human life, and this is explained in Figure 8.1.

8.2 PROTEIN SEPARATION AND IDENTIFICATION

Multiple protein separation techniques to study the proteomes of different fungi are used, but sometimes, their robust cell walls pose challenges due to chitin. Various techniques used in the separation of fungal proteomics are described.

8.2.1 Different Techniques for Protein Extraction

8.2.1.1 Mg/CHAPS Extraction

This is an optimized method used for fungal species containing tough cell walls. This method uses a combination of the detergent CHAPS (3-cholamidopropyl)

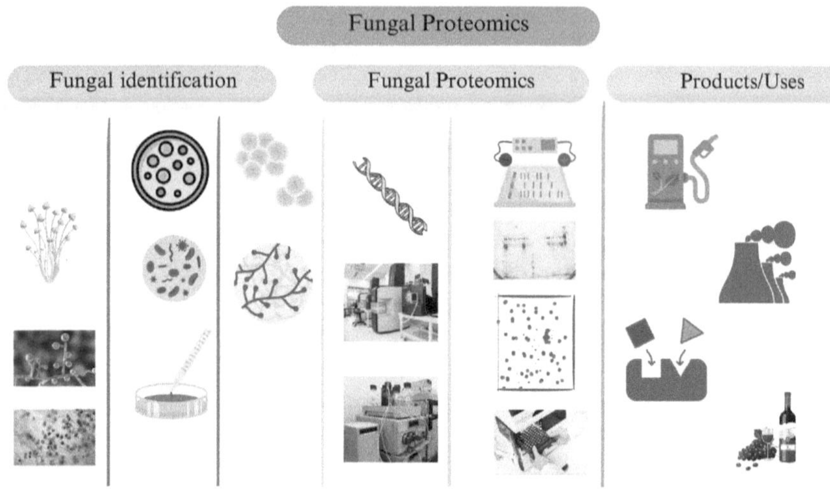

FIGURE 8.1 (A) It shows steps taken for the visual identification of fungi based on the use of growth media and microscopy techniques. (B) It deals with techniques like HPLC and 2D gel electrophoresis for the study of individual proteins under certain conditions or the expression of the whole proteome of fungi under certain influences. (C) It shows uses or products based on fungal proteomics research, which is either used in biofuel production, industrial applications, enzymatic use, or the food industry.

dimethylammonio-1-propanesulfonate and magnesium chloride ($MgCl_2$), which helps with protein extraction from fungal mycelia. $MgCl_2$ stabilizes proteins and helps maintain structural integrity during extraction, while CHAPS enhances overall protein yield and solubilizes the membrane proteins (Bhadauria and Peng, 2010).

8.2.1.2 Liquid Nitrogen Crushing

Liquid nitrogen crushing is a commonly used method for protein extraction from fungi. In this technique, samples are frozen immediately in liquid nitrogen for the preservation of protein integrity. Then, frozen mycelia are crushed in a mortar and pestle, and proteins are released by disrupting cell walls in the surrounding medium (liquid nitrogen). After that, proteins are solubilized in an extraction buffer such as Tris-based buffers. Finally, the mixture is centrifuged to separate soluble proteins from insoluble debris (Krishnaswamy et al., 2019); this is elaborated on briefly in Figure 8.2.

8.2.1.3 Zirconium Bead Lysis

Zirconium beads disrupt fungal cells mechanically and help to release intercellular proteins. It is the most effective technique for the chitinous cell walls of fungi. In this method, samples are prepared by harvesting fungal mycelia and reducing the moisture by lyophilization (freeze-drying). This dried mass is mixed with an appropriate extraction buffer and zirconium beads of 0.1mm diameter. By vigorously shaking, beads collide with fungal cells, effectively break them, and release proteins in the buffer after that mixture is centrifuged to separate soluble proteins and cell debris (Ning et al., 2021; Salem et al., 2016).

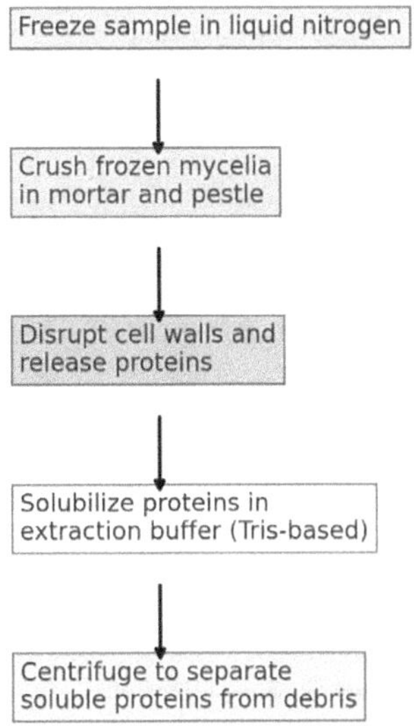

FIGURE 8.2 Overview of the crushing of fungal samples (frozen) to extract soluble proteins from debris by using liquid nitrogen.

8.2.1.4 Osmotic Shock and Enzymatic Treatment

These are two effective methods for protein extraction from fungal cells. In osmotic shock, cells are transferred from a hypertonic to a hypotonic solution for rapid changes in osmotic pressure. This quick and sudden change in internal pressure bursts the cells and releases proteins (Schimek et al., 2020).

In enzymatic treatment, specific enzymes are used to degrade cell walls or membranes to release proteins. Lysozyme breaks the chitinases or cellulases in fungi. This technique leaves the structural components of the cell wall and releases the proteins into the surrounding solution by increasing permeability. This method can enhance the efficacy of protein extraction by conjugating osmotic shock (Klimek-Ochab et al., 2011).

8.2.1.5 Traditional Methods

Traditional methods do not yield optimal results when compared with modern techniques. Tris/EDTA extraction is the most widely used traditional technique for extracting proteins from fungi. This uses buffers containing Tris (tris hydroxymethyl aminomethane) and EDTA (ethylenediaminetetraacetic acid) to extract proteins. Tris buffer maintains pH, while EDTA chelates divalent cations that may stabilize cell walls to facilitate protein release (Bhadauria and Peng, 2010).

8.2.2 Chromatography Techniques

These techniques analyze and separate components in a mixture based on mobile and stationary phases.

8.2.2.1 Types of Chromatography

There are multiple types of chromatography, but the most commonly used are discussed herein.

8.2.2.2 Ion Exchange Chromatography

In this method, proteins are separated based on their charge, allowing for the purification of specific proteins from complex mixtures (Mao and Hearn, 1996).

8.2.2.3 Affinity Chromatography

To isolate proteins, specific binding interactions are utilized, making it especially valuable for studying proteins linked to pathogenic processes and metabolic pathways (Mao and Hearn, 1996).

8.2.3 Electrophoresis

Electrophoresis is a technique used to separate and analyze fungal proteins according to their charge, size, and other properties.

8.2.3.1 Sodium Dodecyl Sulfate Polyacrylamide Gel Electrophoresis (SDS-PAGE)

This technique separates proteins based on their molecular weights. Denatured proteins are coated with SDS, which gives a negative charge, and then separated by using polyacrylamide gel (Nandakumar and Marten, 2002), as shown in Figure 8.3.

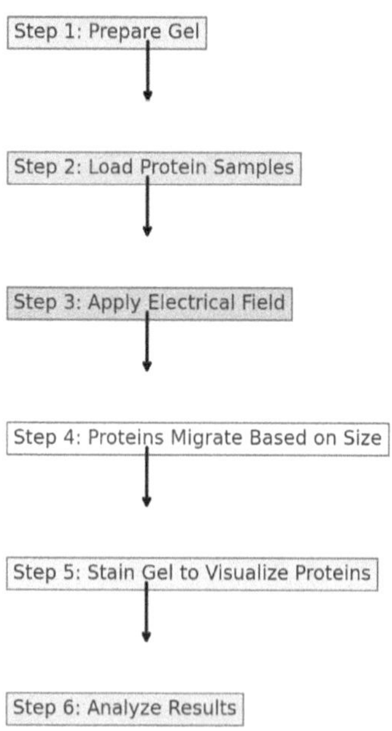

FIGURE 8.3 Schematic overview of multiple steps of SDS-PAGE to analyze results based on the molecular weights of proteins.

8.2.3.2 Isoelectric Focusing (IEF)

This technique separates proteins based on their isoelectric points (pI). Until the charge of proteins is net zero, they migrate in a pH gradient (Kim et al., 2007).

8.2.3.3 Two-Dimensional Gel Electrophoresis (2-DE)

This technique combines both IEF and SDS-PAGE. Primarily, proteins are isolated on a pI basis, and then, they are sorted based on their molecular weight (Fernández and Novo, 2013).

8.3 PROTEIN IDENTIFICATION TECHNIQUES

In fungal proteomics, proteins are identified essentially to understand fungal proteins' interaction and biological functions. Key techniques for protein identification are defined as follows.

8.3.1 Mass Spectrometry (MS)

This technique provides detailed information about protein identity, structure, and post-translational modifications. In the case of fungal proteomics, MS is used in combination with other techniques for the most effective results. The flow chart in Figure 8.4 shows all the steps in this process.

8.3.2 Bottom-Up Proteomics

This technique involves the digestion of proteins into peptides by enzymatic activities and then analysis with MS. This technique provides high throughput identification of a complex mixture of proteins (Söderberg et al., 2006).

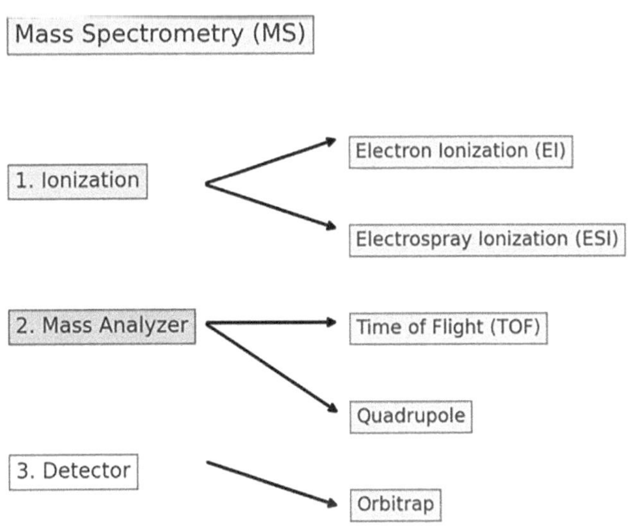

FIGURE 8.4 Flowchart diagram representation of mass spectrometry.

8.3.3 Top-Down Proteomics

This technique usually analyzes intact proteins, which helps to identify proteoforms and post-translational modifications. It is especially valuable for analyzing larger proteins that might be lost during digestion (Söderberg et al., 2006).

8.4 TARGETED PROTEOMICS

This technique follows proteins or peptides of specific interest, which are used to discover and validate biomarkers (Söderberg et al., 2006).

- **Immunoprecipitation**
  This approach uses specific antibodies to isolate particular proteins from a complex mixture. This technique is used in combination with mass spectrometry to identify proteins. This helps to identify translational modifications and protein–protein interactions (Doyle and Owens, 2017).
- **Western Blotting**
  This technique detects particular proteins using antibodies. Following separation by SDS-PAGE, proteins are transferred onto a membrane and then detected using antibodies that recognize the target protein. It helps to confirm the protein presence identified by mass spectrometry (Doyle, 2011).

8.5 BIOINFORMATICS AND DATA ANALYSIS

8.5.1 Databases Used for Protein Identification and Annotation

Fungal proteomics is the large-scale study of proteins, their functions, their structures, and their interactions within the fungal body. Multiple databases have been developed as resources for protein identification, characterization, functional annotation, and comparative analyses. Different databases are presented here that provide a comprehensive overview and information about fungal proteomics, as mentioned in Table 8.1.

TABLE 8.1
Databases Used in Data Retrieval, Functional Annotation, and Comparative Studies in Fungal Proteomics

Databases	Information	References
UniProt (Universal Protein Source)	• Widely used protein database • Provides comprehensive information about protein sequences, their interactions, and functional annotation • Comparative proteomics	Consortium, 2019
FungiDB	• Specialized database for fungal genomics, transcriptomics, and proteomics • Tools may include sequence alignment, domain prediction, and functional annotation	Basenko et al., 2018

TABLE 8.1 (*Continued*)
Databases Used in Data Retrieval, Functional Annotation, and Comparative Studies in Fungal Proteomics

Databases	Information	References
MycoCosm	• Wide range of fungal genomes • Tools for functional annotation and comparative genomics	Ahrendt et al., 2022
InterPro	• Predicts protein models of protein families, sites of function, and domains • Mapping of fungal proteins into conserved motifs and domains	Paysan-Lafosse et al., 2023
PRIDE (Proteomics Identifications Database)	• Based on mass spectrometry-based proteomic data • Enables comparative studies based on given or existing protein datasets • Also explores expression patterns of proteins in fungi	Vizcaíno et al., 2012
PHI-base (Pathogen-Host Interactions Database)	• Provides experimentally verified data on effectors, pathogenicity, and virulence genes of fungi • Useful for host–pathogen interaction-related interactive proteins	Urban et al., 2020
STRING	• Database for predicted and known protein–protein interactions • Predict functional partnerships of fungi (provide insights into fungal cellular processes)	Szklarczyk et al., 2021
Fungal Transcription Factor Database (FTFD)	• Provides roles of transcription factors in the regulation of protein expression in fungi	Park et al., 2008
Swiss-Prot	• Section of the UniProt database • Provides curated data for various fungal proteins • Also provide insights for protein annotation and functional studies	Boeckmann et al., 2003
Phyre2	• Predicts 3D structures and functions of proteins • Useful in studies on novel or less-characterized proteins	Kelley et al., 2015

8.6 PROTEOMICS AND STRUCTURAL PREDICTION OF PROTEINS

Whole-genome sequencing generates a lot of data on various fungi, which could help categorize similar genes based on their function. The arrangement of residues, protein structure, nature of protein, and gene type determine the tag of the proteins in a particular group. It is used to predict the 3D structure of fungal proteins based on their amino acid sequences. Such structures aid in understanding interactions, functions, and their roles in fungal biology, enzyme production, waste decomposition, and in particular pathogenesis in a plant or animal host, as shown in Figure 8.5. Structural predictions can give a better understanding for identifying active domains, receptor sites, and other functional locations in fungal cells (Kelley et al., 2015). Similarly, several online tools and applications help analyze the proteins and provide insight into proteomics, as shown in Table 8.2.

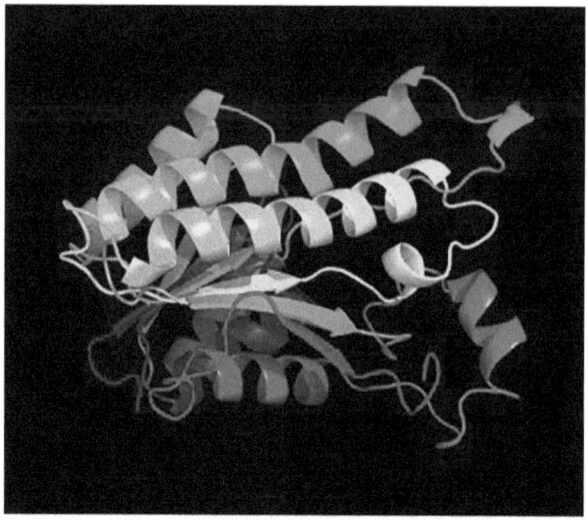

FIGURE 8.5 Structural prediction and visualization of 3-oxoacyl-[acyl-carrier protein] reductase of *Fusarium oxysporum* by using the Phyre2 web tool.

TABLE 8.2
Various Tools and Software to Identify Specific Fungal Proteins and Their Functions and Annotations

Tools and Software	Uses	References
BLAST (Basic Local Alignment Search Tool)	• Most widely used tool to compare amino acid or nucleotide sequences against a database of sequences • Identify unknown fungal proteins from known sequences in the database by screening	Altschul et al., 1990
PANTHER (Protein ANalysis THrough Evolutionary Relationships)	• This site provides functional annotations of fungal proteins and an understanding of their evolutionary relationships • Integrates protein functions by aligning their sequences, protein family trees, and GO annotations	Thomas et al., 2022
InterProScan	• It crucially provides conserved motifs and domains of fungal proteins • This site also integrates various databases of fungal proteins to predict domains, families, and functional sites	Jones et al., 2014
Mascot	• One of the widely used search engines of mass spectrometry data for protein identification • It identifies high-throughput proteins in the fungal proteome from complex samples	Perkins et al., 1999

TABLE 8.2 (*Continued*)
Various Tools and Software to Identify Specific Fungal Proteins and Their Functions and Annotations

Tools and Software	Uses	References
SignalP	• This tool provides the location of signal peptides from the whole proteome • Helps in the identification of secreted proteins to understand fungal pathogenicity • It also determines the role of signal peptides to neutralize the effects of basic antimicrobial peptides	Teufel et al., 2022
Proteome Discoverer	• Software for comprehensive data analysis for mass spectrometry-based proteomics • It performs in-depth characterization of fungal proteins	Orsburn, 2021
PEAKS	• This tool suite is used for de novo sequencing, database search, and PTM analysis • Identifies novel or poorly characterized proteins	Ma et al., 2003

8.6.1 Data Analysis Strategies

In proteomics, data analysis is a crucial process that involves various strategies for the extraction of meaningful insights from complex datasets. In relation to fungal proteomics, typically used strategies are as follows:

- Protein quantification
- Post-translational modification (PTM) analysis
- Pathway analysis

8.6.1.1 Protein Quantification

This includes the following:

8.6.1.1.1 Isotope Labeling

Different techniques like stable isotope labeling by amino acids in cell culture (SILAC) and tandem mass tagging (TMT) work by adding stable isotopes or chemical tags to proteins. These approaches enable accurate quantification by comparing proteins in labeled samples against those in unlabeled ones, offering a clear and precise way to measure differences in protein expression (Thompson et al., 2003).

8.6.1.1.1.1 Label-Free Quantification (LFQ) A popular and widely used technique estimates the relative abundance of proteins from different samples by measuring the intensity of peptide signals in mass spectra. This method eliminates the dire need for chemical labeling, which makes it versatile and applicable to a broad spectrum of samples (Cox and Mann, 2008).

8.6.1.2 Post-Translational Modification (PTM) Analysis

This analysis mainly focuses on detecting and understanding modifications like phosphorylation, glycosylation, and ubiquitination, which are crucial for regulating protein functions and interactions. Researchers often rely on techniques like mass spectrometry and targeted enrichment methods to identify these modifications and explore their impacts on protein behavior (Mann and Jensen, 2003). Databases like Proteome Discoverer provide algorithms specifically designed for detecting PTMs in mass spectrometry data (Cox and Mann, 2008).

8.6.2 Pathway Analysis and Interactions

This includes the following:

8.6.2.1 Gene Ontology (GO) Enrichment Analysis

This helps to identify molecular functions, biological processes, or cellular components that are overexpressed in a set of proteins. It provides deeper insights into the broader biological significance of the proteomic data, allowing researchers to better understand the underlying biological context (Ashburner et al., 2000).

8.6.2.2 Network Analysis

Researchers often construct and analyze protein–protein interaction networks and other molecular networks to gain insights into the functional relationships between proteins. Tools like STRING are commonly used to visualize proteins' functions within a cell (Szklarczyk et al., 2021), as shown in Figure 8.6.

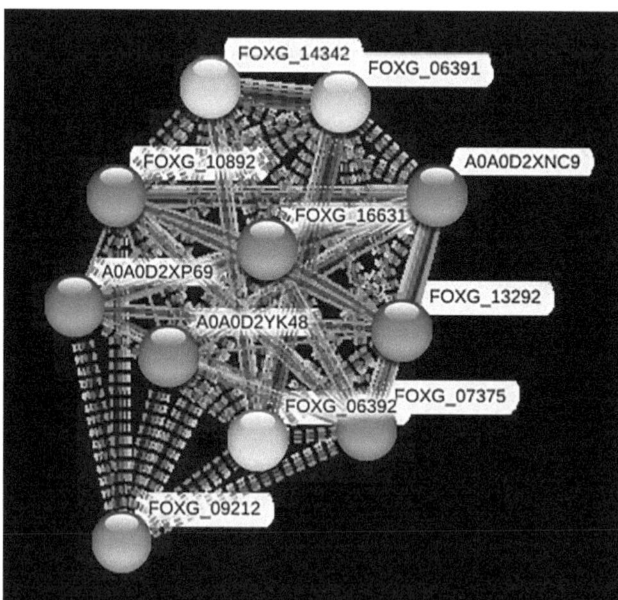

FIGURE 8.6 Protein–protein interaction of 3-oxoacyl-[acyl-carrier protein] reductase in *Fusarium oxysporum* by using the STRING database.

8.6.2.3 Yeast Two-Hybrid (Y2H) Assay

In fungal proteomics, the yeast two-hybrid (Y2H) assay is one of the most important tools for studying fungal effectors and their interactions with plant resistance proteins. Fungal effectors, molecules secreted by pathogenic fungi, have the purpose of manipulating host plant cells. The effectors typically target the immune system through the manipulation of so-called R genes, which the plant either cannot fully disarm or cannot detect at all (Brückner et al., 2009). Researchers use the Y2H assay to spot these interactions with the help of fungal effector proteins (used as bait) fused to a DNA-binding domain and plant resistance proteins (used as prey) fused to a transcriptional activation domain. The activation of a reporter gene in the yeast cell, which is the resulting interaction of the effector and R protein interaction, is the proof and, thus, evidence of the interaction. The method is useful not only for the identification of the molecular mechanisms that are involved in plant–pathogen interactions but also for the discovery of targets that are feasible for the development of disease-resistant crops.

This is crucial for comprehending how plants defend themselves against pathogenic fungi. In the first process, the DBD and AD domains interact, thereby enabling reporter gene transcription. The activation of this gene serves as an indicator of protein interaction within the yeast system, thus providing a readout for the assay. They will interact if the fungal effector and plant resistance proteins are a match. This interaction is crucial for the assay, indicating a potential relationship between the two proteins (McLellan et al., 2021).

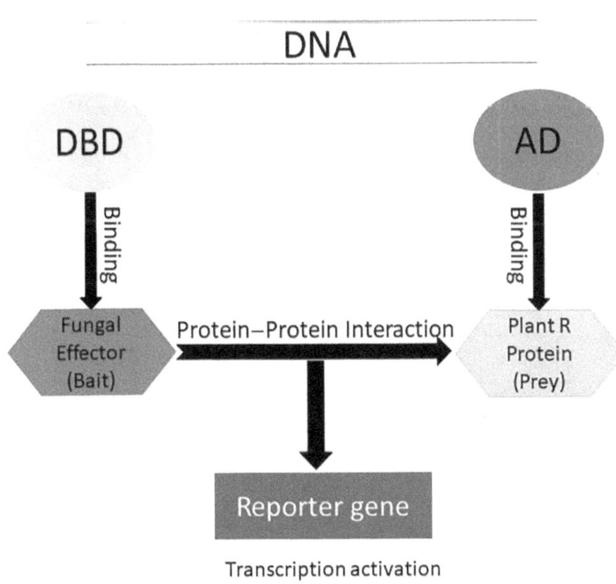

FIGURE 8.7 The diagram visually represents the methodology employed in the yeast two-hybrid (Y2H) assay, which is utilized to elucidate the interactions between fungal effectors and plant resistance proteins.

8.6.2.4 Molecular Docking

This technique helps to predict the interactions among fungal proteins and potential ligands such as drugs and small molecules (Ali et al., 2024). It provides interactive insights into protein–protein interactions among fungal and host proteins. It gives a better understanding of a fungus's invasion within host tissue, evasion of the immune system, and, finally, establishment of infection (Ogunjobi et al., 2024). It also provides predictions to influence the binding strength among fungal proteins and their ligands, just like the binding sites of root-knot nematode effector proteins were predicted for the *Mi-1* resistance gene in tomato (Figure 8.8). This is crucial for understanding how protein function is regulated in fungal cells and how drug resistance might develop (Berthiller et al., 2013). Analyzing fungal or any other proteomics data involves several steps to quantify proteins effectively, identify post-translational modifications, and understand the biological implications of the results through pathway and network analyses. This process relies on different software tools and databases that help convert raw mass spectrometry data into valuable biological information.

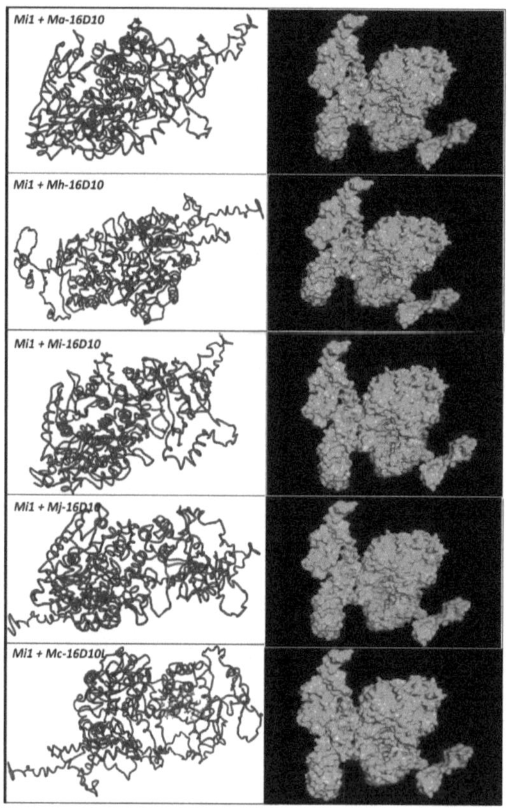

FIGURE 8.8 Protein–protein docking of the Mi-1 protein with different peptides from root-knot nematodes (as a ligand). Each box in the figure corresponds to a distinct peptide–protein interaction, showcasing the top-ranked conformation based on the minimum S score. The peptides are shown in red and blue on the left and right sides of the figure, respectively. *Source:* Adapted from Zahoor et al., 2023, with permission.

REFERENCES

Ahrendt, S. R., Mondo, S. J., Haridas, S., & Grigoriev, I. V. (2022). MycoCosm, the JGI's Fungal genome portal for comparative genomic and multiomics data analyses. In *Microbial Environmental Genomics (MEG)* (pp. 271–291). New York, NY: Springer US.

Ali, M. A., Zahoor, A., Niaz, Z., Jabran, M., Anas, M., Shafique, I., & Abbas, A. (2024). Bioinformatics and computational biology. *Trends in Plant Biotechnology*, 281–334.

Altschul, S. F., Gish, W., Miller, W., Myers, E. W., & Lipman, D. J. (1990). Basic local alignment search tool. *Journal of Molecular Biology*, 215(3), 403–410.

Andersen, M. R., Salazar, M. P., Schaap, P. J., et al. (2011). Comparative genomics of citric-acid-producing Aspergillus niger ATCC 1015 versus enzyme-producing CBS 513.88. *Genome Resources*, 21(6), 885–897.

Ashburner, M., Ball, C. A., Blake, J. A., Botstein, D., Butler, H., Cherry, J. M., . . . & Sherlock, G. (2000). Gene ontology: Tool for the unification of biology. *Nature Genetics*, 25(1), 25–29.

Basenko, E. Y., Pulman, J. A., Shanmugasundram, A., Harb, O. S., Crouch, K., Starns, D., . . . & Hertz-Fowler, C. (2018). FungiDB: An integrated bioinformatic resource for fungi and oomycetes. *Journal of Fungi*, 4(1), 39.

Bergquist, P. L., Te'O, V. S. J., Gibbs, M. D., Cziferszky, A. C. E., De Faria, F. P., Azevedo, M. O., Nevalainen, K. M. H. (2002). Production of recombinant bleaching enzymes from thermophilic microorganisms in fungal hosts. *Applied Biochemistry and Biotechnology*, 98–100, 165–176.

Berka, R. M., Dunn-Coleman, N., & Ward, M. (1992). Industrial enzymes from Aspergillus species. In Bennett, J. W., & Klich, M. A. (eds.), *Aspergillus, Biology and Industrial Applications* (pp. 155–202). Stoneham, MA: Butterworth-Heinemann.

Berthiller, F., Crews, C., Dall'Asta, C., Saeger, S. D., Haesaert, G., Karlovsky, P., & Stroka, J. (2013). Masked mycotoxins: A review. *Molecular Nutrition & Food Research*, 57(1), 165–186.

Bhadauria, V., & Peng, Y. L. (2010). Optimization of a protein extraction technique for fungal proteomics. *Indian Journal of Microbiology*, 50, 127–131.

Boeckmann, B., Bairoch, A., Apweiler, R., Blatter, M. C., Estreicher, A., Gasteiger, E., & Schneider, M. (2003). The SWISS-PROT protein knowledgebase and its supplement TrEMBL in 2003. *Nucleic Acids Research*, 31(1), 365–370.

Brückner, A., Polge, C., Lentze, N., Auerbach, D., & Schlattner, U. (2009). Yeast two-hybrid, a powerful tool for systems biology. *International Journal of Molecular Sciences*, 10(6), 2763–2788. doi: 10.3390/ijms10062763.

Cox, J., & Mann, M. (2008). MaxQuant enables high peptide identification rates, individualized ppb-range mass accuracies and proteome-wide protein quantification. *Nature Biotechnology*, 26(12), 1367–1372.

Dagenais, T. R., & Keller, N. P. (2009). Pathogenesis of Aspergillus fumigatus in invasive aspergillosis. *Clinical Microbiology Reviews*, 22(3), 447–465.

De Vries, R. P., & Visser, J. (2001). Aspergillus enzymes involved in the degradation of plant cell wall polysaccharides. *Microbiology and Molecular Biology Reviews*, 65(4), 497–522.

Doyle, S. (2011). Fungal proteomics: From identification to function. *FEMS Microbiology Letters*, 321, 1–9.

Doyle, S., & Owens, R. A. (2017). Fungal proteomics. *Fungi: Biology and Applications*, 119–146.

Fernández, R. G., & Novo, J. V. J. (2013). Proteomic protocols for the study of filamentous fungi. Laboratory protocols in fungal biology. *Current Methods in Fungal Biology*, 299–308.

Jones, P., Binns, D., Chang, H. Y., Fraser, M., Li, W., McAnulla, C., . . . & Hunter, S. (2014). InterProScan 5: Genome-scale protein function classification. *Bioinformatics*, 30(9), 1236–1240.

Kelley, L. A., Mezulis, S., Yates, C. M., Wass, M. N., & Sternberg, M. J. (2015). The Phyre2 web portal for protein modeling, prediction and analysis. *Nature Protocols*, 10(6), 845–858.

Kim, Y., Nandakumar, M. P., & Marten, M. R. (2007). Proteomics of filamentous fungi. *Trends in Biotechnology*, 25(9), 395–400.

Klimek-Ochab, M., Brzezińska-Rodak, M., Żymańczyk-Duda, E., Lejczak, B., & Kafarski, P. (2011). Comparative study of fungal cell disruption—scope and limitations of the methods. *Folia Microbiologica*, 56, 469–475.

Krishnaswamy, A., Barnes, N., Lotlikar, N. P., & Damare, S. R. (2019). An improved method for protein extraction from minuscule quantities of fungal biomass. *Indian Journal of Microbiology*, 59, 100–104.

Ma, B., Zhang, K., Hendrie, C., Liang, C., Li, M., Doherty-Kirby, A., & Lajoie, G. (2003). PEAKS: Powerful software for peptide de novo sequencing by tandem mass spectrometry. *Rapid Communications in Mass Spectrometry*, 17(20), 2337–2342.

Madhavan, A., Arun, K., Sindhu, R., Jose, A. A., Pugazhendhi, A., Binod, P., Sirohi, R., Reshmy, R., & Awasthi, M. K. (2022). Engineering interventions in industrial filamentous fungal cell factories for biomass valorization. *Bioresource Technology*, 344, 126209.

Mann, M., & Jensen, O. N. (2003). Proteomic analysis of post-translational modifications. *Nature Biotechnology*, 21(3), 255–261.

Mao, Q. M., & Hearn, M. T. W. (1996). Optimization of affinity and ion-exchange chromatographic processes for the purification of proteins. *Biotechnology and Bioengineering*, 52(2), 204–222.

McLellan, H., Armstrong, M. R., Birch, P. R. J. (2021). Yeast two-hybrid screening for identification of protein-protein interactions in solanum tuberosum. *Methods in Molecular Biology*, 2354, 95–110. doi: 10.1007/978-1-0716-1609-3_4.

Meyer, V., Basenko, E. Y., Benz, J. P., et al. (2020). Growing a circular economy with fungal biotechnology: A white paper. *Fungal Biology and Biotechnology*, 7, 5.

Nandakumar, M. P., & Marten, M. R. (2002). Comparison of lysis methods and preparation protocols for one-and two-dimensional electrophoresis of Aspergillus oryzae intracellular proteins. *Electrophoresis*, 23(14), 2216–2222.

Ning, Y. T., Yang, W. H., Zhang, W., Xiao, M., Wang, Y., Zhang, J. J., & Xu, Y. C. (2021). Developing two rapid protein extraction methods using focused-ultrasonication and zirconia-silica beads for filamentous fungi identification by MALDI-TOF MS. *Frontiers in Cellular and Infection Microbiology*, 11, 687240.

Ogunjobi, T. T., Okorie, I. C., Gigam-Ozuzu, C. D., Olorunleke, J. V., Ogunleye, F. I., Irimoren, E. O., & Ojo, E. O. (2024). Bioinformatics tools in protein analysis: structure prediction, interaction modelling, and function relationship. *European Journal of Sustainable Development Research*, 8(1).

Orsburn, B. C. (2021). Proteome discoverer—a community enhanced data processing suite for protein informatics. *Proteomes*, 9(1), 15.

Papagianni, M. (2004). Fungal morphology and metabolite production in submerged mycelial processes. *Biotechnology Advances*, 22, 189–259.

Park, J., Park, J., Jang, S., Kim, S., Kong, S., Choi, J., . . . & Lee, Y. H. (2008). FTFD: An informatics pipeline supporting phylogenomic analysis of fungal transcription factors. *Bioinformatics*, 24(7), 1024–1025.

Paul, S., & Joshi, S. R. (2015). Industrial perspectives of fungi. In Verma, P. (ed.), *Industrial Microbiology and Biotechnology* (pp. 81–105). Springer: Singapore.

Paysan-Lafosse, T., Blum, M., Chuguransky, S., Grego, T., Pinto, B. L., Salazar, G. A., . . . & Bateman, A. (2023). InterPro in 2022. *Nucleic Acids Research*, 51(D1), D418–D427.

Pel, H., de Winde, J., Archer, D., et al. (2007). Genome sequencing and analysis of the versatile cell factory Aspergillus niger CBS 513.88. *Nature Biotechnology*, 25(2), 221–231.

Perkins, D. N., Pappin, D. J., Creasy, D. M., & Cottrell, J. S. (1999). Probability-based protein identification by searching sequence databases using mass spectrometry data. *ELECTROPHORESIS: An International Journal*, 20(18), 3551–3567.

Salem, M. A., Jüppner, J., Bajdzienko, K., & Giavalisco, P. (2016). Protocol: A fast, comprehensive and reproducible one-step extraction method for the rapid preparation of polar and semi-polar metabolites, lipids, proteins, starch and cell wall polymers from a single sample. *Plant Methods*, 12, 1–15.

Schimek, C., Egger, E., Tauer, C., Striedner, G., Brocard, C., Cserjan-Puschmann, M., & Hahn, R. (2020). Extraction of recombinant periplasmic proteins under industrially relevant process conditions: selectivity and yield strongly depend on protein titer and methodology. *Biotechnology Progress*, 36(5), 2999.

Schuster, A., & Schmoll, M. (2010). Biology and biotechnology of Trichoderma. *Applied Microbiology and Biotechnology*, 87(3), 787–799.

Shaw, B., & Hoch, H. (2007). Ions regulate spore attachment, germination, and fungal growth. In Howard, R. J., & Gow, N. A. (eds.), *The Mycota: Biology of the Fungal Cell* (pp. 219–236). Berlin: Springer.

Söderberg, O., Gullberg, M., Jarvius, M., Ridderstråle, K., Leuchowius, K. J., Jarvius, J., & Landegren, U. (2006). Direct observation of individual endogenous protein complexes in situ by proximity ligation. *Nature Methods*, 3(12), 995–1000.

Szklarczyk, D., Gable, A. L., Nastou, K. C., Lyon, D., Kirsch, R., Pyysalo, S., . . . & von Mering, C. (2021). The STRING database in 2021: Customizable protein–protein networks, and functional characterization of user-uploaded gene/measurement sets. *Nucleic Acids Research*, 49(D1), D605–D612.

Teufel, F., Almagro Armenteros, J. J., Johansen, A. R., Gíslason, M. H., Pihl, S. I., Tsirigos, K. D., . . . & Nielsen, H. (2022). SignalP 6.0 predicts all five types of signal peptides using protein language models. *Nature Biotechnology*, 40(7), 1023–1025.

Thomas, P. D., Ebert, D., Muruganujan, A., Mushayahama, T., Albou, L. P., & Mi, H. (2022). PANTHER: Making genome-scale phylogenetics accessible to all. *Protein Science*, 31(1), 8–22.

Thompson, A., Schäfer, J., Kuhn, K., Kienle, S., Schwarz, J., Schmidt, G., . . . & Hamon, C. (2003). Tandem mass tags: A novel quantification strategy for comparative analysis of complex protein mixtures by MS/MS. *Analytical Chemistry*, 75(8), 1895–1904.

UniProt Consortium. (2019). UniProt: A worldwide hub of protein knowledge. *Nucleic Acids Research*, 47(D1), D506–D515.

Urban, M., Cuzick, A., Seager, J., Wood, V., Rutherford, K., Venkatesh, S. Y., . . . & Hammond-Kosack, K. E. (2020). PHI-base: The pathogen–host interactions database. *Nucleic Acids Research*, 48(D1), D613–D620.

Vizcaíno, J. A., Côté, R. G., Csordas, A., Dianes, J. A., Fabregat, A., Foster, J. M., . . . & Hermjakob, H. (2012). The PRoteomics IDEntifications (PRIDE) database and associated tools: Status in 2013. *Nucleic Acids Research*, 41(D1), D1063–D1069.

Zahoor, A., Niaz, Z., Tahir-ul-Qamar, M., & Hawamda, A. I. M. (2023). Computational analysis of 16D10 effector peptides from root-knot nematodes and their interaction with Mi resistance (R) protein through molecular docking. *Integrative Plant Biotechnology*, 1, 1–8.

9 Fungal Signalomics

Zakia Habib, Abdullah, and
Hayssam Mohamad Ali

9.1 INTRODUCTION

In eukaryotic and prokaryotic organisms, intracellular signaling transduction pathways are vital for responding to environmental cues and developmental signals (McLean et al., 2019). Hence, cells and organisms must adapt and be resilient to the fluctuating environment to exist, endure, and evolve for their viability (Bayram and Bayram, 2023). The coevolution of the constituent parts of the corresponding systems and the horizontal gene transfer of bacterial genes resulted in the proliferation of signaling pathways in three kingdoms: Bacteria, Archaea, and Eukarya (Pertseva and Shpakov, 2009). Therefore, signaling systems in unicellular eukaryotes constitute an intermediary step in developing the signaling pathway system between pro- and higher eukaryotic organisms in the context of their functional characteristics and molecular architecture (Jacob et al., 2022).

In all living organisms, the functional organization of growth, metabolism, differentiation, and apoptosis has the following four essential constituents: a signal receptor, which explicitly identifies a signal molecule; a signal transport that is linked to the signal receptor; a signal amplifier, which is typically an ion channel or an enzyme-producing second messenger; and an effector molecule that receives signals. As a result, one or more intracellular signaling events is initiated that ultimately reacts to the external stimuli (Nair et al., 2019). The cellular and molecular responses to stress involve acute and adaptive stages. The acute response often aims to minimize impairment due to the detrimental consequences of stress, such as the preliminary physicochemical pressures induced by osmotic stress. However, the adaptive response involves the refurbishment of cellular homeostasis to facilitate the microbe's development in the changing environment. Therefore, cells must perceive the changes in their surroundings and initiate the corresponding signal transduction pathways to develop acute and adaptive responses (Nikolaou et al., 2009).

Fungi represent one of the most diverse living forms, varying from unicellular yeast to multicellular filamentous fungi and mushrooms (Kosal, 2023). These microorganisms are beneficial to mankind and utilized as food resources, additives, enzyme manufacturers, fermenting agents, and suppliers of various bioactive compounds (Kour et al., 2019). In addition to that, various fungi are pathogenic to plants, animals, and humans. Fungi are vital contributors to the sustainability of the world's resources, as they assist in nutrient recycling and developing associations with plant components (Zanne et al., 2020). Fungal signalomics refers to the study of signaling molecules, signaling pathways, and their impact on fungal physiology. These signaling compounds comprise various small molecules such as peptides, proteins, and

 DOI: 10.1201/9781003452362-9

secondary metabolites, as well as environmental signals, including temperature and light (Patwardhan et al., 2023). So, understanding signaling pathways is essential for unraveling fungi's responses to biotic and abiotic dynamics, including host–pathogen interactions, nutrient accessibility, and stress environments (Jamil et al., 2022).

Recently, fungal signalomics has gained more attention because of its impending applications in various fields, including biotechnology, agriculture, and the pharmaceutical industry (Rana et al., 2019). For instance, understanding fungal signalomics can lead to innovative antifungal approaches, boosted crop protection methods, and developed industrial fermentation techniques (Patwardhan et al., 2023). Furthermore, exploring fungal signaling pathways can provide insight into the evolution of signaling systems in eukaryotes (González-Rubio et al., 2019). The current chapter reviews various transducing molecular signal pathways in fungi to demonstrate how signaling molecules scramble data, manage multiple transmission pathways, and respond to various environmental cues.

9.2 QUORUM SENSING

Microorganisms that are numerous in number and living nearby must communicate with each other. Quorum sensing (QS) is a communication technique utilized by microbes that regulates significant developmental processes and is reliant on cell density. Microorganisms use tiny quorum sensing molecules (QSMs), which can regulate biological activities and behaviors, to initiate and monitor communication (Padder et al., 2018). The concentration of these molecules rises proportionately with the population, setting off regulatory feedback once a critical threshold is attained. Thus, QS-dependent genes are coordinately expressed or repressed (Albuquerque and Casadevall, 2012). Approximately 50 years ago, an exploration of genetic competence in *Streptococcus pneumoniae* and bioluminescence in marine *Vibrio* species demonstrated the existence of quorum sensing in bacteria as a signaling and communication mechanism (Bassler and Losick, 2006). Since then, QS has been found in various bacteria, directing crucial functions like sporulation, the production of antibiotics, the establishment of biofilms, and the release of virulence factors (Tian et al., 2021).

Studies have revealed that in addition to serving important roles in bacteria, quorum sensing (QS) also regulates various population density-related activities, including biofilm formation, pathogenicity, and disease, in fungi (Swetha et al., 2021). The presence of QSMs in fungi was further validated via the discovery of filamentation regulation by the QSM farnesol in the pathogenic fungus *Candida albicans* (Wongsuk et al., 2016). Farnesol was the first identified quorum-sensing molecule in a eukaryotic microbe. Later on, other QSMs, including tryptophol, phenylethanol, and tyrosol, were also discovered (Kebaara et al., 2008; Albuquerque and Casadevall, 2012). Successively, the role of QSMs was well explored in several yeasts and filamentous fungi, including *Aspergillus niger*, *A. nidulans*, *Candida albicans*, *Saccharomyces cerevisiae*, *C. dubliniensis*, and *Fusarium graminearum* (Jacob et al., 2022).

Fungi have been found to retain a huge variety of signaling molecules, which are typically not specific to a fungal strain. More specifically, lipids (oxylipins), alcohols (tyrosol, tryptophol, farnesol, and 1-phenylethanol), peptides (pheromones), acetaldehydes, and several volatile chemicals are notable examples of quorum-sensing molecules (QSMs) in fungi. These molecules directly impact the fungal quorum-sensing mechanism, directing several essential processes, including filamentation, pathogenicity, and morphogenesis (Sarsaiya et al., 2019). Filamentation induction from the cell culture supernatant of a stationary phase of an *S. cerevisiae* strain culture was initially reported in 2006 (Chen and Fink, 2006).

Tryptophol (TrpOH) and phenylethanol (PheOH) are two aromatic alcohols actively causing filamentation through quorum sensing (QS). It has been revealed that cell density potentially impacts the production of these compounds. The production of these aromatic alcohols is driven by the increased expression of two genes, ARO9 and ARO10, particularly when there is high cell density. Therefore, yeast produces abundant aromatic alcohols per cell, corresponding to higher population densities and vice versa (Chen and Fink, 2006; Avbelj et al., 2015). PheOH and TrpOH aromatic alcohols synergistically prompt the up-regulation of the FLO11 gene via the Flo8p transcription factor and Tpk2p, a cAMP-dependent PKA component essential for filamentous growth. It has been reported that *S. cerevisiae* that has either FLO8 or TPK2 inactivated is incapable of producing filaments in PheOH and TrpOH appearance (Chen and Fink, 2006; Tian et al., 2021).

In addition to cell density, nitrogen scarcity serves as a major stimulant for morphogenesis in *S. cerevisiae*, particularly increasing the biosynthesis of PheOH and TrpOH. Research on quorum sensing in fungus lacks a thorough understanding of its signaling sensors and signal transmission mechanisms (Winters et al., 2019). Thus, comprehending the molecular mechanisms underlying microbe interactions is critical for assessing their impact on biofilm synthesis, the etiology of diseases, the quality and security of fermented foodstuffs, and how microbes perform in a biotechnological context.

9.3 MULTISTEP PHOSPHORELAY SYSTEMS

Understanding how signals are interpreted, transduced, and processed in living cells is essential to comprehend the molecular processes that allow microorganisms to adapt to changing environments (Defosse et al., 2015; Hérivaux et al., 2016). Prokaryotic organisms utilize a two-component signaling system; however, eukaryotic individuals depend on an intricate multistep phosphorelay system, or MSP (Kabbara et al., 2019). These systems rely on phosphorylation, the transfer of a phosphoryl group within a signaling cascade to transduce communications in response to their changing environment (Jacob and Thines, 2017). In the prokaryotic two-component system, a response regulator protein perceives information from a histidine protein kinase (HK) that senses external stimuli (Stock et al., 2000; Jacob and Thines, 2017). Contrariwise, the more intricate eukaryotic multistep phosphorelay system, or MSP, uses three additional components that include a phosphotransfer protein, a response regulator protein, and a two-component hybrid histidine kinase (HHK) to perceive and communicate environmental changes (Huo et al., 2020).

Prokaryotic histidine kinase has a more multifaceted structure than eukaryotic histidine kinase, which is regarded as an important distinction between the prokaryotic two-component system and the eukaryotic MSP (Bourret et al., 2021). The phosphoryl group is transferred in the multistep phosphorelay system in the following order: His-Asp-His-Asp (Liao et al., 2021).

In this signaling cascade, hybrid histidine kinases (HHKs) are the primary sensor proteins that have diverse sensory domains at the N-terminus, such as a HK domain with an autophosphorylation site, a histidine-like ATPase catalytic domain (HATPase domain), and HAMP or HAMP-like linker domains (poly-HAMP). The aspartate residue that holds the phosphoryl group is present in the C-terminal response regulator domain (REC) of HHKs (Bersching and Jacob, 2021; Kaur et al., 2021). Afterward, the next phase in the phosphorelay mechanism is relocating the phosphate group from the aspartate residue to the histidine residue in the phosphotransfer domain (HPt) of the phosphotransfer proteins. The last step includes the transfer of the phosphate group from the HPt to the aspartate residue located in the REC domain of the response regulator proteins (Jacob and Thines, 2017). In prokaryotes, HPts are linked to the C-terminal of hybrid histidine kinases (HHKs). However, eukaryotic HPts are distinctive proteins that can traffic between the nucleus and the cytosol and expedite communication between the response regulator (RR) and HK. Consequently, they work as mediator proteins between these units, interacting with other proteins or signaling cascades apart from the multistep phosphorelay pathway (Singh et al., 2021).

It was stated that *S. cerevisiae* was the first to have a multistep phosphorelay system identified in its signaling pathway. This MSP involved HOG pathway components, which included the phosphotransfer protein Ypd1p and the hybrid histidine kinase Sln1p (Kaur et al., 2021). Filamentous pathogenic fungi exhibit several HHK-coding genes, unlike *S. cerevisiae*, which only possesses one HHK-coding gene, SLN1 (Catlett et al., 2003). The former fungi, including *Magnaporthe oryzae*, *Fusarium verticillioides*, *A. nidulans*, *Cochliobolus heterostrophus*, *Botrytis cinerea*, *Neurospora crassa*, and *C. albicans*, have a high prevalence of HHKs (Jacob et al., 2014). Although the HOG pathway's MSP is fungi's most studied signaling pathway, the exact chemical mechanisms underlying phosphotransfer are still imprecise and lack sufficient records. For instance, in *M. oryzae* (a filamentous fungus), MoHog1p is phosphorylated at Y173 and T171 within the dual phosphorylation motif during HOG pathway activation via osmotic stress. MoHog1p drifts into the nucleus due to this phosphorylation occurrence, initiating the cellular stress response (Bohnert et al., 2019a, 2019b).

In addition to osmoregulation, MSPs also play a dynamic role in supervising fungi's significant cellular functions and responses under numerous stressful scenarios like oxidative damage, light, or osmotic stress (Jacob et al., 2022). They also play a proficient role in regulating all facets of fungal physiology. Hence, the exploration of MSP in filamentous fungi is indispensable not only for understanding the molecular mechanisms of fungal signaling but also because of the fungicides that particularly target the HOG pathway. Group III HHKs unique to filamentous fungi include MoHik1p in *M. oryzae*, Drk1p in *Debaryomyces hansenii*, and Nik1p in *C. albicans*. To date, no homologs have been discovered in plants or animals. The efficacious mechanism of the synthetic fungicide fludioxonil is mediated by these HHKs (Kaur et al., 2021). As a result, research on MSPs provides novel avenues for creating innovative antifungal substances with little toxicity to other species (Braicu et al., 2019; Liao et al., 2021).

9.4 LIPID SIGNALING

Lipid signaling networks have become a major area of study in recent years. Lipid signaling is now understood to occur in microbial cells, especially in yeasts and molds, despite being mostly researched in mammalian cells (Singh and Del Poeta, 2011). Lipids function as signaling molecules in pathophysiological processes in mammalian cells, including inflammation, cancer, autoimmune illnesses, cardiovascular disease, and neurological disorders (Haspula and Clark, 2020). Changes in lipid signaling networks may cause disrupted cellular homeostasis, leading to various disorders. Lipids play structural roles in plants but also regulate and transmit signals to different parts of the cell. In particular, sphingolipids function as regulatory signaling molecules in a variety of processes, including mineral ion homeostasis, temperature-induced signal transduction, cell division and growth, salt and drought tolerance, membrane stability, pollen development, cell type differentiation and organogenesis, cell wall formation, stomatal closure, and interactions between plants and microbes (Bersching and Jacob, 2021).

Fungal lipid signaling is important in hypervirulence, which increases fungal resistance to host cells or cell death induced by environmental stressors, including immunological responses. Sphingolipids, oxylipins, and farnesol are fungal pathogens' primary lipid signaling molecules (Singh and Del Poeta, 2011) that mediate certain biological activities, such as cell death, differentiation, and proliferation. A thorough understanding of the molecular mechanisms underlying fungal lipid signaling can allow researchers to present well-explained pathophysiological mechanisms directed by lipids and develop novel therapeutic methods (Rhome et al., 2007; Rhome and Del Poeta, 2010).

Many studies on sphingolipids in *Cryptococcus neoformans* (a yeast-like fungus) have suggested that sphingolipids are essential to control virulence via directing melanin synthesis by forming diacylglycerol, which in turn affects protein kinase C1 (Pkc1p) activity (Luberto et al., 2001; Heung et al., 2004). Sphingolipids affect signaling events by stimulating the transcription factor 2 (Atf2p), which then stimulates the antiphagocytic protein 1 (App1p), promoting phagocytosis (Mare et al., 2005; Tommasino et al., 2008). Furthermore, sphingolipids govern fungal growth in both intracellular and external circumstances via the pharmacological activity of both inositol phosphosphingolipid phospholipase C1 (Isc1p) and GlcCer synthase (Gcs1p) (Luberto et al., 2001; Heung et al., 2004).

These lipid-driven mechanisms majorly impact *C. neoformans* virulence in the lung environment, ultimately affecting how the disease progresses. Remarkably, environment-specific sphingolipids regulate the intra- and extracellular growth of *C. neoformans*, suggesting that the fungus utilizes distinct lipid molecules corresponding to the specific host cell compartment it inhabits. More precisely, the expression pattern of various genes involved in sphingolipid metabolism, including Ipc1p, increases when *C. neoformans* is found in the host's macrophages (Fan et al., 2005). Research corroborating this theory has shown that the enzymes Ipc1p and Isc1p are crucial for fungal cells to withstand the changing environment as they change from neutral or alkaline to acidic pH levels (Garcia et al., 2008). This is validated by the observation that *C. neoformans* gets into the body via the respiratory system, where it encounters alveolar macrophage phagolysosomes having an acidic pH and alveolar spaces with a neutral pH. Therefore, a thorough comprehension of fungal lipid signaling and *C. neoformans* adaptations to the changing environment would enhance our understanding of the pathogen's interactions with its hosts.

9.5 PHEROMONE AND GLUCOSE SIGNALING

Microorganisms have developed intricate molecular mechanisms to direct their surroundings and respond to pheromones and nutritional cues. The most extensive transmembrane receptors, identified as G-protein-coupled receptors (GPCRs), are important for identifying these signals and are likely imperative mediators for host–microbe interactions in eukaryotes (Cohen et al., 2017; Brown et al., 2018). All GPCR family members have a primary architecture comprising seven transmembrane domains and a communal molecular mechanism of signal transduction, regardless of their significant functional and sequence discrepancy. In various fungi, GPCRs are critical for nutrition and pheromone sensing. The function of GPCRs in fungal pheromone detection has been well explored. Fungal mating is known to be triggered by the binding of pheromones to GPCRs on the cell surface (Bahn et al., 2007).

In ascomycetes like *S. cerevisiae*, two pheromones are identified and released for opposing mating types by two independent GPCRs (Ste2p and Ste3p) (Naider and Becker, 2020). These pheromones are known as the α and a sex peptide pheromones, respectively, and they trigger the Ste2p and Ste3p GPCRs, which ultimately activates the Gα protein Gpa1p through GDP–GTP exchange. The MAPK cascade is initiated following Gpa1p's dissociation from the βγ dimer, composed of Ste4p and Ste18p (Velazhahan et al., 2021). The GPCR architecture is substantially preserved in the ascomycete phylum, as demonstrated by species such as *Schizosaccharomyces pombe*, *C. neoformans*, *A. nidulans*, *M. oryzae*, and *N. crassa* (Jacob et al., 2022).

Many species use glucose as their main source of carbon, and it has an immense impact on how carbon metabolism is regulated and other cellular characteristics. Consequently, organisms have evolved sophisticated glucose-sensing mechanisms. Because glucose serves as both a food and a signaling molecule, with overlapping roles, understanding the molecular underpinnings of glucose sensing has proven challenging. To identify and transport glucose, fungi have evolved a variety of techniques (Rolland et al., 2002).

In *S. cerevisiae*, for instance, the GPCR sugar receptor Gpr1p detects both glucose and sucrose, which causes Gpa2p to become activated. Consequently, elevated cAMP levels and PKA activation are caused by stimulating adenylyl cyclase (Lemaire et al., 2004). Additionally, Gpr1p homologs in *C. albicans* (CaGpr1p) and *S. pombe* (Git3p receptor) detect glucose and initiate cAMP signaling, which controls morphogenesis and the change from yeast to hyphal growth. In *N. crassa*, the Gpr1p homolog Gpr4p interacts carbon source-dependently with the Gα subunit Gna1p to regulate asexual development and cAMP production (Brown et al., 2015). One can distinguish between the nutritional role of glucose and its regulatory or signaling role by noting that many glucose-induced actions require glucose metabolism. On the other hand, it appears that most glucose-induced signal transduction pathways function independently of glucose metabolism.

Apart from GPCRs, *S. cerevisiae* has a family of hexose transporters (Hxts) that are involved in transport as well as sugar sensing (Nijland and Driessen, 2020). Cells can adapt their cellular physiology and growth by controlling glucose uptake and metabolism in response to external factors with the help of these varied low- and high-affinity Hxts. Twelve putative transmembrane domains are commonly found in members of the HXT family; prominent examples of these include the genes SNF3, RGT2, and HXT1–17, which are connected to the sugar transporter (Scharff-Poulsen et al., 2018). This demonstrates how important transporters are in the process of glucose sensing. Interestingly, Snf3p and Rgt2p serve as extracellular glucose sensors to control the expression of HXT genes despite having a connection to transporters and are distinguished from other popular Hxts by this feature (Bharatula and Broach, 2018).

REFERENCES

Albuquerque, P. and A. Casadevall. 2012. Quorum sensing in fungi–a review. *Med. Mycol.* 50:337–345.

Avbelj, M., J. Zupan, L. Kranjc and P. Raspor. 2015. Quorum-sensing kinetics in Saccharomyces cerevisiae: A symphony of ARO genes and aromatic alcohols. *J. Agric. Food Chem.* 63:8544–8550.

Bahn, Y.-S., C. Xue, A. Idnurm, J.C. Rutherford, J. Heitman and M.E. Cardenas. 2007. Sensing the environment: Lessons from fungi. *Nat. Rev. Microbiol.* 5:57–69.

Bassler, B.L. and R. Losick. 2006. Bacterially speaking. *Cell* 125:237–246.

Bayram, Ö.S. and Ö. Bayram. 2023. An anatomy of fungal eye: Fungal photoreceptors and signalling mechanisms. *J. Fungi* 9:591.

Bersching, K. and S. Jacob. 2021. The molecular mechanism of fludioxonil action is different to osmotic stress sensing. *J. Fungi* 7:393.

Bharatula, V. and J.R. Broach. 2018. The nutrient stress response in yeast. *Stress response Mech. fungi Theor. Pract. Asp.* 131–159.

Bohnert, S., L. Heck, C. Gruber, H. Neumann, U. Distler, S. Tenzer, A. Yemelin, E. Thines and S. Jacob. 2019a. Fungicide resistance toward fludioxonil conferred by overexpression of the phosphatase gene MoPTP2 in Magnaporthe oryzae. *Mol. Microbiol.* 111:662–677.

Bohnert, S., H. Neumann, E. Thines and S. Jacob. 2019b. Visualizing fungicide action: An in vivo tool for rapid validation of fungicides with target location HOG pathway. *Pest Manag. Sci.* 75:772–778.

Bourret, R.B., C.A. Foster and W.E. Goldman. 2021. Predicted Functional and Structural Diversity of Receiver Domains in Fungal Two-Component Regulatory Systems. *Msphere* 6:10–1128.

Braicu, C., M. Buse, C. Busuioc, R. Drula, D. Gulei, L. Raduly, A. Rusu, A. Irimie, A.G. Atanasov and O. Slaby. 2019. A comprehensive review on MAPK: A promising therapeutic target in cancer. *Cancers (Basel).* 11:1618.

Brown, N.A., T.F. Dos Reis, L.N.A. Ries, C. Caldana, J. Mah, J. Yu, J.M. Macdonald and G.H. Goldman. 2015. G-protein coupled receptor-mediated nutrient sensing and developmental control in A spergillus nidulans. *Mol. Microbiol.* 98:420–439.

Brown, N.A., S. Schrevens, P. Van Dijck and G.H. Goldman. 2018. Fungal G-protein-coupled receptors: Mediators of pathogenesis and targets for disease control. *Nat. Microbiol.* 3:402–414.

Catlett, N.L., O.C. Yoder and B.G. Turgeon. 2003. Whole-genome analysis of two-component signal transduction genes in fungal pathogens. *Eukaryot. Cell* 2:1151–1161.

Chen, H. and G.R. Fink. 2006. Feedback control of morphogenesis in fungi by aromatic alcohols. *Genes Dev.* 20:1150–1161.

Cohen, L.J., D. Esterhazy, S.-H. Kim, C. Lemetre, R.R. Aguilar, E.A. Gordon, A.J. Pickard, J.R. Cross, A.B. Emiliano and S.M. Han. 2017. Commensal bacteria make GPCR ligands that mimic human signalling molecules. *Nature* 549:48–53.

Defosse, T.A., A. Sharma, A.K. Mondal, T. Dugé de Bernonville, J. Latgé, R. Calderone, N. Giglioli-Guivarc'h, V. Courdavault, M. Clastre and N. Papon. 2015. Hybrid histidine kinases in pathogenic fungi. *Mol. Microbiol.* 95:914–924.

Fan, W., P.R. Kraus, M.-J. Boily and J. Heitman. 2005. Cryptococcus neoformans gene expression during murine macrophage infection. *Eukaryot. Cell* 4:1420–1433.

Garcia, J., J. Shea, F. Alvarez-Vasquez, A. Qureshi, C. Luberto, E.O. Voit and M. Del Poeta. 2008. Mathematical modeling of pathogenicity of Cryptococcus neoformans. *Mol. Syst. Biol.* 4:183.

González-Rubio, G., T. Fernández-Acero, H. Martín and M. Molina. 2019. Mitogen-activated protein kinase phosphatases (MKPs) in fungal signaling: Conservation, function, and regulation. *Int. J. Mol. Sci.* 20:1709.

Haspula, D. and M.A. Clark. 2020. Cannabinoid receptors: An update on cell signaling, pathophysiological roles and therapeutic opportunities in neurological, cardiovascular, and inflammatory diseases. *Int. J. Mol. Sci.* 21:7693.

Hérivaux, A., Y.-S. So, A. Gastebois, J.-P. Latgé, J.-P. Bouchara, Y.-S. Bahn and N. Papon. 2016. Major sensing proteins in pathogenic fungi: The hybrid histidine kinase family. *PLoS Pathog.* 12:e1005683.

Heung, L.J., C. Luberto, A. Plowden, Y.A. Hannun and M. Del Poeta. 2004. The sphingolipid pathway regulates Pkc1 through the formation of diacylglycerol in Cryptococcus neoformans. *J. Biol. Chem.* 279:21144–21153.

Huo, R., Z. Liu, X. Yu and Z. Li. 2020. The interaction network and signaling specificity of two-component system in Arabidopsis. *Int. J. Mol. Sci.* 21:4898.

Jacob, S., S. Bühring and K. Bersching. 2022. Recent advances in research on molecular mechanisms of fungal signaling. *Encyclopedia* 2:840–863.

Jacob, S., A.J. Foster, A. Yemelin and E. Thines. 2014. Histidine kinases mediate differentiation, stress response, and pathogenicity in Magnaporthe oryzae. *Microbiologyopen* 3:668–687.

Jacob, S. and E. Thines. 2017. Multistep phosphorelay in fungi: The enigma of multiple signals and a limited number of signaling pathways. *Mycol. Prog.* 16:1007–1013.

Jamil, F., H. Mukhtar, M. Fouillaud and L. Dufossé. 2022. Rhizosphere signaling: Insights into plant–rhizomicrobiome interactions for sustainable agronomy. *Microorganisms* 10:899.

Kabbara, S., A. Hérivaux, T. Dugé de Bernonville, V. Courdavault, M. Clastre, A. Gastebois, M. Osman, M. Hamze, J.M. Cock and P. Schaap. 2019. Diversity and evolution of sensor histidine kinases in eukaryotes. *Genome Biol. Evol.* 11:86–108.

Kaur, H., S.P. Sasan, A. Yadav, Y. Martoliya and A.K. Mondal. 2021. Distinct role of HAMP and HAMP-like linker domains in regulating the activity of Hik1p, a hybrid histidine kinase 3 from Magnaporthe oryzae. *Mol. Genet. Genomics* 296:1135–1145.

Kebaara, B.W., M.L. Langford, D.H. Navarathna, R. Dumitru, K.W. Nickerson and A.L. Atkin. 2008. Candida albicans Tup1 is involved in farnesol-mediated inhibition of filamentous-growth induction. *Eukaryot. Cell* 7:980–987.

Kosal, E. 2023. The Fungi Kingdom. *Introd. Biol. Ecol. Evol. Biodivers.*

Kour, D., K.L. Rana, N. Yadav, A.N. Yadav, J. Singh, A.A. Rastegari and A.K. Saxena. 2019. Agriculturally and industrially important fungi: Current developments and potential biotechnological applications. *Recent Adv. white Biotechnol. through fungi Vol. 2 Perspect. value-added Prod. Environ.* 1–64.

Lemaire, K., S. Van de Velde, P. Van Dijck and J.M. Thevelein. 2004. Glucose and sucrose act as agonist and mannose as antagonist ligands of the G protein-coupled receptor Gpr1 in the yeast Saccharomyces cerevisiae. *Mol. Cell* 16:293–299.

Liao, B., X. Ye, X. Chen, Y. Zhou, L. Cheng, X. Zhou and B. Ren. 2021. The two-component signal transduction system and its regulation in Candida albicans. *Virulence* 12:1884–1899.

Luberto, C., D.L. Toffaletti, E.A. Wills, S.C. Tucker, A. Casadevall, J.R. Perfect, Y.A. Hannun and M. Del Poeta. 2001. Roles for inositol-phosphoryl ceramide synthase 1 (IPC1) in pathogenesis of C. neoformans. *Genes Dev.* 15:201–212.

Mare, L., R. Iatta, M.T. Montagna, C. Luberto and M. Del Poeta. 2005. APP1 transcription is regulated by inositol-phosphorylceramide synthase 1-diacylglycerol pathway and is controlled by ATF2 transcription factor in Cryptococcus neoformans. *J. Biol. Chem.* 280:36055–36064.

McLean, T.C., R. Lo, N. Tschowri, P.A. Hoskisson, M.M. Al Bassam, M.I. Hutchings and N.F. Som. 2019. Sensing and responding to diverse extracellular signals: An updated analysis of the sensor kinases and response regulators of Streptomyces species. *Microbiology* 165:929–952.

Naider, F. and J.M. Becker. 2020. A paradigm for peptide hormone-GPCR analyses. *Molecules* 25:4272.

Nair, A., P. Chauhan, B. Saha and K.F. Kubatzky. 2019. Conceptual evolution of cell signaling. *Int. J. Mol. Sci.* 20:3292.

Nijland, J.G. and A.J.M. Driessen. 2020. Engineering of pentose transport in Saccharomyces cerevisiae for biotechnological applications. *Front. Bioeng. Biotechnol.* 7:464.

Nikolaou, E., I. Agrafioti, M. Stumpf, J. Quinn, I. Stansfield and A.J.P. Brown. 2009. Phylogenetic diversity of stress signalling pathways in fungi. *BMC Evol. Biol.* 9:1–18.

Padder, S.A., R. Prasad and A.H. Shah. 2018. Quorum sensing: A less known mode of communication among fungi. *Microbiol. Res.* 210:51–58.

Patwardhan, S.B., C. Pandit, S. Pandit, D. Verma, D. Lahiri, M. Nag, R.R. Ray, P. Jha and R. Prasad. 2023. Illuminating the signalomics of microbial biofilm on plant surfaces. *Biocatal. Agric. Biotechnol.* 47:102537.

Pertseva, M.N. and A.O. Shpakov. 2009. The prokaryotic origin and evolution of eukaryotic chemosignaling systems. *Neurosci. Behav. Physiol.* 39:793–804.

Rana, K.L., D. Kour, I. Sheikh, A. Dhiman, N. Yadav, A.N. Yadav, A.A. Rastegari, K. Singh and A.K. Saxena. 2019. Endophytic fungi: Biodiversity, ecological significance, and potential industrial applications. *Recent Adv. white Biotechnol. through fungi Vol. 1 Divers. Enzym. Perspect.* 1–62.

Rhome, R. and M. Del Poeta. 2010. Sphingolipid signaling in fungal pathogens. *Sphingolipids as Signal. Regul. Mol.* 232–237.

Rhome, R., T. McQuiston, T. Kechichian, A. Bielawska, M. Hennig, M. Drago, G. Morace, C. Luberto and M. Del Poeta. 2007. Biosynthesis and immunogenicity of glucosylceramide in Cryptococcus neoformans and other human pathogens. *Eukaryot. Cell* 6:1715–1726.

Rolland, F., J. Winderickx and J.M. Thevelein. 2002. Glucose-sensing and-signalling mechanisms in yeast. *FEMS Yeast Res.* 2:183–201.

Sarsaiya, S., J. Shi and J. Chen. 2019. A comprehensive review on fungal endophytes and its dynamics on Orchidaceae plants: Current research, challenges, and future possibilities. *Bioengineered* 10:316–334.

Scharff-Poulsen, P., H. Moriya and M. Johnston. 2018. Genetic analysis of signal generation by the Rgt2 glucose sensor of Saccharomyces cerevisiae. *G3 Genes Genomes Genet.* 8:2685–2696.

Singh, A. and M. Del Poeta. 2011. Lipid signalling in pathogenic fungi. *Cell. Microbiol.* 13:177–185.

Singh, D., P. Gupta, S.L. Singla-Pareek, K.H.M. Siddique and A. Pareek. 2021. The journey from two-step to multi-step phosphorelay signaling systems. *Curr. Genom.* 22:59–74.

Stock, A.M., V.L. Robinson and P.N. Goudreau. 2000. Two-component signal transduction. *Annu. Rev. Biochem.* 69:183–215.

Swetha, T.K., A. Priya and S.K. Pandian. 2021. Natural molecules against QS-associated biofilm formation of pathogens. In *Microbial and Natural Macromolecules*. Elsevier, pp. 317–348.

Tian, X., H. Ding, W. Ke and L. Wang. 2021. Quorum sensing in fungal species. *Annu. Rev. Microbiol.* 75:449–469.

Tommasino, N., M. Villani, A. Qureshi, J. Henry, C. Luberto and M. Del Poeta. 2008. Atf2 transcription factor binds to the APP1 promoter in Cryptococcus neoformans: Stimulatory effect of diacylglycerol. *Eukaryot. Cell* 7:294–301.

Velazhahan, V., N. Ma, G. Pándy-Szekeres, A.J. Kooistra, Y. Lee, D.E. Gloriam, N. Vaidehi and C.G. Tate. 2021. Structure of the class D GPCR Ste2 dimer coupled to two G proteins. *Nature* 589:148–153.

Winters, M., N. Arneborg, R. Appels and K. Howell. 2019. Can community-based signalling behaviour in Saccharomyces cerevisiae be called quorum sensing? A critical review of the literature. *FEMS Yeast Res.* 19:foz046.

Wongsuk, T., P. Pumeesat and N. Luplertlop. 2016. Fungal quorum sensing molecules: Role in fungal morphogenesis and pathogenicity. *J. Basic Microbiol.* 56:440–447.

Zanne, A.E., K. Abarenkov, M.E. Afkhami, C.A. Aguilar-Trigueros, S. Bates, J.M. Bhatnagar, P.E. Busby, N. Christian, W.K. Cornwell and T.W. Crowther. 2020. Fungal functional ecology: Bringing a trait-based approach to plant-associated fungi. *Biol. Rev.* 95:409–433.

10 Industrial Applications of Fungi

Mohamed Mohany, Faiz Ur Rahman, Ayesha Ahmed, Salim S. Al-Rejaie, Riffat Malik, Munir Ahmad, Shahzad Munir, and Jinhao Zhang

10.1 INTRODUCTION

Fungi are eukaryotes that commonly undergo both asexual and sexual reproduction, featuring a distinctive sturdy cell wall composed of cellulose, chitin, or a combination of both. As osmotrophic chemoheterotrophs, they generally metabolize various substrates, from simple sugars to complex carbohydrates. Although most fungi thrive in aerobic conditions, some exhibit obligate or facultative anaerobic characteristics. Fungi release extracellular enzymes such as xylanases, pectinases, and cellulases that facilitate the breakdown of insoluble materials into soluble components. With hundreds of species, fungi hold significant economic importance for humanity, as our daily lives are intricately intertwined with their beneficial or detrimental effects (Little et al., 2012).

Mycology, originating from the Greek words "mykes," meaning mushroom, and "logos" (discourse), refers to the exploration of fungi. Fungi exhibit a global distribution and thrive in diverse habitats, including harsh environments. The collective assembly of all fungi within a specific area or geographical region is called mycobiota. Taxonomists have documented approximately 120,000 fungal species. Fungi play crucial roles in various domains, such as providing antibiotics in medicine, contributing to soil fertility, and causing fruit and crop diseases in agriculture, serving as the foundation for numerous industries and as essential components of the food chain. Additionally, some fungi are useful resources for studying basic biological processes (Hawksworth and Lücking, 2017).

Fungi are organisms that rely on chemical sources for nutrition and are found in subsoil and subaerial environments. They are essential decomposers and are significant symbiotic partners for plants and animals. They can act as pathogens and contribute to the deterioration of both synthetic and natural materials. They are essential to preserving soil structure because of their frequent exopolymer synthesis and filamentous branching development. Most fungi exhibit a filamentous growth pattern, and some display polymorphism, existing in both unicellular yeast or yeast-like forms and filamentous mycelia. Endophytic fungi serve as valuable sources of natural bioactive compounds with potential uses in various fields, including the food, agriculture, and pharmaceutical industries (Solomon et al., 2019)

DOI: 10.1201/9781003452362-10

Some fungi, especially yeasts and mold, have detrimental effects by causing the deterioration of stored commodities like rubber, textiles, food, timber, plastic, glass, and even leather. Conversely, they are useful sources of industrial chemicals such as enzymes, acetone, alcohols, and antibiotics. Additionally, these fungi are essential in the fermentation process, contributing to the creation of vinegar, bread dough, alcoholic beverages, and cheese. Fungi serve as the foundation for numerous vital industries. The transition from chemical to biological processing, utilizing enzymes from fungi instead of chemicals in various industries like leather, paper, pulp, and textiles, has considerably mitigated adverse environmental impacts (Hawksworth and Lücking, 2017).

Numerous microbial enzymes, particularly those derived from fungi, are recognized for their crucial roles in the food processing sector, and their significance is expected to grow significantly in the years ahead. Fungal enzymes are widely employed to optimize overall processes or introduce additional flavors. Additionally, enzyme-mediated processes are favored for their high specificity and minimal production of by-products. The use of enzymes in the food processing industry has historical roots, with alcoholic fermentation involving yeasts being the oldest known enzyme-mediated process. Today, enzymes find applications in various sectors such as sugar, meat, dairy, brewing, baking, fruit processing, and more. Among the diverse enzymes utilized are tannases, lipases, amylases, gluco-oxidases, proteases, pectinases, and others (Saxena et al., 2001).

Incorporating microbial enzymes into detergents has led to effective laundry cleaning even at lower temperatures, resulting in a notable decrease in CO_2 emissions. Various industrial processes leverage the biochemical capabilities of specific fungi to achieve positive outcomes. Fungi play a crucial part in tackling significant global challenges, and using fungal processes and products can enhance sustainability by optimizing the utilization of natural resources. The applications span from enhancing biowaste to create enriched products to utilizing renewable biomass from plants as a viable alternative to oil-based products like fuel, plastics, biochemicals, and fertilizers. Introducing fungal inocula with seeds in soil can enhance the growth of plants by improving water and nutrient absorption, which is crucial for sustaining crop yields in the changing climate conditions. Fungal enzymes produce prebiotic food ingredients, contributing to a more robust human gut microbiota and increased resistance against lifestyle-related diseases. Likewise, the utilization of fungi presents a convenient approach to enhance the quality of animal feed and reduce reliance on antibiotics, particularly in sectors such as meat production, which currently contributes significantly to the emergence of bacterial resistance to multiple drugs. Fungi represent a promising reservoir for discovering novel drug candidates and antimicrobial agents. Additionally, fungi hold considerable potential for revolutionizing the production of biological medicines and various bio-based products, offering a fresh avenue for manufacturing (Solomon et al., 2019).

10.2 APPLICATIONS OF FUNGI IN THE PHARMACEUTICAL INDUSTRY

The pharmaceutical sector plays a crucial role in healthcare by creating, producing, and promoting medications. It is an essential component of the healthcare system, and each year, countless patients with life-threatening illnesses benefit from medications derived from fungi. Exploring the vast array of secondary metabolites produced by fungi, like antibiotics and cholesterol-lowering drugs, has spurred efforts to understand the chemical potential of fungal strains. This involves the discovery of new strategies to activate associated biosynthetic genes. Fungi exist in diverse habitats and can develop various survival mechanisms, making them a valuable resource for the pharmaceutical industry. Their presence in various environments enhances their potential for contributing to the development of medicinal drugs. Furthermore, the relatively straightforward cultivation and extraction processes make fungi an important and cost-effective source for the pharmaceutical industry in producing therapeutic drugs (Dutta et al., 2022).

10.3 ANTIBIOTIC PRODUCTION

Sir Alexander Fleming initially identified fungi's involvement in synthesizing antibiotic compounds in 1929. One such organic substance with antimicrobial properties is penicillin, derived from a fungus belonging to the green mildew *Penicillium* genus that belongs to Deuteromycetes. Various species within the *Penicillium* genus, including *Penicillium chrysogenum* and *Cenococcum* spp., are utilized in antibiotic production. Fungi are important in the development of significant and innovative medications. One of them is cyclosporine, an anti-rejection substance that has substantially contributed to the advancement of organ transplant procedures in recent years. The term "antibiotic" refers to an organic substance produced by microorganisms, inhibiting the growth of specific other organisms. Streptomycin, derived from *Streptomyces griseus*, holds significant medical value, particularly for its efficacy against gram-negative penicillin-resistant organisms. Furthermore, *Aspergillus* cultures have been a source of several antibiotics (Neely and Ghannoum, 2000).

10.4 SECONDARY METABOLITES

Aspergillus nidulans is employed to isolate secondary metabolites. Drugs are derived from these secondary metabolites. Among these secondary metabolites are ergot alkaloids (Ergotamine and ergometrine) and lovastatin, a well-known drug that reduces cholesterol levels. Fungal metabolites exhibit diverse therapeutic properties, including antibacterial, antiviral, immunosuppressive, and antitumor activities.

10.4.1 ANTIFUNGAL

Human fungal infections range from surface-level conditions affecting the skin and nails to widespread and potentially life-threatening diseases. Human health is increasingly at risk from serious invasive fungal infections from pathogens such as *Aspergillus* species, *Candida* spp., and *Cryptococcus neoformans*. Griseofulvin, derived from the *Penicillium griseofulvum* mycelium and various other species, exhibits antifungal characteristics. It disrupts hyphal growth by interrupting the formation of cells. The fungal cell wall is one of the targets for antifungal agents. The polysaccharide glucan is an important cell wall component, so inhibitors of glucan exhibit antifungal activities (Pela, 2004).

Echinocandins represent a category of antifungal medications with fungicidal effects on various fungi, including *Candida* species, while exerting fungistatic properties on the *Aspergillus* genus. This drug class has been observed to induce less severe side effects in comparison to polyenes and azoles. Since the initial identification of the first echinocandin in the early 1970s, various fungi have yielded numerous members of this family (Vicente et al., 2003). Pneumocandins, a subgroup originating from the fungus *Glarea lozoyensis* through fermentation, have proven effective in creating a new recently introduced antifungal medication. Clinical trials have shown that caspofungin originating from the fungus *Glarea lozoyensis* is well received and effective in treating aspergillosis, candidiasis, invasive candidiasis, and esophageal infection (Mora-Duarte et al., 2002).

Even though sphingolipids are only found in small amounts within the cytoplasmic membrane of fungi, they play a crucial role in cellular functions. Inhibiting sphingolipid synthesis leads to the inhibition of growth and the death of cells. Since several enzymes are unique to fungi, sphingolipid synthesis can be a target for antifungal therapy. In the quest for novel antifungals, three pivotal enzymes, namely, serine palmitoyltransferase, inositol phosphoceramide synthase, and ceramide synthase, in this pathway have been targeted, and inhibitors of these enzymes have been identified in fungi. Serine palmitoyltransferase is inhibited by viridiofungins, sphingofungins, and sphingosine-like metabolites obtained from *Colletotrichum acutatum*. Ceramide synthase is inhibited by fumonisin B1 and australifungin, while inositol phosphoceramide synthase is inhibited by khafrefungin and aureobasidins (Mandala and Harris, 2000).

Targeting protein synthesis has long been an appealing strategy for developing antimicrobial agents. However, because fungi are eukaryotic and their protein synthesis machinery is similar to that of mammals, applying this idea to antifungal medication presents difficulties. Despite this challenge, elongation factor 2 (EF-2) has become a viable target for antifungals among the soluble translation factors now in use because of its functional differences from its mammalian counterpart. The sordarin family of antifungal drugs, which was isolated from the fermentations of *Sordaria araneosa*, is the main group functioning at this level. Early findings revealed that sordarin targeted a protein that participates in mRNA translation. Strong and specific translation inhibitors in fungi, sordarins work by maintaining the stability of the fungal EF-2–ribosome complex. Compounds in this category demonstrated inhibitory effects on in vitro translation in various *Candida* species such as *C. kefyr*, *C. neoformans*, *C. albicans*, and *C. tropicalis* to varying extents, while showing no activity against *C. parapsilosis*, *C. krusei*, or *C. glabrata* (Domínguez and Martín, 1998).

10.4.2 Antitumor

The large puffball mushroom *Clavatia* contains calvacin, an anticancer substance, and consuming this fungus has been associated with a decreased risk of stomach tumors. Illudins, a class of sesquiterpene antibiotics, are frequently recognized for their antibacterial and antitumor properties. They are obtained from basidiomycetes belonging to the genera *Coprinus* and *Omphalotus* (Del Val et al., 2003). A derivative of illudin S, known as irofulvene, has demonstrated significant effectiveness in animal models of cancer as a standalone treatment or in conjunction with other anticancer agents. While the exact mechanism of action remains unclear in tumor cells, irofulvene has been demonstrated to cause caspase-mediated apoptosis (Wang et al., 2002).

Another antitumor compound derived from fungi that has undergone human testing is rhizoxin, obtained from *Rhizopus chinensis*. This compound functions as an antimitotic agent, inhibiting tubulin polymerization. Similar mechanisms of action have been described for other fungal compounds, including phomopsin A produced by *Phomopsis leptostromiformis* and the related compound ustiloxin A produced by *Ustilaginoidea virens*. However, these compounds have not yet been tested in human beings (Carter, 2018).

In recent years, one of the most promising targets for antitumor research has been histone deacetylase (HDAC). HDAC inhibitors exhibit de-transforming activity and can cause oncogene-transformed cells to revert to their normal morphology. A reversible HDAC inhibitor is obtained from *Streptomyces hygroscopicus* (Ververis and Karagiannis, 2012). Hypocrellins, pigments isolated from *Shiraia bambusicola* and *Hypocrella bambusae*, are utilized in traditional Chinese medicine and have been extensively investigated for their capacity to sensitize cells to ultrasonic waves and light. These polycyclic quinones affect cells by producing hydrogen peroxide after photoactivation, which triggers caspase-3 activation and apoptosis (Ali et al., 2002). Brefeldin A is a fungal metabolite, originally identified as an antibiotic in 1958. This compound is obtained from *Penicillium* and *Ascochyta* species. Subsequently, it was revealed to possess various activities, including antitumor, antiviral, antifungal, and nematocidal effects (Fox et al., 2001).

10.4.3 Antibacterial

The first biological action of fungal metabolites with therapeutic significance was antibiosis against bacteria, with the landmark discovery and commercial development of penicillin from *Penicillium chrysogenum*. After the discovery of penicillin, other fungus-derived antibiotics have been successfully brought to the market, with fusidic acid and cephalosporin particularly noteworthy. Cephalosporin, similar to penicillin, functions as a β-lactam antibiotic and inhibits the formation of bacterial cell walls. In the 1940s, it was originally isolated from the *Acremonium chrysogenum* species. Cephalosporin has served as the foundation for various generations of semisynthetic derivatives (such as ceftriaxone, cefixime, cephalexin, cefotaxime, ceftazidime, cefuroxime, and cefaclor), with some of them remaining among the top-selling antibiotics (Vilvanathan, 2021).

The bacteriostatic antibiotic fusidic acid is derived from *Fusidium coccineum* and other species. In bacteria, it hinders protein synthesis by targeting elongation factor G. Fusidic acid is effective against gram-positive bacteria; it does not exhibit activity against gram-negative bacteria. Its primary applications include topical treatment for skin and ocular infections. Additionally, it is prescribed for intestinal infections caused by *Clostridium difficile* and infections in various organs caused by *Staphylococcus aureus* and other gram-positive bacteria (Sharma et al., 2022). Another metabolite, clerocidin, obtained from *Fusidium viride* and similar species, demonstrates a wide range of antibacterial activities, proving to be more potent than ciprofloxacin against anaerobic and gram-positive bacteria (Misiek and Hoffmeister, 2007).

10.4.4 Atherosclerotic

The considerable recognition of fungi as a compelling source of pharmaceutically valuable compounds is largely because of statins, which are inhibitors of 3-hydroxymethylglutaryl-CoA reductase and represent extensively utilized cholesterol-lowering drugs. The initial natural statin, lovastatin, was pivotal in establishing this reputation. An enzyme frequently targeted for developing cholesterol-lowering drugs is acyl-CoA cholesterol acyltransferase (ACAT), known for its role in cholesterol accumulation. Numerous compounds derived from fungi, including those isolated from *Aspergillus fumigatus*, have been identified as inhibitors of ACAT (Rho et al., 2002).

Cholesterol ester transfer protein (CETP) has been targeted to develop agents to enhance high-density lipoproteins (HDLs) that are anticipated to offer protective effects against atherosclerosis. CETP is involved in the reverse cholesterol transport process from HDL to LDL. Fungal inhibitors of CETP encompass a range of derivatives of rosenonolactone derived from *Trichothecium roseum*, an apiosporamide-related metabolite reported from *Cytospora* spp. (Zhao et al., 2014).

10.4.5 Antiparasitic

Fungi are recognized for producing numerous compounds with antiparasitic properties. Apicidin, a cyclic tetrapeptide obtained from a species of *Fusarium*, was effective against various apicomplexan parasites. It is a strong inhibitor of the parasites' histone deacetylases (HDACs). In mice, picaridin demonstrates extensive in vitro efficacy against *Apicomplexa* and exhibits in vivo effectiveness against *Plasmodium berghei* malaria (Colletti et al., 2001).

Paraherquamide is a versatile anthelminthic agent generated by *Penicillium paraherquei*. This compound exhibits significant activity against various nematode strains. Another class of anthelminthic agents, the aspergillimides, obtained from *Aspergillus* spp., demonstrates even greater potency than paraherquamide. Nodulisporic acids represent innovative indole diterpenes with potent insecticidal properties, and *Nodulisporium* species synthesize these acids. Their mechanism of action involves the activation of insect glutamate-gated chloride channels (Polishook et al., 2001). Chitinases, enzymes capable of breaking down linear polymers of chitin, are potential targets for developing insecticidal and antimalarial drugs. Inhibitors, such as argidin and argifin, cyclic pentapeptides obtained from *Clonostachys* and *Gliocladium* spp., respectively, have recently been identified as effective in killing *Lucilia cuprina* blowfly larvae (Houston et al., 2002).

10.4.6 Neurologic

In neurological disorders, ergot alkaloids are the most significant clinically beneficial fungal compounds, particularly ergotamine. This is derived from the ergot fungus *Claviceps purpurea*, which was responsible for widespread poisoning incidents in Europe, as the consumption of rye bread contaminated with the fungus's sclerotia led to adverse effects. Owing to its potent vasoconstrictor properties, ergotamine found applications in conventional Regime to induce childbirth and treat hemorrhage after delivery (Aloe et al., 2015).

Several fungal metabolites have been identified for their neurotrophic or neuritogenic activities, suggesting potential applications in various neurological disorders, such as dementia. For example, stachybotrin C, produced by *Stachybotrys parvispora*, has been shown to induce PC12 cell differentiation to neuron-like cells, similar to the effects of nerve growth factor (NGF), at micromolar concentrations. Furthermore, in primary neuronal cultures, this compound protects against neuronal damage. NG-061, obtained from *Penicillium minioluteum*, demonstrated similar effects on PC12 cells. Its effects were found to be additive to those of NGF. However, it failed to exhibit any protective impacts on primary cultures of neurons (Aloe et al., 2015).

10.4.7 Antiviral

Several research groups have invested significant efforts in exploring natural products for potential antiviral agents. The need for antiviral drugs has increased due to the HIV pandemic. Researchers have focused on identifying compounds that target two viral genome-encoded enzymes, namely protease and reverse transcriptase, in the quest for effective anti-HIV agents. Despite the successful development of clinically useful drugs targeting these enzymes, no such drug has originated from natural sources. Various fungal metabolites, such as mniopetals from the basidiomycetes *Mniopetalum* spp., which are weak inhibitors of HIV reverse transcriptase, and semicochliodinols related to asterriquinones from *Chrysosporium merdarium*, exhibit inhibition against protease of HIV (Pela, 2004).

Flutimide, a substituted 2,6-diketopiperazine, shares structural similarities with aspergillic acid. This compound is derived from *Delitschia confertaspora*, a recently discovered fungal species obtained from the dassie dung in Namibia. Flutimide demonstrates specificity in targeting the endonuclease activity of the cap-dependent transcriptase found in A and B influenza viruses while showing ineffectiveness against other viral polymerases (Hsieh and Hsu, 2007).

10.5 ROLE OF FUNGI IN THE FOOD INDUSTRY

Since ancient times, fungi have been used to manufacture many foods and drinks, including breads and alcoholic beverages. Moreover, current industrial fermentation for the synthesis of food products involves both yeasts and molds. For the large-scale production of fatty acids, organic acids, enzymes, pigments, and vitamins, among other materials, they are employed as biotransformation agents.

10.5.1 Alcoholic Fermentation

Fermentation processes leverage fungi to transform liquid and solid substrates into various products. Any material that promotes microbial growth can be used as a substrate. The goal is to create the required bioproduct as soon as feasible with a high rate and yield economically and straightforwardly. Fungi are important in fermentative industries, particularly in alcoholic fermentation. They are utilized to produce antibiotics, organic acids, ethanol, and various enzymes, including fungal glycosidase, gluconase, and cellulases. Notably, *S. cerevisiae* and *Monilia* spp. are employed in ethanol production. Fungi also contribute to industrial processes such as cheese ripening and the processing of various products. This utilization forms the basis of two crucial global industries: brewing and baking. Both industries rely on the fermentation of sugar solutions through yeasts, resulting in the production of carbon dioxide and ethyl alcohol. In brewing or winemaking, the primary product is alcohol, while carbon dioxide, once considered a waste product, is now considered valuable. It is collected, congealed, and marketed as "dry ice." In the baking or bread-making industry, carbon dioxide serves two essential purposes: causing the rising of dough and making light bread (Chisti, 2014).

Alcohol is the incidental byproduct in this process, as yeasts release an enzyme complex known as zymase, facilitating the transformation of sugar into alcohol. Numerous high-quality yeast strains are currently available, although they do not have diastase and cannot convert starch into sugar. Molds, a type of fungi, are famous for secreting a diverse range of enzymes, leading to complex carbohydrate fermentation. During alcohol synthesis in industries, molds are used as starters to initiate starch saccharification; in the subsequent step, yeast is applied to the sugar. Commonly used molds for scarification include *Mucor rouxii*, *M. racemosus*, and certain species of *Rhizopus*; for producing African native beer, *Aspergillus flavusis* is employed.

10.5.2 Organic Acids

Organic acids play a crucial role in food processing, serving primarily as acidulants, buffers, preservatives, flavoring agents, and technological adjuvants. Currently, the predominant fungal organic acids are gluconic and citric acids, which are synthesized through sucrose or glucose fermentation by *Aspergillus niger*. Other fungus-derived organic acids, such as malic, oxalic, itaconic, succinic, fumaric, tartaric, and lactic acids, can also be produced to a lesser extent. The substantial accumulation of organic acids occurs under specific and defined conditions, with some instances demonstrating a nearly complete transformation of the carbon substrate into organic acids. In producing citric acid, crucial parameters include maintaining high dissolved oxygen and carbohydrate concentrations, a low manganese concentration and pH, and consistent agitation (Moresi and Parente, 2014).

10.5.2.1 Citric Acid

Citric acid constitutes over 95% of global production and is the most versatile. It plays a significant role in the food industry, contributing to approximately 80% of food acidulant usage. Its versatility is evident in its pH and tartness control used in jellies and jams. Furthermore, it is an acid flavor adjuster in various beverages, including wine, soft drinks, and fruit juices. Citric acid can form complexes with heavy metals, making it a common treatment for frozen foods to preserve vitamin C and minimize enzymatic browning. While *A. niger* is mostly utilized in citric acid fermentation, other molds and yeasts, such as *Penicillium citrinum*, *Aspergillus wentii*, and *Candida lipolytica*, are also able to accumulate citric acid (Porro and Branduardi, 2017).

10.5.2.2 Gluconic Acid

Gluconic acid is generated through a bioprocess involving *A. niger*. In modern applications, this process can occur without fungal cells, utilizing enzymes like catalase and glucose oxidase obtained from *A. niger*. This enzymatic approach achieves almost complete conversion of glucose to gluconic acid, eliminating the need for additional product purification steps. While gluconic acid production has historically relied on fungal fermentation, the use of enzymes has gained prominence due to its cost-effectiveness. Filamentous fungi serve as primary enzyme sources in food processing to improve the simplicity and efficiency of processing steps, ultimately enhancing the quality of the end product. Enzyme processing plays an important role in influencing the appearance, texture, aroma, stability, color, flavor, and nutritional aspects of food materials (Magnuson and Lasure, 2004).

10.5.3 Enzyme Synthesis

As for amylases, their commercial production involves the use of *Aspergillus* species such as *Aspergillus oryzae*, *A. awamori*, and *A. niger*. Alpha-amylase specifically breaks down α-1,4-glycosidic bonds in starch, yielding α-limit dextrins and oligosaccharides. Subsequently, glucoamylase and beta-amylase complete the conversion to maltose, glucose, and β-limit dextrins. Amylases can hydrolyze up to 75 to 90% of amylose and 50 to 60% of amylopectin into glucose. The synthesis of glucose syrups accounts for the majority of α-amylase consumption in food processing, including bakery products. Similar to amylases, proteases find extensive applications in the food industry, contributing to various processes such as enhancing bread quality, meat tenderization, milk coagulation, and chill proofing of beer, among other uses. *A. oryzae* plays a crucial role in the production of proteases. At the same time, *Rhizomucor miehei* is significant for rennet production. In commercial fungal pectinolytic enzymes, the primary components are typically polygalacturonases (pectinases), although the enzymatic complex may also include pectin methylesterases (pectases). *Aspergillus* spp. stands out as the most prevalent source of these enzymes. Pectinases find widespread application in juice clarification, particularly for grape and apple juices. Moreover, these enzymes can be employed during the grape crushing process in winemaking, serving not only for clarification but also to enhance juice yield by approximately 10% and to eliminate coffee beans' mucilage coat. Most fungi generate invertase, but the invertase used in commercial applications is typically derived from *Saccharomyces* spp. cells by using a proteolytic enzyme. Invertase plays a crucial role in producing artificial honey, confections, and soft-centered candies, where the high concentration of sucrose will be crystallized until converted by invertase. Commercial α-galactosidase is derived from *Mortierella vinaceae*. This enzyme functions by hydrolyzing raffinose into galactose and sucrose. In the industrial context, α-galactosidase is employed in refining beet sugar to enhance the crystal formation of sucrose (Yada, 2015).

In milk, lactose is found, and whey can undergo hydrolysis through the action of lactase (β-galactosidase), forming galactose and glucose. Industrially, lactases are primarily derived from molds such as *A. oryzae* and *A. niger*, as well as yeasts like *Kluyveromyces marxianus*. Lactase obtained from filamentous fungi exhibits greater heat stability than yeast lactases, with a wider pH range and increased acid stability. This makes it particularly suitable for acid whey processing. Lactase is the most significant ingredient in producing dairy products and lactose-free milk tailored for individuals with lactose intolerance (Copetti, 2019).

10.5.4 Production of Vitamins and Pigments

Filamentous fungi are recognized for generating diverse pigments, encompassing various chemical classes like carotenoids, quinones, flavins, phenazines, melanins, indigo, violacein, and monastics. Regarding natural coloring, producing colorants using fungi offers a significant advantage by freeing manufacturers from reliance on seasonal supplies of raw vegetable materials like carrots, beetroot, and grapes. Additionally, it helps to reduce the batch-to-batch differences. Concerning artificial colorings, there has been a recent reduction in the number of dyes in food use due to toxicological studies. Furthermore, consumers are increasingly advocating for more natural food ingredients. Filamentous fungi are utilized in the industrial production of fermentative food-grade pigments, including *Monascus* pigments, riboflavin from *Ashbya gossypii*, β-carotene, natural red pigment from *Penicillium oxalicum*, and lycopene from *Blakeslea trispora*. All these pigments, except that from *Monascus*, are approved for European food products. Fungi hold great potential as a natural pigment source to address the growing need for safe and healthful substitutes for artificial colorants (Dufosse et al., 2014).

In the extensively studied *Monascus* spp., eight kinds of chemical structures of pigments have been recognized. The structures are categorized into red pigments, yellow pigments (monascin, ankaflavin, yellow II, and xanthomonascin A), and orange pigments (monascorubrin and rubropunctatin). Several of the pigments have demonstrated antifungal and antibiotic activities, as well as cytotoxic effects on various kinds of cancer cells. They exhibit other biological effects, including antioxidant activity and antiobesity potential. In Southeast Asia, pigments derived from *Monascus* spp. have been utilized as food colorants for over a millennium. Still, they lack approval for such usage in the United States and the European Union because of mycotoxin citrinin contamination. To address this safety issue, an alternative approach involves the generation of hydroxy-anthraquinoid and polyketide-*Monascus*-like pigments from carefully chosen non-toxic fungal strains, aiming to broaden the spectrum of colors available in modern natural food colorants (Mapari et al., 2010).

Carotenoids are naturally occurring orange–red food coloring agents with about 750 structural variations. Fungus-produced carotenoids share structural similarities with their plant-derived counterparts. In the food industry, β-carotene, a noteworthy carotenoid, is an antioxidant and vitamin A precursor and functions as an orange–red pigment. Within industrial fermentation, *Blakeslea trispora* is a pivotal fungus essential for the economical fermentation of lycopene, an otherwise costly pigment (Copetti, 2019).

Riboflavin is increasingly used as a yellow biocolor in addition to its function as vitamin B2. Yeasts such as *Debaryomyces subglobosus* or *Candida guilliermondii* are responsible for its production, while at large-scale fermentation, the filamentous fungus *A. gossypii* is a primary producer of this compound. Vitamin B2 is predominantly manufactured through fermentation processes. The excessive production of riboflavin, typical in wild strains, can be boosted by adding glycine to the growth medium. Several fungi can accumulate over 20% of their biomass as lipids (Stahmann, 2011).

10.5.5 Production of Polyunsaturated Fatty Acids

The filamentous fungus *Mortierella alpina* tends to create longer polyunsaturated fatty acids (PUFAs), while wild yeasts can accumulate more oil, up to 85% of their dry weight. PUFAs function as vitamin-like substances, having structural functions in cell membranes and acting as precursors in hormone manufacturing; prominent examples include docosahexaenoic and arachidonic acids. While not essential, these compounds are commercially significant and are manufactured by fungal synthesis. Molecular engineering approaches, particularly in yeasts such as *Candida lipolytica* and *Saccharomyces cerevisiae*, have enhanced the microbial synthesis of PUFAs. Volatile flavors are obtained from fungi (Ochsenreither et al., 2016).

10.5.6 Role of Fungi in Flavors

The bioprocess for generating the blue cheese flavor, characterized by a high content of volatile fatty acids, aldehydes, alkanols, and 2-alkanones, was pioneered using *Penicillium roqueforti*. This process marked one of the initial industrial endeavors in flavor production, resulting in a 7–12-fold higher ketone concentration than high-quality blue cheese while being four times more efficient (Fraatz and Zorn, 2011).

White-rot fungi, including *Ischnoderma* spp., can convert L-phenylalanine into benzaldehyde, imparting the bitter almond flavor, a significant flavor compound, after vanillin. Additionally, *Kluyveromyces* spp. or *Saccharomyces* spp. yeasts can produce 2-phenylethanol, which is an alcohol having a roselike odor. Basidiomycetes submerged cultures can produce various volatiles, including cinnamic acid esters, cinnamaldehyde, eugenol (contributing to a clove aroma), methyl benzoate (imparting a dry fruit aroma), and benzyl acetate (yielding a jasmine aroma). *Pichia ohmeri* and *C. lipolytica* can accumulate 5-alkanolide and 4-alkanolide lactones with carbons 8C to 12C, adding a fatty, nutty, and fruity aroma to food (Copetti, 2019).

10.6 USES OF FUNGI IN THE TEXTILE INDUSTRY

The textile industry's environmental hazards arise from wastewater discharge, characterized by its dark hue and synthetic dyes. The textile sector extensively uses water in various dyeing and finishing processes, consequently discharging a significant volume of wastewater into natural water bodies. This discharged effluent, characterized by its resistance to conventional biological treatment, contributes to environmental issues by blocking sunlight and adversely affecting aquatic life in ecosystems. Fungi have demonstrated their potential for degrading dyes, offering a promising solution. The key advantage of utilizing fungi is their ease of cultivation and rapid growth. Of 7000 enzymes, 75 have proven effective and efficient in textile processing and production. Several fungal enzymes have demonstrated their unique roles in the textile industry (Nigam, 2013).

10.6.1 Amylases

Amylases break down starch into smaller glucose polymers, including dextrins. The warp threads in cotton blends are typically coated with an adhesive called "size," which protects and lubricates yarn threads from abrasion in the weaving process. Although various materials have been employed for sizing fabrics, the most widely utilized are starch and its derivatives because of its remarkable film-forming capabilities, affordability, and accessibility. For dyeing the fabrics, sizing agents are removed by treating them with oxidizing agents, acids, and bases at high temperatures before using amylases, which often results in imperfect dyeing and the degradation of cotton fiber, affecting its natural softness. Amylases are now commercially utilized for resizing due to their high specificity and efficiency, ensuring minimal impact on the fabric. Amylases work by randomly cleaving starch into water-soluble dextrins, which can be easily washed away, reducing the discharge of harmful compounds into the environment (Feitkenhauer et al., 2003).

10.6.2 Cellulase

The hydrolytic enzyme cellulase breaks down cellulose into oligosaccharides and, ultimately, glucose, serving as a catalyst. The widespread application of cellulase began in the late 1980s, particularly in denim staining. It plays a significant role in textile wet processing and is a surface modifier for fibers. Its importance grew notably in achieving a stone-washed appearance for denim without using stones. Cellulase exhibits various effects on cellulose-based artificial fabrics, such as lyocell (known as TENCEL), where cellulase activity alters drape ability and removes surface fuzz. Similar effects are observed on viscose, where cellulase reduces viscose's pilling and minimizes lyocell's fibrillate tendency. When pretreated with cellulase, fabrics show a substantial increase in the uptake of natural dyes like carmine and chlorophyll without compromising fastness. Beyond imparting a stone-washed appearance to denim, cellulase enhances the appearance and handling of fabrics like linen, knits, and cotton. Cellulase finds its best and most advantageous uses in aging fabric surfaces, creating a stone-washed look in denim garments, and renewing and cleaning fabric surfaces damaged by fuzz, loose fibers, and micro-fibrils (Sukharnikov et al., 2011).

10.6.3 Catalases

Catalases, more accurately termed hydroperoxides, are involved in the breakdown of H_2O_2 into O_2 and H_2O. Various microorganisms, including bacteria and fungi, produce these enzymes. After desizing and scouring, bleaching with H_2O_2 can occur before the dyeing phase. Catalases are employed to decompose any excess H_2O_2, eliminating the requirement for a reducing agent and reducing the necessity for rinsing water. This results in reduced polluted wastewater and decreased water usage. Introducing immobilized enzymes offers a practical means of reducing the enzyme cost of degrading hydrogen peroxide in bleaching effluents. Additionally, this approach enables enzyme recovery and allows for the reuse of bleaching effluents treated in the dyeing process (Zámocký et al., 2012).

10.6.4 Laccases

Laccases, classified as multi-copper enzymes, operate extracellularly and utilize molecular oxygen to oxidize phenols and a diverse array of non-aromatic and aromatic compounds through a reaction mechanism catalyzed by radicals. While laccases can be found in bacteria, plants, and insects, they are most prevalent in fungi, with over 60 fungal strains exhibiting laccase activity. These enzymes typically thrive in acidic pH conditions. One notable characteristic of laccase enzymes is their absence of substrate specificity, allowing them to react with a wide range of substrates. For denim garments, bleaching is employed to achieve various shades. Specifically, bleaching powder (sodium hypochlorite) is used for denim garments. Different wash techniques, such as ice washing, are controlled, transformed, and manipulated to create distinctive effects, such as the Bata wash (Jeon et al., 2012).

Nonetheless, this method has a notable drawback due to chlorine-based bleaches, which are environmentally unfriendly and create an unsafe working environment. Laccase, used in indigo decolorization, is important in treating and finishing denim fabrics. Laccases are widely used in textile processing because of their capacity to decolorize effluents and their effectiveness in degrading various dye structures, including synthetic dyes, making them an environmentally friendly option for treating wastewater containing dyes. Cotton whitening through flavonoid oxidation is achieved with the assistance of laccase. Less fiber and water conservation damage is achieved when enzymatic bleaching is used in conjunction with or instead of chemical bleaching (Haq et al., 2015).

10.6.5 Pectinases

Pectin and pectic substances, which are polysaccharides, are present in plants' cell walls and middle lamellae. Enzymes called pectinases are responsible for breaking down these pectic substances. Plant pathogens (fungi and bacteria) and saprophytes primarily produce this enzyme to degrade the plant cell wall. Raw cotton contains various noncellulosic impurities like hemicellulose, waxes, and mineral salts, which are present in the fibers' cuticles and cell walls. These contaminants make raw cotton more hydrophobic, which makes chemical processing difficult, especially when dyeing and finishing cotton (Yadav et al., 2018).

Therefore, a pretreatment process known as scouring is necessary to eliminate substances that hinder dye binding before dyeing cotton yarn. Scouring, which utilizes sodium hydroxide, enhances fabric wettability. While these chemicals are known to decrease cellulose strength, they also reduce fabric weight. Moreover, this treatment raises COD (chemical oxygen demand), BOD (biochemical oxygen demand), and salt in the wastewater. Bioscouring addresses these issues associated with chemicals. In the enzymatic process, cellulose remains intact, preventing the loss of material weight and strength. Bioscouring presents numerous benefits when compared to conventional scouring techniques. It is typically conducted at a neutral pH, significantly reducing total water usage compared to traditional procedures and preserving the softness of the cotton fiber. Various enzymes, including lipases, pectinases, cellulases, and cutinases, have been individually and collectively studied for cotton bioscouring, with pectinases proving highly effective (Karnwal et al., 2019).

10.7 USES OF FUNGI IN THE PAPER AND PULP INDUSTRY

The paper and pulp industry stands out as one of the most environmentally harmful sectors, employing numerous chemicals in various stages like bleaching, deinking, and pulping of waste papers. Utilizing these chemicals results in the release of multiple pollutants into the effluent. To address this issue, cost-effective and environmentally friendly approaches, such as using different microbes or their enzymes, including xylanase, lignin, and cellulase degrading enzymes, are employed in the paper and pulp industry. White-rot fungi, for example, can modify or degrade lignin, reducing chemical and energy requirements during pulping. Similarly, laccases and xylanases play a role in decreasing chemical consumption during pulp bleaching. Microbial xylanases and cellulases are used in the enzymatic deinking of recycled fiber and pulp refining processes. The integration of enzymes in deinking lowers chemical consumption and enhances pulp's drainage rate and strength properties. Additionally, pulp treated with enzymes effectively reduces mechanical energy consumption during refining (Gil et al., 2009).

10.8 BIOPULPING

Pulping refers to creating fibrous raw materials for paper production by disrupting the bonds in wood or other raw materials like grasses, agro-residues, and non-wood. The goal during papermaking is to remove lignin as much as possible. In chemical pulping, different chemicals are employed to obtain cellulosic fibers by dissolving lignin from the cell walls of plants. Raw materials are treated with specific chemicals in an aqueous solution at elevated pressures and temperatures, transforming them into a fibrous mass or pulp. Three chemical pulping processes, sulfite, kraft, and soda pulping, are commonly used. It is worth noting that pulping constitutes a crucial stage in papermaking, contributing significantly to generating pollutants throughout the process. Biopulping is an eco-friendly alternative to conventional pulping methods, offering a cost-effective approach to delignify cellulosic raw material. The biopulping process refers to the pretreatment of raw materials with white-rot fungi before pulping. White-rot fungi play a crucial role by producing lignin-degrading enzymes like peroxidase laccase, manganese peroxidase, and lignin during biopulping. The primary objective of biopulping is to selectively remove or loosen lignin within the wood matrix without compromising the integrity of cellulose (Kumar et al., 2016a).

In the pulping process, fungal enzymes facilitate the removal of hemicellulose and lignin. Various white-rot fungi, including *Schizophyllum commune*, *Corioulus versicolor*, *Trametes versicolor*, *Phanerochaete chrysosporium*, *Ganoderma colossus*, *Physisporinus rivulosus*, *Phlebiopsis gigantea*, and *Ceriporiopsis subvermispora*, have proven to be highly suitable for biopulping. It enhances pulp quality, improves paper properties, and lowers environmental impact and energy expenses compared to traditional pulping processes (Saini, 2017).

In the initial phase of biopulping, wood is either transformed into chips, or alternative raw materials like grasses and agro-residues are finely chopped. Wood chips are typically contaminated with various fungal and bacterial spores. These undesirable microorganisms can generate cellulases, leading to the degradation of cellulose and adversely impacting strength properties. Therefore, it becomes imperative to decontaminate wood chips and other raw materials, with steam sterilization being the preferred method. In the subsequent step, the selected fungus is introduced to the substrate through inoculation and then incubated in aerated conditions at an appropriate temperature for a period ranging from 1 to 4 weeks. The incubation duration depends on the substrate's specific properties and the fungus involved. Solid-state fermentation for the fungal pretreatment of raw materials involves maintaining a substrate moisture level between 50% and 80%. In warm and moist conditions, fungi that break down lignin colonize the surface and penetrate the interior of wood chips. As a result, the treated materials can be pulped more efficiently, requiring reduced pulping chemicals and electrical energy (Singh et al., 2016).

10.9 BIOBLEACHING

Utilizing microbial enzymes, including laccases and xylanases, is considered a cost-effective and environmentally friendly alternative for bleaching pulps, known as biobleaching. In the paper industry, biobleaching is the process of bleaching pulp by adding biological agents, like enzymes or whole microorganisms. When xylanase is used in the pretreatment of pulp, it is termed "prebleaching" or "bleach boosting." This process enhances the penetration of bleaching agents by disrupting the network of xylan, facilitating the removal of trapped lignin from pulp fibers rather than directly removing or attacking lignin-based chromophores (Kumar et al., 2016b). Utilizing xylanase in the bleaching process presents a cost-effective alternative for the paper industry, offering several advantages. When employed alongside ozone and hydrogen peroxide in total chlorine-free bleaching, xylanase bleaching significantly enhances the final brightness of the pulp while concurrently reducing bleaching costs (Gangwar et al., 2014).

Xylanase proves highly effective in minimizing bleaching agents like chlorine dioxide and chlorine, leading to a potential 25% reduction in effluent discharge of adsorbable organic halides (AOXs). In the pulp bleaching process, cellulase-free xylanases play a crucial role. Xylanase preparations with cellulase contamination can enhance cellulose degradation, adversely impacting pulp strength and viscosity. Hence, cellulase-free xylanases are preferable for prebleaching applications. Xylanase must exhibit activity at high temperatures and within a broad pH range, particularly in the alkaline spectrum, to ensure its optimal functionality in bleaching during incubation periods of up to 3 to 5 hours (Nie et al., 2015).

Xylanases devoid of cellulases, derived from various microorganisms such as fungi and bacteria, have proven effective in the biobleaching process of pulp. These xylanases contribute to lignin removal by breaking down xylan, promoting the release of lignin from pulp. This results in improved brightness, enhanced pulp properties, a reduced kappa number, and decreased reliance on chlorine as a bleaching agent. The pretreatment of pulp with xylanase significantly reduces the chlorine requirement during the bleaching stages, ranging from 10% to 50% (Gautam et al., 2015).

10.9.1 Deinking

Deinking is a vital step in the recycling process of waste paper because it removes dirt and ink particulates from waste paper. Recycling waste paper is good for the environment and the economy, as it diminishes the need for raw materials sourced from forests and decreases the volume of used paper entering landfills. Substituting virgin pulp with recycled fibers not only conserves green plants but also results in savings in energy and water consumption per ton of pulp. The primary goal of paper recycling is to remove ink and other impurities and maintain the strength and characteristics of the fiber. The energy consumption for recycling waste paper is notably lower, ranging from 23% to 70%, compared to virgin fiber production (Mäkinen et al., 2013).

Additionally, traditional deinking methods typically involve the use of substantial quantities of chemical agents like sodium carbonate, hydrogen peroxide, sodium silicate, sodium hydroxide, diethylenetriaminepentacetic acid (DTPA), and surfactants, leading to expensive wastewater treatment processes. As an alternative, employing enzymes to remove toner particles from fibers has been recognized as an efficient and less environmentally damaging solution to address disposal challenges. Cellulases are often preferred in many cases, but other enzymes like lipases, amylases, and hemicellulases can be considered for optimizing the process based on the specific characteristics of the ink and paper (Pathak et al., 2011).

10.9.2 Ink-Degrading Enzymes

Esterase and lipase enzymes can break down ink derived from vegetable oil, proving to be efficient in the deinking process of waste paper printed with soybean oil-based ink. Additionally, due to their amphiphilic nature, lipases may exhibit a surfactant effect, which aids in waste paper deinking (Verma et al., 2012).

10.10 FIBER-ATTACKING ENZYMES

Xylanases, pectinases, cellulases, and laccase can interact with the fiber's surface, modifying either the surface or bonds in the proximity of ink molecules. This alteration helps to remove the ink through flotation and washing processes. The enzymatic treatment of pulp weakens bonds, likely by enhancing fibrillation or removing layers of the surface of individual fibers. Among these enzymes, xylanases and cellulases have been extensively studied for deinking. Cellulases remove the fibrils from fiber surfaces, thereby liberating ink particles for removal. In the case of starch-degrading enzymes, amylases are essential for acting on starch and helping cellulases separate ink particles from the fiber surface (Dutt et al., 2012).

10.10.1 Pulp Refining

Unrefined or unbeaten pulp produced in the pulp mill is unsuitable for producing most paper grades. Paper from unrefined virgin pulp exhibits a rough surface, bulkiness, and low strength. Therefore, mechanical treatment of the pulp is necessary to form it into a uniform sheet and establish strong bonds at contact points. During refining, pulp fibers undergo compression and shear forces, thus inducing various morphological changes. This includes the removal of outer fiber layers, allowing for fiber swelling through hydration, a reduction in length because of cutting (fiber shortening), external fibrillation, internal fibrillation, and enhanced flexibility. Consequently, there is an expansion of the fiber surface area, facilitating better bond formation and enhancing the strength characteristics of the paper, except tear strength (Gharehkhani et al., 2015).

Refining is a process that involves shortening excessively long fibers to enhance paper formation, ensuring a more uniform sheet, and improving properties such as porosity, opacity, and absorbance. However, a significant drawback of current refining technologies is their high energy consumption. One ton of refining pulp typically needs 150–500 kWh of energy, constituting 30–50% of the total energy utilized in paper manufacturing. The enzymatic pretreatment of pulp before mechanical refining has proven effective in reducing energy consumption during refining. Xylanase and cellulase enzymes are mostly employed in the refining process. When cellulose fibers are treated with xylanase, xylans or hemicelluloses are broken down, which makes it easier for water molecules to enter the spaces inside the fibers. This action results in the disruption of hydrogen bonds connecting cellulose chains. As a consequence of these bond breakages, the three-dimensional structure of the fibers becomes more flexible. As a result, the paper structure becomes more compact and has improved physical strength characteristics (Przybysz Buzała et al., 2016).

Enzymes have demonstrated effectiveness in modifying the surface of fibers, influencing fiber bonding, and subsequently impacting the strength properties of paper. Enzymatic treatment enhances collapsibility, flexibility, and fibrillation, enhancing network packing and sheet consolidation. Xylanases and cellulases are commonly employed individually or in combination for enzyme-assisted pulp refining, aiming to decrease energy consumption in mechanical refining processes. Enzymes derived from various fungal sources, particularly cellulases and xylanases, effectively reduce energy consumption while enhancing drainage efficiency (Singh and Chen, 2008).

Applying xylanase treatment resulted in the partial hydrolysis of xylan, a component found in small amounts associated with cellulose fibers. This process led to a slight loosening of the fiber structure. Consequently, xylanase treatment played a significant role in reducing energy and enhancing the static strength characteristics of the resulting paper. Enzymatic pretreatment, performed before refining, resulted in an increased pulp drainage rate and fiber hydration. This effect was more pronounced in the case of cellulases, possibly because these enzymes primarily degrade the cell wall at the fiber surface. This degradation provides a larger surface area for interacting with water molecules, which increases fiber hydration. Despite some fiber degradation observed in the case of cellulases, the overall impact of enzymatic treatment was positive on the strength qualities of the pulp (Kumar et al., 2017).

10.10.2 Role of Fungi in the Agricultural Industry

Fungi can operate effectively in diverse soil environments, from acidic to basic conditions. Additionally, fungi are essential for several physiological functions, including absorbing minerals and water, photosynthesis, stomatal regulation, and synthesizing chemicals like ethylene, lignan, auxins, and biostimulants. These enhance plants' adaptability to recognize and withstand environmental stressors such as salinity, drought, cold, heat, and heavy metals.

10.10.3 Role of Soil Fungi

Fungi thrive in slightly acidic or low-pH soils, particularly in undisturbed conditions. Their roles involve:

- Breaking down organic residues.
- Facilitating the decomposition process.
- Enabling other microbial types to transform residues into beneficial products.

Around 90% of all plants engage in symbiotic relationships with mycorrhizal fungi, forming hyphal networks. Through these mycorrhizae, plants primarily acquire phosphate and other minerals like copper and zinc from the soil, while the fungus receives nutrients such as sugars from plant roots. This beneficial mutual association is referred to as a mycorrhizal network. Although soil fungi can adapt to a broad pH range of soil, they are more abundant under acidic environments because they face strong competition from bacteria that prefer neutral pH values. Mostly fungi are aerobes and prefer optimal soil moisture levels for growth (Smith and Read, 2010).

The rhizosphere, located near the plant's roots, is primarily inhabited by soil microbes that engage in various chemical and biochemical processes. Soil fungi constitute a significant portion of the rhizosphere, ranging from 10% to 30%. With their capacity to produce diverse extracellular enzymes, fungi can effectively break down organic matter, decomposing soil components and maintaining the nutrient and carbon balance for optimal soil health. This ability enables fungi to create connections within the soil, facilitating the transportation of nutrients back to the plants.

Fungi rely on the soil as a fundamental source for their growth and are closely linked to plants' roots. They also generate various bioactive metabolites that can enhance the growth of plants. Additionally, fungi provide inorganic nutrients like phosphate ammonium and nitrate to plants that are utilized as biofertilizers. Rhizosphere microorganisms exhibit resilience against competing soil factors and can thrive in varying environmental conditions (Waqas et al., 2014).

10.10.4 The Economic Advantages of Fungi

Saprophytic fungi involved in decomposition play a crucial role in the perpetual cycle of greenhouse gas emissions and are essential to the chemical processes of plants in the natural world. These fungi also contribute to the decomposition and decay of plant and animal remnants, providing nutrients in an accessible form to plants. Certain types of fungi act as antagonists, suppressing pathogenic fungi responsible for seedling diseases, thereby positively influencing crop growth. Some fungi, such as *Metarhizium anisopliae*, and *Empusa sepulchralis*, can manage specific insect pests. Other insects are parasitic, particularly those that form spores. Fungal spores can be sprayed on crops to control pests like spittlebugs, citrus rust mites, and Colorado potato beetles. These fungi are characterized by loops in their mycelia, which strangle and trap nematodes as they try to get through, subsequently absorbing nutrients from the nematodes (Gianinazzi et al., 2002).

10.10.5 Vesicular Arbuscular Mycorrhizae

Vesicular arbuscular mycorrhizal (VAM) fungi, members of Glomeromycota, are considered fundamental fungi situated at the base of the hierarchy for higher fungi like basidiomycetes. These fungi produce tiny structures or relatively small sporocarps. Although over 200 described species exist, they can form mycorrhizal associations with most plants. During mycorrhizal association, the underground mycelium connects with the roots of plants without causing harm to the plants. Mycorrhizal fungi are important in promoting the growth of plants by enhancing nutrient uptake, producing growth-promoting chemicals, providing tolerance to salinity and drought, and promoting collaborative relationships with other beneficial microbes (Alqarawi et al., 2014).

In sustainable agriculture, the soil conditions are likely more conducive to arbuscular mycorrhizal fungi than in conventional agriculture. These fungi are distributed in agricultural and natural environments and are associated with over 80% of land plants, including grasses, woody gymnosperms, angiosperms, and ferns. Arbuscular mycorrhizal fungi (AMF) form beneficial symbiotic relationships with numerous land plants, and they can enhance soil characteristics, thus supporting plant growth in both normal and challenging environments. Plant colonization by arbuscular mycorrhizal fungi improves plant growth, leading to nutritional, physiological, and morphological changes that enhance resistance to various abiotic stresses. In instances of salinity stress, the inoculation of arbuscular mycorrhizal fungi has been observed to protect *Ocimum basilicum* by enhancing chlorophyll synthesis, water usage efficiency, and mineral uptake. Similarly, tomato plants inoculated with these fungi exhibit increased leaf area and elevated levels of phosphorus, calcium, potassium, and nitrogen, contributing to enhanced plant growth (Balliu et al., 2015).

10.10.6 Plant Responses to AM Fungal Inoculation

Soil phosphorus plays a crucial role in the response of plants, and generally, plants exhibit better responses to low levels of phosphorus. The interaction between strains of fungi and host genotypes influences how plants respond to inoculation. A global field experiment has indicated that the absence of effective AM fungal endophytes in soils with marginal phosphorus deficiency increases yields for crops such as barley, maize, wheat, cowpea, and potatoes. Furthermore, AM fungus inoculation enhances zinc uptake in wheat, maize, potato, and peach plants in zinc-deficient soils. AM associations are involved in increased absorption of calcium, sulfur, and water, and tolerance against drought in avocado and citrus seedlings.

Additionally, plants inoculated with AM fungi exhibit increased levels of chlorophyll and cytokinins. Hence, numerous researchers are exploring alternative strategies involving manipulating or adding microorganisms to boost plants' defense mechanisms against various pathogens. Beneficial microbes, such as antagonistic bacteria (e.g., *Bacillus subtilis*) and fungi (e.g., arbuscular mycorrhizal fungi, *Trichoderma*), engage in competition with plant pathogens for habitat and nutrients. They achieve this by entangling pathogens and antibiotics (Landis et al., 2004).

The effective administration of AM fungus is particularly crucial for organic and low-input agricultural systems where phosphorus levels are generally low. However, all agricultural systems can obtain benefits by promoting the establishment of arbuscular mycorrhizae. Some crops, like flax, exhibit limited chemotaxis ability and rely heavily on AM-mediated phosphorus uptake, especially in soils with low to intermediate phosphorus concentrations. Properly managing AMF in agroecosystems can enhance land productivity and soil quality. Agricultural practices like reduced tillage, limited phosphorus fertilizer use, and perennialized cropping systems foster functional mycorrhizal symbiosis (Mo et al., 2016).

10.10.7 Role of AM Fungi in Phytoremediation and Soil Quality

The utilization of arbuscular mycorrhizal fungus in the restoration of the environment, particularly in phytoremediation, has been demonstrated to alter the establishment of host plants in damaged soil, improving soil quality. Glomalin, a soil protein, is produced and is responsible for improving soil aggregation stability. AM fungi are important for agriculture, particularly in soils low in phosphorus, where phosphorus in the vesicle diffuses into the cytoplasm and is subsequently absorbed by the plants. Fungi from *Glomus* and *Endogone* genera are involved in forming this association (Liu and Wang, 2003).

10.10.8 Function of AM Fungi in Salinity Tolerance

Mycorrhizae assist plants in overcoming extreme conditions of the environment (e.g., saline environments), and numerous AM species have been identified in saline habitats. The root systems of 50% of plants found close to shorelines are estimated to have mycorrhizal associations. Similarly, plants in salt marshes have been found to have a variety of AM species. AM species can survive in extremely salinized environments with electrical conductivities above 150 dS/m. AM fungi employ different mechanisms to help plants manage salt stress. They may enhance plants' ability to absorb nutrients from the soil. When an AM fungus was introduced to lettuce and onion plants under salt stress, the plants accumulated more phosphorus. In plants, arbuscular mycorrhizal (AM) fungi can influence the ionic balance, particularly concerning Na^+ and Cl– ions (Garg and Manchanda, 2009).

Additionally, introducing AM to *Solanum lycopersicum* (tomato) in a saline environment enhanced the generation of antioxidant enzymes, protecting cell membranes from harm. These fungi also contribute to the increased secretion of various hormones, including abscisic acid. The impact of mycorrhizae on hormones is significant, as these substances can assist plants in overcoming various environmental stresses.

For instance, under saline conditions, adding *Glomus intraradices* to lettuce resulted in high levels of hormones in these plants, influencing the regulation of stomatal closure. Salinity can induce drought stress in plants. Hence, AM fungi may aid in enhancing water absorption. The inoculation of mycorrhizae to *Allium porrum* (leek) resulted in enhanced root surface area, thereby improving water uptake. Under salt stress, in lettuce plants, water use efficiency is notably enhanced by including mycorrhizae (Husband et al., 2002).

10.10.9 Role of AM Fungi in Drought Conditions

Rice is predominantly grown in rain-fed environments, and its yield is significantly diminished due to inadequate water supply, making drought a major limitation for rice production. While rice possesses inherent mechanisms to cope with drought stress, living soil organisms, such as AM fungi, assist rice in boosting its resistance to drought. Under drought stress, AM fungi aid in water and nutrient absorption by plants and contribute to regulating plant hormones and stomatal behavior. Intercropping can boost AM fungal colonization and improve rice root morphology, which is helpful for drought resistance (Berg and Smalla, 2009).

10.10.10 Mycoinsecticides

Fungi have been employed to manage insect pests. Fungi infect insects through the surface of their bodies, distinct from infections resulting from protozoa, viruses, and bacteria. These fungi are termed entomogenous fungi. These fungi that prey on insects attach their conidia to the insect's integument, where they grow. Under optimal humidity and temperature conditions, germ tubes penetrate the insect body. The fungus then multiplies within the insect, covering its body with conidia and mycelia. Newly formed conidia disperse, leading to further infections, and this cycle continues. Myconematicides are named according to the type of fungal biocontrol agent, including endoparasites (e.g., *Hirsutella*), nematode-trapping fungi (e.g., *Arthrobotrys*, *Dactylella*), and highly specific egg parasites. Examples of common and commercially available myco-nematicides include Royal 300 R (*Arthrobotrys robata*) and Royal 350 R (*Arthrobotrys suporba*) (Yuvaraj and Ramasamy, 2020).

10.11 ROLE OF FUNGI IN THE PETROLEUM INDUSTRY

The growing global need for fuel and the decline in traditional crude oil reserves have sparked significant interest in exploring unconventional oil reservoirs globally. Unlike conventional methods, these alternative fossil fuel sources are unsuitable for extraction, transportation, and refinement using standard approaches. The extensive reservoirs of unconventional crude oils harbor elevated levels of harmful pollutants, heavy metals, and complex high-molecular-weight compounds. Therefore, there is a significant need to advance technologies designed to mitigate the environmental consequences of extracting, producing, and refining unconventional crude oils. Achieving sustainable development in the oil industry necessitates the creation of innovative, environmentally friendly technologies that can simultaneously yield higher economic returns while remediating the environmental impact on ecosystems affected by human activities. Integrating modern biotechnology into the oil industry is a complementary approach, introducing new tools to enhance processes and products throughout the production chain. This, in turn, reduces operational costs and enhances productivity while minimizing the environmental footprint (Naranjo-Briceño et al., 2013).

Prokaryotes, particularly those belonging to the bacterial domain, can metabolize various hydrocarbons through various metabolic pathways. They can accomplish this under varying electron acceptors and environmental factors, employing adaptation strategies to manage the toxic effects associated with hydrocarbons (Kleindienst and Knittel, 2020).

Bacteria are absorptive heterotrophs, absorbing soluble substances from their surroundings. As a result, biodegradation can face obstacles when substrates are not readily available for biological activity. This limitation may arise when substrates cannot traverse the cell wall, potentially due to low water solubility or strong adsorption onto surfaces within the soil matrix. Additionally, the metabolic rates of many hydrocarbonoclastic bacteria can experience adverse effects under these growth conditions. Fungi exhibit distinct physiological and metabolic advantages compared to bacteria, as they are not dependent solely on soluble and easily absorbed organic compounds for their nutritional needs. Instead, they release diverse enzymes into their surroundings that effectively break down polymeric substances (Harms et al., 2011). These enzymes cleave complex molecules, and the resulting soluble breakdown products are subsequently absorbed and further broken down by internal fungal enzymes. Beyond benefiting the primary fungal oxidizers, the released substrates also provide advantages to various other microorganisms. This dynamic interplay creates complex interactions, prompting fungi to co-evolve intricate strategies (Kohlmeier et al., 2005). In the natural environment, fungi are crucial for recycling resistant polymeric organic substances like cellulose, chitin, lignin, keratin, and melanin. Due to their diverse biochemistry, fungi can also break down certain challenging-to-degrade xenobiotics through co-metabolism. In these scenarios, fungi can degrade parent hydrocarbons to varying extents, ranging from partial conversions to total biodegradation (Boer et al., 2005).

The oil sector, referred to as the petroleum industry or oil patch, encompasses worldwide activities such as exploring, extracting, refining, transporting (commonly through pipelines and oil tankers), and marketing petroleum products. In addressing intricate problems that involve breaking down complex organic substrates into complicated products, in the future, microbial and fungal consortia may offer a more efficient solution. *Onygena* species, which are non-pathogenic, specialize in degrading keratin present in horns, feathers, and hooves. Keratin consists of proteins arranged in a form that is not easily accessible for biological processes. In *O. corvina*, numerous proteases are produced, which can break down pig bristles and feathers. The breakdown of substances by living systems is termed biodegradation. Primary biodegradation involves transforming original substances into new products with different properties. In contrast, ultimate biodegradation refers to the whole conversion of the original material into new microbial biomass, water, and carbon (IV) oxide by mineralization. Bioremediation, a form of environmental biotechnology, involves using organisms or biological agents to facilitate the breakdown of environmental pollutants through biodegradation processes (Demain et al., 2004).

Oil-degrading microorganisms, including bacteria and fungi, are not limited to specific ecosystems and have been identified in temperate regions, the Antarctic, and the Arctic. However, there is limited research on temperate ecosystems. Some fungi have shown higher hydrocarbon biodegradation ability than bacteria, so their capacities to use crude oil and n-alkane vary. Bioremediation, introducing microorganisms to impacted areas, is an environmentally friendly and economically viable way to deal with marine oil spills. Fungi thrive on crude oil, breaking down hydrocarbons. However, their capacity to grow on pure n-alkane does not reliably predict their ability to thrive on crude oil. Fungi exhibit remarkable adaptability and diversity in utilizing various organic molecules as a carbon source. However, their proficiency in degrading a particular hydrocarbon for energy and biomass may vary. Its chemical composition can also determine the kinds of fungi that can grow on crude oil. Fungi typically use n-alkanes and crude oil as carbon and energy sources, converting these molecules into CO_2 and biomass. It was observed that certain hydrocarbons exhibit lethality to some fungi; other hydrocarbons are not as poisonous (Elshafie et al., 2007).

Kerosene possesses a complex and multi-component structure, so various microbe cultures with enzymatic systems are applied to break it down. Biodegradation is done mostly by using fungi. Most filamentous fungi cannot completely mineralize aromatic hydrocarbons; instead, they convert them into indirect products characterized by reduced toxicity and enhanced susceptibility to decomposition facilitated by bacteria. Filamentous fungi like *Aspergillus* and *Cladosporium* are involved in the biodegradation of aliphatic hydrocarbons, while *Cunninghamella*, *Aspergillus*, *Fusarium*, and *Penicillium* can participate in the decomposition of aromatic hydrocarbons. For the mineralization of monoaromatic hydrocarbons such as benzene, toluene, ethylbenzene, and xylene (BTEX), *Cladophialophora* species of fungi are used (Prenafeta-Boldú et al., 2002).

There is a more active decomposition process for toluene, ethylbenzene, and m-xylene than for benzene. The addition of *Rhodococcus* sp. bacteria led to a significant enhancement in benzene mineralization. The same monooxygenase enzyme broke down toluene, ethylbenzene, and m-xylene in a lateral chain. Rot fungi, such as *Pleurotus ostreatus* and *Phanerochaete chrysosporium*, synthesize extracellular enzymes like lignin peroxidase, which are involved in degrading the lignin cell wall and oxygenating different xenobiotics (Husaini et al., 2008).

Microorganisms convert polycyclic aromatic hydrocarbons (PAHs) into quinone derivatives, and subsequently, the aromatic ring of quinones undergoes sequential stages of degradation, leading to complete mineralization. The effectiveness of PAH biodegradation significantly improves when *Pseudomonas putida* bacteria are enriched with *Irpex lacetus* and *Pleurotus ostreatus* fungi. This enhancement is particularly notable in 5–6-chain aromatic hydrocarbons, with a reduction in their content ranging from 16% to 19%. Certain fungi, such as *Phanerochaete chrysosporium*, can biodegrade aromatic compounds like BTEX and PAH. The collaboration between fungi and bacteria proves beneficial for mineralizing petroleum hydrocarbons. Notably, promising outcomes in PAH mineralization, particularly for 5–6-chain PAH, have been observed (Steliga, 2012).

A vast array of fungal biodiversity exists, exhibiting a variety of enzymatic mechanisms capable of transforming diverse hydrocarbon chemical structures. These structures include heavy-weight polycyclic aromatics and short-chain aliphatics. Hydrocarbonoclastic fungi can break down hydrocarbons as their exclusive energy and carbon sources (Prenafeta-Boldú et al., 2018). These fungi exhibit the remarkable capability to use hydrocarbons despite the challenging hydrophobicity properties and high toxicity. As a result, extremophilic hydrocarbonoclastic fungi are essential to the breakdown and conversion of petroleum. The overall fungal diversity isolated is primarily composed of the phyla (83%) Ascomycota, (10%) Zygomycota, (1%) Basidiomycota, and (6%) other phyla. Within the Ascomycota phylum, the most prevalent orders include Eurotiales, Hypocreales, Microascales, and Saccharomycetales. The dominant genera within this phylum are *Penicillium*, *Aspergillus*, and *Fusarium* (Arulazhagan et al., 2017).

REFERENCES

Ali, S. M.; Chee, S. K.; Yuen, G. Y.; Olivo, M. Hypocrellins and Hypericin Induced Apoptosis in Human Tumor Cells: A Possible Role of Hydrogen Peroxide. *Int. J. Mol. Med.*, 2002, *9* (5), 461–472.

Aloe, L.; Rocco, M. L.; Balzamino, B. O.; Micera, A. Nerve Growth Factor: A Focus on Neuroscience and Therapy. *Curr Neuropharmacol.*, 2015, *13* (3), 294–303. doi: 10.2174/1570159x13666150403231920

Alqarawi, A. A.; Abd Allah, E. F.; Hashem, A. Alleviation of Salt-Induced Adverse Impact via Mycorrhizal Fungi in Ephedra Aphylla Forssk. *J. Plant Interact.*, 2014, *9* (1), 802–810.

Arulazhagan, P.; Mnif, S.; Rajesh Banu, J.; Huda, Q.; Jalal, M. A. B. HC-0B-01: Biodegradation of Hydrocarbons by Extremophiles. *Biodegrad. Bioconversion Hydrocarb.*, 2017, 137–162.

Balliu, A.; Sallaku, G.; Rewald, B. AMF Inoculation Enhances Growth and Improves the Nutrient Uptake Rates of Transplanted, Salt-Stressed Tomato Seedlings. *Sustainability*, 2015, *7* (12), 15967–15981.

Berg, G.; Smalla, K. Plant Species and Soil Type Cooperatively Shape the Structure and Function of Microbial Communities in the Rhizosphere. *FEMS Microbiol. Ecol.*, 2009, *68* (1), 1–13.

Boer, W. de; Folman, L. B.; Summerbell, R. C.; Boddy, L. Living in a Fungal World: Impact of Fungi on Soil Bacterial Niche Development. *FEMS Microbiol. Rev.*, 2005, *29* (4), 795–811.

Carter, A. Investigations of Secondary Metabolites from Fungi: Isolation, Structure Elucidation, and Computational Studies. *Shareok.org.*, 2018, 1–200.

Chisti, Y. Fermentation (Industrial)| Basic Considerations. *Food Microbiol.* 2014, 1, 751–761.

Colletti, S. L.; Myers, R. W.; Darkin-Rattray, S. J.; Gurnett, A. M.; Dulski, P. M.; Galuska, S.; Allocco, J. J.; Ayer, M. B.; Li, C.; Lim, J. Broad Spectrum Antiprotozoal Agents That Inhibit Histone Deacetylase: Structure–Activity Relationships of Apicidin. Part 1. *Bioorg. Med. Chem. Lett.*, 2001, *11* (2), 107–111.

Copetti, M. V. Fungi as Industrial Producers of Food Ingredients. *Curr. Opin. food Sci.*, 2019, *25*, 52–56.

Del Val, A. G.; Platas, G.; Arenal, F.; Orihuela, J. C.; Garcia, M.; Hernandez, P.; Royo, I.; De Pedro, N.; Silver, L. L.; Young, K. Novel Illudins from Coprinopsis Episcopalis (Syn. Coprinus Episcopalis), and the Distribution of Illudin-like Compounds among Filamentous Fungi. *Mycol. Res.*, 2003, *107* (10), 1201–1209.

Demain, A. L.; Velasco, J.; Adrio, J. L. Industrial Mycology: Past, Present and Future. *Handb. Ind. Mycol.*, 2004, 30 (2), 187–214.

Domínguez, J. M.; Martín, J. J. Identification of Elongation Factor 2 as the Essential Protein Targeted by Sordarins in Candida Albicans. *Antimicrob. Agents Chemother.*, 1998, *42* (9), 2279–2283.

Dufosse, L.; Fouillaud, M.; Caro, Y.; Mapari, S. A. S.; Sutthiwong, N. Filamentous Fungi Are Large-Scale Producers of Pigments and Colorants for the Food Industry. *Curr. Opin. Biotechnol.*, 2014, *26*, 56–61.

Dutt, D.; Tyagi, C. H.; Singh, R. P.; Kumar, A. Effect of Enzyme Concoctions on Fiber Surface Roughness and Deinking Efficiency of Sorted Office Paper. *Cellul. Chem. Technol*, 2012, *46* (9–10), 611–623.

Dutta, B.; Lahiri, D.; Nag, M.; Ghosh, S.; Dey, A.; Ray, R. R. Fungi in Pharmaceuticals and Production of Antibiotics. In *Applied Mycology: Entrepreneurship with Fungi*; Springer, 2022; pp. 233–257.

Elshafie, A.; AlKindi, A. Y.; Al-Busaidi, S.; Bakheit, C.; Albahry, S. N. Biodegradation of Crude Oil and N-Alkanes by Fungi Isolated from Oman. *Mar. Pollut. Bull.*, 2007, *54* (11), 1692–1696.

Feitkenhauer, H.; Fischer, D.; Fäh, D. Microbial Desizing Using Starch as Model Compound: Enzyme Properties and Desizing Efficiency. *Biotechnol. Prog.*, 2003, *19* (3), 874–879.

Fox, B. M.; Vroman, J. A.; Fanwick, P. E.; Cushman, M. Preparation and Evaluation of Sulfide Derivatives of the Antibiotic Brefeldin a as Potential Prodrug Candidates with Enhanced Aqueous Solubilities. *J. Med. Chem.*, 2001, *44* (23), 3915–3924.

Fraatz, M. A.; Zorn, H. Fungal Flavours. *Ind. Appl.*, 2011, 249–268.

Gangwar, A. K.; Prakash, N. T.; Prakash, R. Applicability of Microbial Xylanases in Paper Pulp Bleaching: A Review. *BioResources*, 2014, *9* (2), 3733–3754.

Garg, N.; Manchanda, G. Role of Arbuscular Mycorrhizae in the Alleviation of Ionic, Osmotic and Oxidative Stresses Induced by Salinity in Cajanus Cajan (L.) Millsp. (Pigeonpea). *J. Agron. Crop Sci.*, 2009, *195* (2), 110–123.

Gautam, A.; Kumar, A.; Dutt, D. Production of Cellulase-Free Xylanase by Aspergillus Flavus ARC-12 Using Pearl Millet Stover as the Substrate under Solid-State Fermentation. *J. Adv. Enzym. Res.*, 2015, *1*, 1–9.

Gharehkhani, S.; Sadeghinezhad, E.; Kazi, S. N.; Yarmand, H.; Badarudin, A.; Safaei, M. R.; Zubir, M. N. M. Basic Effects of Pulp Refining on Fiber Properties—A Review. *Carbohydr. Polym.*, 2015, *115*, 785–803.

Gianinazzi, S.; Schüepp, H.; Barea, J. M.; Haselwandter, K. *Mycorrhizal Technology in Agriculture: From Genes to Bioproducts*; Vol. 01. Springer Science & Business Media, 2002; p. 296.

Gil, N.; Gil, C.; Amaral, M. E.; Costa, A. P.; Duarte, A. P. Use of Enzymes to Improve the Refining of a Bleached Eucalyptus Globulus Kraft Pulp. *Biochem. Eng. J.*, 2009, *46* (2), 89–95.

Haq, U. N.; Khan, M. M. R.; Khan, M. M. R. Investigation of the Bulk, Surface and Transfer Properties of Chlorine Bleached Denim Apparel at Different Condition. *Eur. Sci. J.*, 2015, *11* (12), 213–226.

Harms, H.; Schlosser, D.; Wick, L. Y. Untapped Potential: Exploiting Fungi in Bioremediation of Hazardous Chemicals. *Nat. Rev. Microbiol.*, 2011, *9* (3), 177–192.

Hawksworth, D. L.; Lücking, R. Fungal Diversity Revisited: 2.2 to 3.8 Million Species. *Microbiol. Spectr.*, 2017, *5* (4), 10–1128.

Houston, D. R.; Shiomi, K.; Arai, N.; Ōmura, S.; Peter, M. G.; Turberg, A.; Synstad, B.; Eijsink, V. G. H.; Van Aalten, D. M. F. High-Resolution Structures of a Chitinase Complexed with Natural Product Cyclopentapeptide Inhibitors: Mimicry of Carbohydrate Substrate. *Proc. Natl. Acad. Sci.*, 2002, *99* (14), 9127–9132.

Hsieh, H.-P.; Hsu, J. T.-A. Strategies of Development of Antiviral Agents Directed against Influenza Virus Replication. *Curr. Pharm. Des.*, 2007, *13* (34), 3531–3542.

Husaini, A.; Roslan, H. A.; Hii, K. S. Y.; Ang, C. H. Biodegradation of Aliphatic Hydrocarbon by Indigenous Fungi Isolated from Used Motor Oil Contaminated Sites. *World J. Microbiol. Biotechnol.*, 2008, *24 (12)*, 2789–2797.

Husband, R.; Herre, E. A.; Turner, S. L.; Gallery, R.; Young, J. P. W. Molecular Diversity of Arbuscular Mycorrhizal Fungi and Patterns of Host Association over Time and Space in a Tropical Forest. *Mol. Ecol.*, 2002, *11* (12), 2669–2678.

Jeon, J.; Baldrian, P.; Murugesan, K.; Chang, Y. Laccase-catalysed Oxidations of Naturally Occurring Phenols: From in Vivo Biosynthetic Pathways to Green Synthetic Applications. *Microb. Biotechnol.*, 2012, *5* (3), 318–332.

Karnwal, A.; Singh, S.; Kumar, V.; Sidhu, G. K.; Dhanjal, D. S.; Datta, S.; Amin, D. S.; Saini, M.; Singh, J. Fungal Enzymes for the Textile Industry. *Recent Adv. White Biotechnol. Through Fungi Vol. 1 Divers. Enzym. Perspect.*, 2019, 459–482.

Kleindienst, S.; Knittel, K. Anaerobic Hydrocarbon-Degrading Sulfate-Reducing Bacteria at Marine Gas and Oil Seeps. *Mar. Hydrocarb. Seeps Microbiol. Biogeochem. Glob. Mar. Habitat*, 2020, 21–41.

Kohlmeier, S.; Smits, T. H. M.; Ford, R. M.; Keel, C.; Harms, H.; Wick, L. Y. Taking the Fungal Highway: Mobilization of Pollutant-Degrading Bacteria by Fungi. *Environ. Sci. Technol.*, 2005, *39* (12), 4640–4646.

Kumar, A.; Dutt, D.; Gautam, A. Production of Crude Enzyme from Aspergillus Nidulans AKB-25 Using Black Gram Residue as the Substrate and Its Industrial Applications. *J. Genet. Eng. Biotechnol.*, 2016a, *14* (1), 107–118.

Kumar, A.; Gautam, A.; Dutt, D.; Yadav, M.; Sehrawat, N.; Kumar, P. Applications of Microbial Technology in the Pulp and Paper Industry. *Microbiol. Biotechnol. Sustain. Environ.*, 2017, 185–206.

Kumar, V.; Kaur, S.; Singh, S.; Upadhyay, N. Unexpected Formation of N′-Phenyl-Thiophosphorohydrazidic Acid O, S-Dimethyl Ester from Acephate: Chemical, Biotechnical and Computational Study. *3 Biotech*, 2016b, *6*, 1–11.

Landis, F. C.; Gargas, A.; Givnish, T. J. Relationships among Arbuscular Mycorrhizal Fungi, Vascular Plants and Environmental Conditions in Oak Savannas. *New Phytol.*, 2004, *164* (3), 493–504.

Little, C. R.; Carris, L. M.; Stiles, C. M. Introduction to Fungi. American Phytopathological Society. *APS*. 2012, 1–25.

Liu, R.; Wang, F. Selection of Appropriate Host Plants Used in Trap Culture of Arbuscular Mycorrhizal Fungi. *Mycorrhiza*, 2003, *13* (3), 123–127.

Magnuson, J. K.; Lasure, L. L. Organic Acid Production by Filamentous Fungi. In *Advances in Fungal Biotechnology for Industry, Agriculture, and Medicine*; Springer, 2004; pp. 307–340.

Mäkinen, L.; Ämmälä, A.; Körkkö, M.; Niinimäki, J. The Effects of Recovering Fibre and Fine Materials on Sludge Dewatering Properties at a Deinked Pulp Mill. *Resour. Conserv. Recycl.*, 2013, *73*, 11–16.

Mandala, S. M.; Harris, G. H. Isolation and Characterization of Novel Inhibitors of Sphingolipid Synthesis: Australifungin, Viridiofungins, Rustmicin, and Khafrefungin. In *Methods in Enzymology*, Vol. 311; Elsevier, 2000; pp. 335–348.

Mapari, S. A. S.; Thrane, U.; Meyer, A. S. Fungal Polyketide Azaphilone Pigments as Future Natural Food Colorants? *Trends Biotechnol.*, 2010, *28* (6), 300–307.

Misiek, M.; Hoffmeister, D. Fungal Genetics, Genomics, and Secondary Metabolites in Pharmaceutical Sciences. *Planta Med.*, 2007, *73* (02), 103–115.

Mo, Y.; Wang, Y.; Yang, R.; Zheng, J.; Liu, C.; Li, H.; Ma, J.; Zhang, Y.; Wei, C.; Zhang, X. Regulation of Plant Growth, Photosynthesis, Antioxidation and Osmosis by an Arbuscular Mycorrhizal Fungus in Watermelon Seedlings under Well-Watered and Drought Conditions. *Front. Plant Sci.*, 2016, *7*, 644–661.

Mora-Duarte, J.; Betts, R.; Rotstein, C.; Colombo, A. L.; Thompson-Moya, L.; Smietana, J.; Lupinacci, R.; Sable, C.; Kartsonis, N.; Perfect, J. Comparison of Caspofungin and Amphotericin B for Invasive Candidiasis. *N. Engl. J. Med.*, **2002**, *347* (25), 2020–2029.

Moresi, M.; Parente, E. Fermentation (Industrial) | Production of Some Organic Acids (Citric, Gluconic, Lactic, and Propionic); Batt, C. A., Tortorello, M. L. B. T.-E. of F. M. (Second E., Eds.; Academic Press: Oxford, 2014; pp. 804–815. doi: 10.1016/B978-0-12-384730-0.00111-7.

Naranjo-Briceño, L.; Pernía, B.; Guerra, M.; Demey, J. R.; De Sisto, Á.; Inojosa, Y.; González, M.; Fusella, E.; Freites, M.; Yegres, F. Potential Role of Oxidative Exoenzymes of the Extremophilic Fungus P Estalotiopsis Palmarum BM-04 in Biotransformation of Extra-heavy Crude Oil. *Microb. Biotechnol.*, 2013, *6* (6), 720–730.

Neely, M. N.; Ghannoum, M. A. The Exciting Future of Antifungal Therapy. *Eur. J. Clin. Microbiol. Infect. Dis.*, 2000, *19*, 897–914.

Nie, S.; Wang, S.; Qin, C.; Yao, S.; Ebonka, J. F.; Song, X.; Li, K. Removal of Hexenuronic Acid by Xylanase to Reduce Adsorbable Organic Halides Formation in Chlorine Dioxide Bleaching of Bagasse Pulp. *Bioresour. Technol.*, 2015, *196*, 413–417.

Nigam, P. S. Microbial Enzymes with Special Characteristics for Biotechnological Applications. *Biomolecules*, 2013, *3* (3), 597–611.

Ochsenreither, K.; Glück, C.; Stressler, T.; Fischer, L.; Syldatk, C. Production Strategies and Applications of Microbial Single Cell Oils. *Front. Microbiol.*, 2016, *7*, 1539.

Pathak, P.; Bhardwaj, N. K.; Singh, A. K. Optimization of Chemical and Enzymatic Deinking of Photocopier Waste Paper. *BioResources*, 2011, *6* (1), 447–463.

Pela, F. Biological Activities of Fungal Metabolites. In *Handbook of Industrial Mycology*; CRC Press, 2004; pp. 68–111.

Polishook, J. D.; Ondeyka, J. G.; Dombrowski, A. W.; Peláez, F.; Platas, G.; Teran, A. M. Biogeography and Relatedness of Nodulisporium Strains Producing Nodulisporic Acid. *Mycologia*, 2001, 93 (6), 1125–1137.

Porro, D.; Branduardi, P. Production of Organic Acids by Yeasts and Filamentous Fungi. *Biotechnol. Yeasts Filamentous Fungi*, 2017, 205–223. doi: 10.1007/978-3-319-58829-2_7.

Prenafeta-Boldú, F. X.; De Hoog, G. S.; Summerbell, R. C. Fungal Communities in Hydrocarbon Degradation. In *Microbial Communities Utilizing Hydrocarbons and Lipids: Members, Metagenomics and Ecophysiology*, McGenity, T. J. (Ed.); Springer Nature, 2018; pp. 1–36.

Prenafeta-Boldú, F. X.; Vervoort, J.; Grotenhuis, J. T. C.; Van Groenestijn, J. W. Substrate Interactions during the Biodegradation of Benzene, Toluene, Ethylbenzene, and Xylene (BTEX) Hydrocarbons by the Fungus Cladophialophora Sp. Strain T1. *Appl. Environ. Microbiol.*, 2002, *68* (6), 2660–2665.

Przybysz Buzała, K.; Przybysz, P.; Kalinowska, H.; Derkowska, M. Effect of Cellulases and Xylanases on Refining Process and Kraft Pulp Properties. *PLoS One*, 2016, *11* (8), 0161575–0161589.

Rho, M.-C.; Lee, H. S.; Chang, K.-T.; Song, H. Y.; Kwon, O. E.; Lee, S. W.; Ko, J. S.; Hong, S. G.; Kim, Y.-K. Phenylpyropene C, a New Inhibitor of Acyl-CoA: Cholesterol Acyltransferase Produced by Penicillium Griseofulvum F1959. *J. Antibiot.*, 2002, *55* (2), 211–214.

Saini, R. D. Textile Organic Dyes: Polluting Effects and Elimination Methods from Textile Waste Water. *Int J Chem Eng Res*, 2017, *9* (1), 121–136.

Saxena, R. K.; Gupta, R.; Saxena, S.; Gulati, R. Role of Fungal Enzymes in Food Processing. In *Applied Mycology and Biotechnology*, Vol. 1; Elsevier, 2001; pp. 353–386.

Sharma, A.; Dhiman, K.; Sharma, A.; Goyal, K.; Pandit, V.; Ashawat, M. S.; Jindal, S. Fusidic Acid: A Therapeutic Review. *AJRC* 2022, 15 (5), 372.

Singh, D.; Chen, S. The White-Rot Fungus Phanerochaete Chrysosporium: Conditions for the Production of Lignin-Degrading Enzymes. *Appl. Microbiol. Biotechnol.*, 2008, *81*, 399–417.

Singh, R.; Kumar, M.; Mittal, A.; Mehta, P. K. Microbial Enzymes: Industrial Progress in 21st Century. *3 Biotech*, 2016, *6*, 1–15.

Smith, S. E.; Read, D. J. *Mycorrhizal Symbiosis*, Vol. 3; Academic Press, Elsevier, 2010; pp. 11–145.

Solomon, L.; Tomii, V. P.; Dick, A. A. Importance of Fungi in the Petroleum, Agro-Allied, Agriculture and Pharmaceutical Industries. *N.Y. Sci. J.*, 2019, *12*, 8–15.

Stahmann, K.-P. Production of Vitamin B 2 and a Polyunsaturated Fatty Acid by Fungi. *Ind. Appl.*, 2011, 8, 235–247.

Steliga, T. Role of Fungi in Biodegradation of Petroleum Hydrocarbons in Drill Waste. *Polish J. Environ. Stud.*, 2012, *21* (2), 471–479.

Sukharnikov, L. O.; Cantwell, B. J.; Podar, M.; Zhulin, I. B. Cellulases: Ambiguous Non-homologous Enzymes in a Genomic Perspective. *Trends Biotechnol.*, 2011, *29* (10), 473–479.

Verma, S.; Saxena, J.; Prasanna, R.; Sharma, V.; Nain, L. Medium Optimization for a Novel Crude-Oil Degrading Lipase from Pseudomonas Aeruginosa SL-72 Using Statistical Approaches for Bioremediation of Crude-Oil. *Biocatal. Agric. Biotechnol.*, 2012, *1* (4), 321–329.

Ververis, K.; Karagiannis, T. C. Overview of the Classical Histone Deacetylase Enzymes and Histone Deacetylase Inhibitors. *Int. Sch. Res. Not.*, 2012, *2012*, 1–12.

Vicente, M. F.; Basilio, A.; Cabello, A.; Peláez, F. Microbial Natural Products as a Source of Antifungals. *Clin. Microbiol. Infect.*, 2003, *9* (1), 15–32.

Vilvanathan, S. Penicillins, Cephalosporins, and Other β-Lactam Antibiotics. *Introd. to Basics Pharmacol. Toxicol. Vol. 2 Essentials Syst. Pharmacol. From Princ. to Pract.*, 2021, 821–834.

Wang, A.; Xu, Y.; Gao, Y.; Huang, Q.; Luo, X.; An, H.; Dong, J. Chemical and Bioactive Diversities of the Genera Stachybotrys and Memnoniella Secondary Metabolites. *Phytochem. Rev.*, 2015, *14*, 623–655.

Wang, W.; Waters, S. J.; MacDonald, J. R.; Roth, C.; Shentu, S.; Freeman, J.; Von Hoff, D. D.; Miller, A. R. Irofulven (6-Hydroxymethylacylfulvene, MGI 114)-Induced Apoptosis in Human Pancreatic Cancer Cells Is Mediated by ERK and JNK Kinases. *Anticancer Res.*, 2002, *22* (2A), 559–564.

Waqas, M.; Khan, A. L.; Lee, I.-J. Bioactive Chemical Constituents Produced by Endophytes and Effects on Rice Plant Growth. *J. plant Interact.*, 2014, *9* (1), 478–487.

Yada, R. Y. Factors affecting the enzyme activity. In *Improving and Tailoring Enzymes for Food Quality and Functionality*; Elsevier, 2015; pp. 11–55.

Yadav, A. N.; Verma, P.; Kumar, V.; Sangwan, P.; Mishra, S.; Panjiar, N.; Gupta, V. K.; Saxena, A. K. Biodiversity of the Genus Penicillium in Different Habitats. In *New and Future Developments in Microbial Biotechnology and Bioengineering*; Elsevier, 2018; pp. 3–18.
Yuvaraj, M.; Ramasamy, M. Role of Fungi in Agriculture. *Biostimulants Plant Sci.*, 2020, 1–12.
Zámocký, M.; Gasselhuber, B.; Furtmüller, P. G.; Obinger, C. Molecular Evolution of Hydrogen Peroxide Degrading Enzymes. *Arch. Biochem. Biophys.*, 2012, *525* (2), 131–144.
Zhao, D.-M.; Li, W.-Y.; Shi, Y.-F.; Xiong, X.-Q.; Song, S.; Hao, C.-Z.; Cheng, M.-S.; Shen, J.-K. Pharmacophore-Based Design, Synthesis, and Biological Evaluation of Novel 3-((3, 4-Dichlorophenyl)(4-Substituted Benzyl) Amino) Propanamides as Cholesteryl Ester Transfer Protein (CETP) Inhibitors. *Chinese Chem. Lett.*, 2014, *25* (2), 299–304.

Index

Note: Page numbers in *italics* indicate a figure and page numbers in **bold** indicate a table on the corresponding page.

D

E

F

K

L

M

FSC
www.fsc.org
MIX
Papier | Fördert gute Waldnutzung
FSC® C083411

Kolibri 360 GmbH
Ferdinand-Jühlke-Straße 7
99095 Erfurt, Deutschland
produktsicherheit@kolibri360.de